Acta Numerica 2009

Managing editor

A. Iserles
DAMTP, University of Cambridge,
Centre for Mathematical Sciences, Wilberforce Road,
Cambridge CB3 0WA, England

Editorial Board

D. N. Arnold, *University of Minnesota, Minneapolis, USA*
C. de Boor, *University of Wisconsin, Madison, USA*
F. Brezzi, *Instituto di Analisi Numerica del CNR, Italy*
J. C. Butcher, *University of Auckland, New Zealand*
P. G. Ciarlet, *City University of Hong Kong, China*
W. Dahmen, *RTWH Aachen, Germany*
B. Engquist, *University of Texas, Austin, USA*
H.-O. Kreiss, *University of California, Los Angeles, USA*
M. J. D. Powell, *University of Cambridge, England*
E. Tadmor, *University of Maryland, College Park, USA*
R. Temam, *Université Paris-Sud, France*
L. N. Trefethen, *University of Oxford, England*

Acta Numerica

Volume 18   2009

CAMBRIDGE UNIVERSITY PRESS
Cambridge, New York, Melbourne, Madrid, Cape Town,
Singapore, São Paulo, Delhi, Tokyo, Mexico City

Cambridge University Press
The Edinburgh Building, Cambridge CB2 8RU, UK

Published in the United States of America by Cambridge University Press, New York

www.cambridge.org
Information on this title: www.cambridge.org/9780521290661

© Cambridge University Press 2009

This publication is in copyright. Subject to statutory exception
and to the provisions of relevant collective licensing agreements,
no reproduction of any part may take place without the written
permission of Cambridge University Press.

First published 2009
First paperback edition 2011

A catalogue record for this publication is available from the British Library

ISBN 978-0-521-19211-8 Hardback
ISBN 978-0-521-29066-1 Paperback

Cambridge University Press has no responsibility for the persistence or
accuracy of URLs for external or third-party internet websites referred to in
this publication, and does not guarantee that any content on such websites is,
or will remain, accurate or appropriate.

Contents

**Recent trends in the numerical solution
of retarded functional differential equations** 1
  A. Bellen, N. Guglielmi, S. Maset and M. Zennaro

Adaptivity with moving grids 111
  Chris J. Budd, Weizhang Huang and Robert D. Russell

**Fast direct solvers for integral equations in complex
three-dimensional domains** 243
  L. Greengard, D. Gueyffier, P.-G. Martinsson and V. Rokhlin

**Blow-up or no blow-up? A unified computational
and analytic approach to 3D incompressible Euler
and Navier–Stokes equations** 277
  Thomas Y. Hou

Recent trends in the numerical solution of retarded functional differential equations*

Alfredo Bellen, Stefano Maset and Marino Zennaro
Dipartimento di Matematica e Informatica,
Università degli Studi di Trieste, I-34100 Trieste, Italy
E-mail: {bellen}{maset}{zennaro}@units.it

Nicola Guglielmi
Dipartimento di Matematica Pura e Applicata,
Università degli Studi di L'Aquila, I-67100 L'Aquila, Italy
E-mail: guglielm@univaq.it

Retarded functional differential equations (RFDEs) form a wide class of evolution equations which share the property that, at any point, the rate of the solution depends on a discrete or distributed set of values attained by the solution itself in the past. Thus the initial problem for RFDEs is an infinite-dimensional problem, taking its theoretical and numerical analysis beyond the classical schemes developed for differential equations with no functional elements. In particular, numerically solving initial problems for RFDEs is a difficult task that cannot be founded on the mere adaptation of well-known methods for ordinary, partial or integro-differential equations to the presence of retarded arguments. Indeed, efficient codes for their numerical integration need specific approaches designed according to the nature of the equation and the behaviour of the solution.

By defining the numerical method as a suitable approximation of the solution map of the given equation, we present an original and unifying theory for the convergence and accuracy analysis of the approximate solution. Two particular approaches, both inspired by Runge–Kutta methods, are described. Despite being apparently similar, they are intrinsically different. Indeed, in the presence of specific types of functionals on the right-hand side, only one of them can have an explicit character, whereas the other gives rise to an overall procedure which is implicit in any case, even for non-stiff problems.

In the panorama of numerical RFDEs, some critical situations have been recently investigated in connection to specific classes of equations, such as the accurate location of discontinuity points, the termination and bifurcation of

---

* This work was supported by INdAM–GNCS.

the solutions of neutral equations, with state-dependent delays, the regularization of the equation and the generalization of the solution behind possible termination points, and the treatment of equations stated in the implicit form, which include singularly perturbed problems and delay differential-algebraic equations as well. All these issues are tackled in the last three sections.

In this paper we have not considered the important issue of stability, for which we refer the interested reader to the comprehensive book by Bellen and Zennaro (2003).

CONTENTS

1	Introduction	2
2	Some particular RFDEs	7
3	Discontinuity points and vanishing delays	11
4	Existence and uniqueness	15
5	Numerical methods for RFDEs	21
6	Functional continuous Runge–Kutta methods	32
7	Order conditions for FCRK methods	39
8	The standard approach	62
9	Implementation issues in the standard approach	80
10	Neutral problems with state-dependent delays	88
11	Implicit problems with state-dependent delays	102
	References	108

1. Introduction

In this paper, we present methods for numerically solving the *Cauchy problem*, or *initial problem* (IP), for the very general *retarded functional differential equation* (RFDE)

$$y'(t) = F(t, y_t), \qquad (1.1)$$

where y is an $\mathbb{R}^d$-valued function of a real variable, $F : \mathbb{R} \times X \to \mathbb{R}^d$, X being a subset of the set $\mathcal{C}$ of the continuous functions $(-\infty, 0] \to \mathbb{R}^d$, and, according to the Hale–Krasovski notation, $y_t \in X$ is given by

$$y_t(\vartheta) = y(t + \vartheta), \quad \vartheta \in (-\infty, 0].$$

The set X is called the *data set* of the RFDE (1.1) and the function y_t is called the *state* at time t, since, under minimal assumptions, it uniquely determines the future evolution $y(s)$, $s \geq t$.

In order to define the IP for RFDEs, we must associate to (1.1) an initial point $t_0 \in \mathbb{R}$ and initial data $\phi \in X$. The resulting problem takes the form

$$y'(t) = F(t, y_t), \quad t \geq t_0, \qquad (1.2)$$
$$y_{t_0} = \phi,$$

where the function $\phi \in X$ represents the initial state of the system.

Equation (1.1), also called the *Volterra functional differential equation*, provides a powerful tool for modelling many phenomena in applied mathematics and, in the literature, is often referred to by different terminology, such as *time delay system, hereditary system, system with memory, system with after-effect*, etc.

There are many kinds of RFDEs characterized by the action of the functional F on the state y_t. In particular, we include in (1.1) the *neutral functional differential equations*, where F also acts on the derivative of the state y_t, i.e.,

$$F(t, y_t) = G(t, y_t, y_t').$$

In any case, they are all evolution systems which share the property that, at any t, the dynamic depends not only on the current value $y(t)$, but also on a discrete or distributed set of values of the solution y in the past. This fact, together with the need for an initial function rather than an initial value, makes the theoretical analysis, as well as the numerical approximation of the IP (1.2), much more complicated than the initial value problem for ordinary differential equations it formally resembles.

The general theory of RFDEs is widely developed, and we refer the reader to the classical books by Bellman and Cooke (1963), El'sgol'ts and Norkin (1973), Hale (1977), Driver (1977), Kolmanovskii and Nosov (1986), Kolmanovskii and Myshkis (1992), Hale and Verduyn Lunel (1993), Kuang (1993) and Diekmann, van Gils, Verduyn Lunel and Walther (1995), which also include many real-life examples of RFDEs and more general retarded functional differential equations.

As for the numerics, apart from some isolated earlier papers, the analysis of numerical methods for RFDEs started in the early 1960s. Since then, specific methods have been separately developed by adapting the well-known methods for ordinary differential equations to the presence of delays. An exhaustive collection of methods and related references up to the beginning of 1970s was given by Cryer (1972). Other papers reporting the state of the art for general or particular classes of RFDEs appeared from time to time in the subsequent decades. In particular, Bellen (1985) and Meinardus and Nürnberger (1985) surveyed papers up to the 1980s, followed by Zennaro (1995), Baker, Paul and Willé (1995*a*, 1995*b*) and Baker (1996, 2000) up to the publication of the monograph by Bellen and Zennaro (2003), which was the first book completely devoted to the numerical analysis of the Cauchy problem for differential equations with delays. After some historical remarks on theoretical and numerical methods for RFDEs, the book provides a detailed analysis of continuous Runge–Kutta (RK) methods $(A, b(\theta), c)$, in view of their application in the following general procedure, called the *standard approach*, for IPs of the form

$$y'(t) = \hat{F}(t, y(t), y_t, y'_t), \quad t \geq t_0,$$
$$y_{t_0} = \phi, \tag{1.3}$$

where $\hat{F}$ explicitly separates the dependence on $y(t)$ from that on y_t.

Given a mesh $\Delta = \{t_0, t_1, \ldots, t_n, \ldots\}$, the standard approach for (1.3) consists in solving step by step, by means of a continuous RK method, the local problems

$$w'_{n+1}(t) = \hat{F}(t, w_{n+1}(t), x_t, x'_t), \quad t_n \leq t \leq t_{n+1},$$
$$w_{n+1}(t_n) = y_n, \tag{1.4}$$

where

$$x(s) = \begin{cases} \phi(s - t_0) & \text{for } s \leq t_0, \\ \eta(s) & \text{for } t_0 \leq s \leq t_n, \\ w_{n+1}(s) & \text{for } t_n \leq s \leq t_{n+1}, \end{cases}$$

and $\eta(s)$ is the continuous approximate solution computed by the method itself up to t_n.

The philosophy underlying the standard approach consists in considering (1.3) as an ODE, where the states x_t and x'_t, acting as forcing terms, are virtually known and given by the approximate solution itself, either having been or to be computed. It is clear that this approach relies on the availability of a *continuous numerical method*, that is, a numerical method which provides a continuous approximate solution. It is also clear that, whenever the right-hand side functional in (1.4) requires values of the functions x and x' at some points lying in the current interval, the method becomes implicit even if the underlying RK method is explicit. This makes the procedure more suited for stiff problems, for which the RK method is itself expected to be implicit. The standard approach, even before being so-named, was the most widely adopted method for RFDEs in the literature from the 1970s to the 1990s, also using continuous approximations other than continuous RK methods. In particular, for continuous RK methods, significant results on convergence, variable step-size implementation and stability analysis were achieved. A detailed presentation of such results up to 2002 is available in Bellen and Zennaro (2003), along with an exhaustive bibliography.

Since the publication of Bellen and Zennaro (2003), important work on the numerical solution of RFDEs has appeared. Part of it was devoted to further analyses of specific problems using well-known and consolidated techniques, especially as far as stability issues are concerned (see Wang and Li (2004) and the book by Kuang and Cong (2005)). Another part was devoted to developing new methods and to tackling some topics that, due to their more intrinsic difficulty, had not yet been well investigated, namely equations of neutral type and equations with state-dependent delays.

This paper, rather than reporting the state of the art, aims at providing, in an original and unifying approach, results on well-posedness and the error analysis of different numerical schemes based on continuous RK methods for as large as possible classes of RFDEs. The paper also reports some recent results on specific issues that needed, and still need, further investigation. In particular, we consider the termination and bifurcation of the solution at some critical point for RFDEs of neutral type, with state-dependent delays and the possible generalization of the solution beyond such points as well.

The paper is organized as follows. In Section 2 we provide some particular classes of RFDEs together with suitable data sets X, where they are naturally defined and where the true and the approximate solutions will be sought. For other classes of RFDEs and bibliographic references, see also Brunner (2004).

In Section 3 we analyse a specific phenomenon, typical of RFDEs, which is not present in ordinary differential equations, namely the appearance of so-called *discontinuity points*, often called *breaking points*. These originate in the possible lack of continuity in the derivative of the solution of (1.1) at the initial point t_0, that is,

$$\phi'^-(0) \neq y'(t_0)^+ = F(t_0, \phi).$$

This event implies that the solution, as defined in the forthcoming Definition 1.1, must be considered in the 'almost everywhere' sense.

In Section 4 we provide a short review of existence and uniqueness results for the solutions of the various classes of equations considered in Section 2. In order to do that, it is essential to establish what we actually mean by a solution of (1.2) in a right neighbourhood of t_0.

Definition 1.1. Let $T > 0$. A solution of (1.2) on $(-\infty, t_0 + T]$ is a continuous function $y : (-\infty, t_0 + T] \to \mathbb{R}^d$ such that:

- $y_t \in X$ for all $t \in [t_0, t_0 + T]$;
- the function $t \mapsto F(t, y_t)$, $t \in [t_0, t_0 + T]$, is measurable and bounded;
- for all $t \in (-\infty, t_0 + T]$, we have

$$y(t) = \begin{cases} \phi(0) + \int_{t_0}^t F(s, y_s)\,ds & \text{if } t \in [t_0, t_0 + T], \\ \phi(t - t_0) & \text{if } t \in (-\infty, t_0]. \end{cases}$$

Note that the third condition in the above definition is equivalent to requiring that y is differentiable almost everywhere in $[t_0, t_0 + T]$, $y'(t) = F(t, y_t)$ for almost all $t \in [t_0, t_0 + T]$ and $y_{t_0} = \phi$.

In Section 5 we develop an original and unifying approach which allows us to analyse well-posedness and convergence for most of the methods developed so far for the whole class of equations stated in the form (1.2).

Sections 6, 7 and 8 are devoted to the construction and accuracy analysis of two classes of methods, both based on suitable continuous RK methods $(A, b(\theta), c)$, for which the continuous approximation $\eta(s)$ is given step by step, by

$$\eta(t_n+\theta h_{n+1}) = y_n + h_{n+1} \sum_{i=1}^{\nu} b_i(\theta) F\left(t_{n+1}^i, Y_{t_{n+1}^i}, Y'_{t_{n+1}^i}\right), \quad 0 \le \theta \le 1, \quad (1.5)$$

for the problem (1.2), and by

$$\eta(t_n+\theta h_{n+1}) = y_n + h_{n+1} \sum_{i=1}^{\nu} b_i(\theta) \hat{F}\left(t_{n+1}^i, Y_{n+1}^i, Y_{t_{n+1}^i}, Y'_{t_{n+1}^i}\right), \quad 0 \le \theta \le 1, \quad (1.6)$$

for the class of problems in the form (1.3).

The stages Y_{n+1}^i in (1.6) are the classical stage values of the RK method, whereas the stages $Y_{t_{n+1}^i}$ in (1.5) and (1.6) are *states* and the two methods differ from each other in how they are defined.

In Section 6 we consider the first class of methods, called *functional continuous Runge–Kutta* (FCRK) methods and denoted by $(A(\theta), b(\theta), c)$, where, for each i, the stage $Y_{t_{n+1}^i}$ is a polynomial determined by the coefficients $a_{ij}(\theta)$. A particular class of FCRK methods based on a predictor–corrector version of the collocation method, proposed by Tavernini (1971), is reported as a prototype of the class. Although FCRK methods seem the most natural and direct way to extend RK formulas to RFDEs (1.1), they were neglected for a long time and have been investigated, in their general form, only recently; see the error analysis by Maset, Torelli and Vermiglio (2005). The merit of such a class of methods is that, contrary to the standard approach, they are available in explicit form when values of y_t or y'_t are required at points of the current integration step. In this section the methods are introduced and their well-posedness, proved by Maset (2009), is reported.

In Section 7 the error analysis and order conditions are developed for the class of FCRK methods and explicit schemes of order up to four are constructed.

In Section 8 we consider the second class of methods, namely the standard approach based on continuous RK methods as described in (1.4) for initial problems of the form (1.3). In this case all the states $Y_{t_{n+1}^i}$ are given by the same function η, i.e., $Y_{t_{n+1}^i} = \eta_{t_{n+1}^i}$ for all i. The numerical analysis of this approach is now consolidated and available in the cited book by Bellen and Zennaro (2003). Here we report some general results on the discrete and uniform order of continuous RK methods and on the corresponding methods for the solution of (1.3). These results also serve as a background for the subsequent sections, where only the standard approach is considered.

In Section 9 we address some implementation problems arising in the use of the standard approach, in connection with the accurate computation of the breaking points. In particular, we face the paradoxical situation caused by the state-dependent delays, where the accuracy in the calculation of a breaking point depends on the accuracy of the approximate solution used to detect it, which, in turn, depends on the accuracy of the same breaking point we are trying to locate. The strategy adopted in the earlier and in the last releases of the code RADAR5 by Guglielmi and Hairer (2001, 2008) is described with details and numerical comparisons.

In Section 10 we consider RFDEs of neutral type with state-dependent delays for which the derivative of the solution is discontinuous at the initial point t_0, that is,
$$\phi'^{-}(0) \neq y'(t_0)^{+} = G(t_0, \phi, \phi').$$
Such an inequality produces a sequence of breaking points where, the delay being state-dependent, the solution may either cease to exist or bifurcate. These occurrences are investigated from both the theoretical and numerical point of view. Possible regularizations of the equation, leading to weak (or generalized) solutions defined beyond such termination and bifurcation points, are proposed and compared.

Finally, in Section 11, we address our attention to a special class of state-dependent problems in the implicit form
$$M u'(t) = F\bigl(u(t), u(\alpha(u(t)))\bigr),$$
where the matrix M is constant, and possibly singular. Besides including neutral state-dependent RFDEs, such problems also include singularly perturbed problems and a variety of delay differential-algebraic equations. Since these problems often have a stiff character, they usually need to be integrated by an implicit method, and therefore a Newton or quasi-Newton iterative process is needed. The efficient implementation of such iterations is investigated in detail in the case of overlapping.

2. Some particular RFDEs

Now we introduce some particular and important RFDEs (1.1). In our presentation, we divide the RFDEs into two classes defined by different data sets and corresponding to non-neutral and neutral types.

Let us first consider RFDEs (1.1) with data set $X = \mathcal{C}$.

- *Delay differential equations* (DDEs):
$$y'(t) = f\bigl(t, y(t), y(t - \tau_1(t)), \ldots, y(t - \tau_s(t))\bigr), \qquad (2.1)$$

where $f : \mathbb{R} \times \mathbb{R}^d \times (\mathbb{R}^d)^s \to \mathbb{R}^d$ and $\tau_i : \mathbb{R} \to [0, +\infty)$, $i = 1, \ldots, s$. The functions τ_i, $i = 1, \ldots, s$, are called *delays*. For such equations, the functional F in (1.1) is given by

$$F(t, \varphi) = f(t, \varphi(0), \varphi(-\tau_1(t)), \ldots, \varphi(-\tau_s(t))), \quad (t, \varphi) \in \mathbb{R} \times \mathcal{C}. \quad (2.2)$$

- *Delay integro-differential equations* (DIDEs):

$$y'(t) = f\left(t, y(t), \int_{t-\tau_1(t)}^{t-\tau_2(t)} k(t, t-s, y(s)) \, ds\right), \quad (2.3)$$

where $f : \mathbb{R} \times \mathbb{R}^d \times \mathbb{R}^d \to \mathbb{R}^d$, $\tau_1, \tau_2 : \mathbb{R} \to [0, +\infty)$ and $k : \mathbb{R} \times (0, +\infty) \times \mathbb{R}^d \to \mathbb{R}^d$. The functions τ_i, $i = 1, 2$, are the delays and the function k is called the *kernel*. The functional F takes the form

$$F(t, \varphi) = f\left(t, \varphi(0), \int_{-\tau_1(t)}^{-\tau_2(t)} k(t, -\vartheta, \varphi(\vartheta)) \, d\vartheta\right), \quad (t, \varphi) \in \mathbb{R} \times \mathcal{C}. \quad (2.4)$$

We assume that:

(K) The kernel k is measurable and, for some norm $|\cdot|$ on $\mathbb{R}^d$, for any bounded subset B of $\mathbb{R} \times \mathbb{R}^d$, the function M_B given by

$$M_B(\vartheta) = \sup_{(t,y) \in B} |k(t, -\vartheta, y)|, \quad \vartheta \in (-\infty, 0),$$

is locally integrable.

Under assumption (K), the integral in (2.4) exists and is finite for any $(t, \varphi) \in \mathbb{R} \times \mathcal{C}$. Such an assumption is satisfied if the kernel is continuous or weakly singular, *i.e.*,

$$k(t, x, y) = x^{-\alpha} \cdot a(t, x, y), \quad (t, x, y) \in \mathbb{R} \times (0, +\infty) \times \mathbb{R}^d,$$

or

$$k(t, x, y) = \log x \cdot a(t, x, y), \quad (t, x, y) \in \mathbb{R} \times (0, +\infty) \times \mathbb{R}^d,$$

where $\alpha \in [0, 1)$ and $a : \mathbb{R} \times [0, +\infty) \times \mathbb{R}^d \to \mathbb{R}^d$ is continuous.

Other RFDEs (1.1) with data set $\mathcal{C}$ are DDEs and DIDEs, where the delays also depend on the value $y(t)$.

- *State-dependent delay differential equations* (SDDDEs):

$$y'(t) = f(t, y(t), y(t - \tau_1(t, y(t))), \ldots, y(t - \tau_s(t, y(t)))), \quad (2.5)$$

where $\tau_i : \mathbb{R} \times \mathbb{R}^d \to [0, +\infty)$, $i = 1, \ldots, s$, and

$$F(t, \varphi) = f(t, \varphi(0), \varphi(-\tau_1(t, \varphi(0))), \ldots, \varphi(-\tau_s(t, \varphi(0)))), \quad (2.6)$$
$$\text{for } (t, \varphi) \in \mathbb{R} \times \mathcal{C}.$$

- *State-dependent delay integro-differential equations* (SDDIDEs):

$$y'(t) = f\left(t, y(t), \int_{t-\tau_1(t,y(t))}^{t-\tau_2(t,y(t))} k(t, t-s, y(s))\, ds\right), \qquad (2.7)$$

where $\tau_1, \tau_2 : \mathbb{R} \times \mathbb{R}^d \to [0, +\infty)$ and

$$F(t, \varphi) = f\left(t, \varphi(0), \int_{-\tau_1(t,\varphi(0))}^{-\tau_2(t,\varphi(0))} k(t, -\vartheta, \varphi(\vartheta))\, d\vartheta\right), \quad (t, \varphi) \in \mathbb{R} \times \mathcal{C}.$$

Another possible choice of the data set X is the set $\mathcal{LC}$ of the locally Lipschitz-continuous functions $(-\infty, 0] \to \mathbb{R}^d$, which are known to be differentiable *almost everywhere*. For $\varphi \in \mathcal{LC}$, by defining

$$\varphi'(\vartheta) = \frac{1}{2}\left(\limsup_{h \to 0} \frac{\varphi(\vartheta + h) - \varphi(\vartheta)}{h} + \liminf_{h \to 0} \frac{\varphi(\vartheta + h) - \varphi(\vartheta)}{h}\right)$$

at any point $\vartheta \in (-\infty, 0]$, the derivative φ' belongs to the set $\mathcal{B}$ of the measurable and locally bounded functions $(-\infty, 0] \to \mathbb{R}^d$.

Particular RFDEs (1.1) with the data set $\mathcal{LC}$ are the *neutral functional differential equations* (NFDEs),

$$y'(t) = G(t, y_t, y'_t), \qquad (2.8)$$

where $G : \mathbb{R} \times \mathcal{C} \times \mathcal{B} \to \mathbb{R}^d$ and the functional F in (1.1) is given by

$$F(t, \varphi) = G(t, \varphi, \varphi'), \quad (t, \varphi) \in \mathbb{R} \times \mathcal{LC}.$$

Particular examples of NFDEs are as follows.

- *Neutral delay differential equations* (NDDEs):

$$y'(t) = f\Big(t, y(t), y(t - \tau_1(t)), \ldots, y(t - \tau_s(t)),$$
$$y'(t - \tau_1^*(t)), \ldots, y'(t - \tau_{s^*}^*(t))\Big), \qquad (2.9)$$

where $f : \mathbb{R} \times \mathbb{R}^d \times (\mathbb{R}^d)^s \times (\mathbb{R}^d)^{s^*} \to \mathbb{R}^d$, $\tau_i : \mathbb{R} \to [0, +\infty)$, $i = 1, \ldots, s$, and $\tau_i^* : \mathbb{R} \to [0, +\infty)$, $i = 1, \ldots, s^*$. The functional G in (2.8) is given by

$$G(t, \varphi, \psi) = f\Big(t, \varphi(0), \varphi(-\tau_1(t)), \ldots, \varphi(-\tau_s(t)),$$
$$\psi(-\tau_1^*(t)), \ldots, \psi(-\tau_{s^*}^*(t))\Big),$$
$$\text{for } (t, \varphi, \psi) \in \mathbb{R} \times \mathcal{C} \times \mathcal{B}.$$

- *Neutral delay integro-differential equations* (NDIDEs):

$$y'(t) = f\left(t, y(t), \int_{t-\tau_1(t)}^{t-\tau_2(t)} k(t, t-s, y(s), y'(s))\, ds\right), \qquad (2.10)$$

where $f: \mathbb{R} \times \mathbb{R}^d \times \mathbb{R}^d \to \mathbb{R}^d$, $\tau_1, \tau_2 : \mathbb{R} \to [0, +\infty)$ and $k : \mathbb{R} \times (0, +\infty) \times \mathbb{R}^d \times \mathbb{R}^d \to \mathbb{R}^d$. The functional G takes the form

$$G(t, \varphi, \psi) = f\left(t, \varphi(0), \int_{-\tau_1(t)}^{-\tau_2(t)} k(t, -\vartheta, \varphi(\vartheta), \psi(\vartheta)) \, d\vartheta\right),$$

for $(t, \varphi, \psi) \in \mathbb{R} \times \mathcal{C} \times \mathcal{B}$.

As for the non-neutral case, we assume that:

(NK) The kernel k is measurable, and, for some norm $|\cdot|$ on $\mathbb{R}^d$, for any bounded subset B of $\mathbb{R} \times \mathbb{R}^d \times \mathbb{R}^d$, the function M_B given by

$$M_B(\vartheta) = \sup_{(t,y,z) \in B} |k(t, -\vartheta, y, z)|, \quad \vartheta \in (-\infty, 0),$$

is locally integrable.

Other RFDEs with data set $\mathcal{LC}$ are NDDEs and NDIDEs with state-dependent delays.

- *Neutral state-dependent delay differential equations* (NSDDDEs):

$$y'(t) = f\Big(t, y(t), y(t - \tau_1(t, y(t))), \ldots, y(t - \tau_s(t, y(t))),$$
$$y'(t - \tau_1^*(t, y(t))), \ldots, y'(t - \tau_{s^*}^*(t, y(t)))\Big), \quad (2.11)$$

where $\tau_i : \mathbb{R} \times \mathbb{R}^d \to [0, +\infty)$, $i = 1, \ldots, s$, $\tau_i^* : \mathbb{R} \times \mathbb{R}^d \to [0, +\infty)$, $i = 1, \ldots, s^*$, and

$$G(t, \varphi, \psi) = f\Big(t, \varphi(0), \varphi(-\tau_1(t, \varphi(0))), \ldots, \varphi(-\tau_s(t, \varphi(0))),$$
$$\psi(-\tau_1^*(t, \varphi(0))), \ldots, \psi(-\tau_{s^*}^*(t, \varphi(0)))\Big),$$

for $(t, \varphi, \psi) \in \mathbb{R} \times \mathcal{C} \times \mathcal{B}$.

- *Neutral state-dependent delay integro-differential equations* (NSDDIDEs):

$$y'(t) = f\left(t, y(t), \int_{t-\tau_1(t,y(t))}^{t-\tau_2(t,y(t))} k(t, t-s, y(s), y'(s)) \, ds\right), \quad (2.12)$$

where $\tau_1, \tau_2 : \mathbb{R} \times \mathbb{R}^d \to [0, +\infty)$ and

$$G(t, \varphi, \psi) = f\left(t, \varphi(0), \int_{-\tau_1(t,\varphi(0))}^{-\tau_2(t,\varphi(0))} k(t, -\vartheta, \varphi(\vartheta), \psi(\vartheta)) \, d\vartheta\right), \quad (2.13)$$

for $(t, \varphi, \psi) \in \mathbb{R} \times \mathcal{C} \times \mathcal{B}$.

3. Discontinuity points and vanishing delays

In this section we describe two particular situations caused by the presence of two kinds of points: those where some derivative of the solution is not continuous and those where the delay vanishes.

3.1. Discontinuity points

In this section we briefly analyse the propagation of *discontinuity points*, also called *breaking points*, for the derivatives of the solution of (1.2) along the integration interval.

In order to illustrate this phenomenon, for the sake of simplicity we confine ourselves to the particular class of DDEs,

$$y'(t) = f(t, y(t), y(t - \tau(t, y(t)))), \quad t_0 \leq t \leq t_0 + T,$$
$$y(t) = \phi(t), \quad t \leq t_0, \tag{3.1}$$

and NDDEs,

$$y'(t) = f(t, y(t), y(t - \tau(t, y(t)))), y'(t - \tau(t, y(t)))), \quad t_0 \leq t \leq t_0 + T,$$
$$y(t) = \phi(t), \quad t \leq t_0. \tag{3.2}$$

First consider equation (3.1) and assume that the *deviated argument*

$$\alpha(t) = t - \tau(t, y(t))$$

satisfies $\alpha(t) < t_0$ for some points $t \in [t_0, t_0 + T]$. Moreover, assume that the solution $y(t)$ does not link smoothly to the initial function $\phi(t)$ at t_0, that is,

$$\phi'(t_0)^- \neq y'(t_0)^+ = f(t_0, \phi(t_0), \phi(\alpha(t_0))).$$

If the functions f, ϕ and α are continuous, then it is obvious that $y'(t)$ is also continuous for any $t > t_0$. On the other hand, if f, ϕ and α are differentiable, then $y''(t)$ exists for any t except for the points $\xi_{1,i} (> t_0)$ such that

$$\alpha(\xi_{1,i}) = t_0$$

and

$$\alpha'(\xi_{1,i}) \neq 0,$$

i.e., for the simple roots, if any, of the equation

$$\alpha(t) = t_0.$$

In fact, for any smooth function $f(t, y, x)$ we can formally write

$$y''(t)^{\pm} = \frac{\partial f}{\partial t}(t, y(t), y(\alpha(t))) + \frac{\partial f}{\partial y}(t, y(t), y(\alpha(t)))y'(t)$$
$$+ \frac{\partial f}{\partial x}(t, y(t), y(\alpha(t)))y'(\alpha(t))^{\pm}\alpha'(t), \tag{3.3}$$

and hence
$$y''(\xi_{1,i})^+ = \frac{\partial f}{\partial t}(\xi_{1,i}, y(\xi_{1,i}), y(t_0)) + \frac{\partial f}{\partial y}(\xi_{1,i}, y(\xi_{1,i}), y(t_0))y'(\xi_{1,i})$$
$$+ \frac{\partial f}{\partial x}(\xi_{1,i}, y(\xi_{1,i}), y(t_0))y'(t_0)^+ \alpha'(\xi_{1,i}) \tag{3.4}$$

and
$$y''(\xi_{1,i})^- = \frac{\partial f}{\partial t}(\xi_{1,i}, y(\xi_{1,i}), y(t_0)) + \frac{\partial f}{\partial y}(\xi_{1,i}, y(\xi_{1,i}), y(t_0))y'(\xi_{1,i})$$
$$+ \frac{\partial f}{\partial x}(\xi_{1,i}, y(\xi_{1,i}), y(t_0))\phi'(t_0)^- \alpha'(\xi_{1,i}). \tag{3.5}$$

Since $\alpha'(\xi_{1,i}) \neq 0$ and $\phi'(t_0)^-$ is assumed to be different from $y'(t_0)^+$, y'' does not exist at $\xi_{1,i}$ and its prolongation by $y''(\xi_{1,i}) = y''(\xi_{1,i})^+$ has a jump discontinuity at $\xi_{1,i}$.

These jump discontinuities in y'' are called 1-*level primary discontinuities*. By differentiating (3.3), one easily checks that each 1-level primary discontinuity point $\xi_{1,i}$ gives rise in turn to 2-*level primary discontinuities* in y''' at any point $\xi_{2,j}(>\xi_{1,i})$ which is a simple root of

$$\alpha(t) = \xi_{1,i} \quad \text{for some } i.$$

In general, any k-*level primary discontinuity* point $\xi_{k,i}$ gives rise to $(k+1)$-*level primary discontinuities* in $y^{(k+2)}$ at subsequent points $\xi_{k+1,j}$, where the solution of (3.1) becomes increasingly smooth as the primary discontinuity level increases. This increase in the regularity of $y(t)$ will be referred to as *smoothing of the solution*.

Definition 3.1. Every point where some derivative $y^{(s)}$ jumps will be called a *discontinuity point* or *breaking point*. We also say that a breaking point ξ has order k if the solution is C^k-continuous at ξ. In particular, by $k = -1$ we mean that the solution is discontinuous at ξ.

On the contrary, the same argument applied to (3.2) reveals that, for neutral DDEs, smoothing does not occur and, in general, the solution remains C^0-continuous at any primary discontinuity point where the derivative y' jumps. This motivates the weaker definition of the solution in the 'almost everywhere' sense given in Definition 1.1. Obviously, if the splicing condition
$$\phi'(t_0)^- = y'(t_0)^+ = f(t_0, \phi(t_0), \phi(\alpha(t_0)))$$
holds, no discontinuities propagate from t_0 and the solution is meant in the classical sense.

It is also remarkable that, if $\alpha(t) \geq t_0$ for all $t \geq t_0$, then no values of y and/or y' are needed in (3.1) and (3.2) behind t_0 and, therefore, no primary discontinuities propagate from t_0.

Other discontinuities can appear if the functions f, τ and ϕ in (3.1) and (3.2) have some discontinuities with respect to t in some of their derivatives. Then such discontinuities are also propagated by the deviated argument $\alpha(t)$, according to the primary discontinuity propagation rule, and are called *secondary discontinuities*.

From the numerical point of view, it is important to analyse how the discontinuity points propagate through the integration interval $[t_0, t_0+T]$, and how smoothness possibly increases at any discontinuity point with respect to its *ancestor*, the discontinuity point from which it originates. In fact, it is known that every step-by-step numerical method for initial value problems (IVPs) achieves its own accuracy order provided that the solution is sufficiently smooth at each step interval $[t_n, t_{n+1}]$. More precisely, for a method to be of order p, we usually ask the solution to be at least C^{p+1}-continuous on $[t_n, t_{n+1}]$. Therefore, discontinuity points of a suitable level ought to be included in the mesh.

To finish with, we briefly consider the difficulties related to the case when the delay $\tau(t, y(t))$ is state-dependent. In order to locate the discontinuities, one should in principle apply the general propagation rule

$$\xi_{k,j} - \tau(\xi_{k,j}, y(\xi_{k,j})) = \xi_{k-1,i} \quad \text{for some } i, \tag{3.6}$$

and solve it for $\xi_{k,j}$. Because the delay is dependent on $y(t)$, this cannot be done *a priori* without any knowledge of the solution. Moreover, it is evident that, even assuming some approximation of $y(t)$ is available, we must be satisfied with an approximation of the discontinuity point $\xi_{k,j}$.

In conclusion, the impossibility of locating the discontinuity points *a priori* makes the implementation and convergence analysis of numerical methods for (3.1) and (3.2) a rather complicated task, which will be further investigated in Section 9.

For a more detailed analysis of the propagation of discontinuity points in systems of DDEs with different delays we refer the reader to Bellen and Zennaro (2003) and the references therein.

3.2. Non-vanishing delays

The theoretical analysis of (1.2) for the classes of RFDEs considered in Section 2, as well as the development of numerical methods for its approximate solution, is considerably simplified if the problem reduces to a finite sequence of IVPs for ordinary differential equations (ODEs) on any interval $[t_0, t_0+T]$. In order to characterize such an occurrence, let us consider the following condition on a delay τ.

(H_1^*) There exists a constant $\tau_0 > 0$ such that $\tau(t) \geq \tau_0$ for all $t \in \mathbb{R}$ or, for state-dependent delays, $\tau(t, z) \geq \tau_0$ for all $t \in \mathbb{R}$ and $z \in \mathbb{R}^d$.

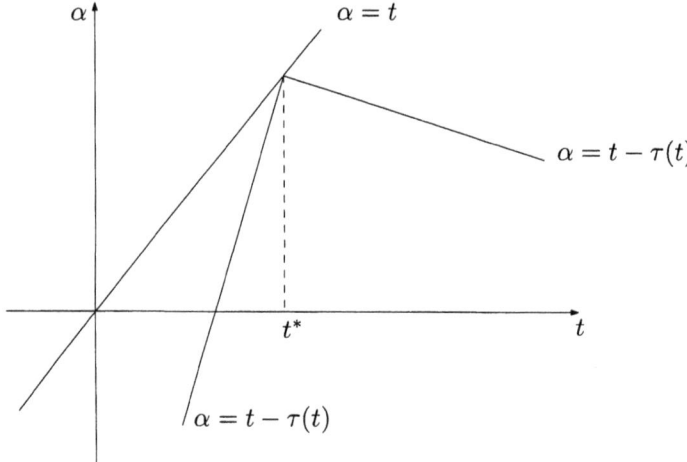

Figure 3.1. An example of a vanishing delay satisfying condition (HH_1^*).

A delay satisfying the condition (H_1^*) is said to be *non-vanishing*; otherwise it is called *vanishing*. It is evident that if all the delays in the given equation are non-vanishing, then the IP reduces on any interval to a finite sequence of IVPs for ODEs.

However, to be non-vanishing is not a necessary condition for the delays in order to reduce the problem to a finite sequence of ordinary ones. In fact, this occurs if and only if any delay τ fulfils this weaker condition.

(HH_1^*) For any $\sigma \in \mathbb{R}$ and $T > 0$, there exist $\sigma_0, \sigma_1, \ldots, \sigma_{K-1}, \sigma_K$ such that
$$\sigma = \sigma_0 < \sigma_1 < \cdots < \sigma_{K-1} < \sigma_K = \sigma + T$$
and
$$t - \tau(t) \leq \sigma_k, \quad t \in [\sigma_k, \sigma_{k+1}], \quad \text{and} \quad k = 0, 1, \ldots, K-1,$$
or
$$t - \tau(t, z) \leq \sigma_k, \quad t \in [\sigma_k, \sigma_{k+1}], \quad z \in \mathbb{R}^d \quad \text{and} \quad k = 0, 1, \ldots, K-1,$$
in the case of a state-dependent delay.

A delay τ satisfying (HH_1^*) is called *weakly non-vanishing*. A delay which is not weakly non-vanishing is called *strongly vanishing*. An example of a vanishing delay which is weakly non-vanishing is $\tau(t) = t - [t]$, where $[\cdot]$ denotes the greatest integer function. Such a delay occurs in some retarded differential real-life models (see Cooke and Wiener (1984)). On the other hand, an example of a vanishing delay τ which is strongly vanishing is depicted in Figure 3.1, where $\tau(t^*) = 0$. For an equation involving such a delay, an IP (1.2) with $t_0 < t^*$ may reduce to a finite sequence of IVPs for ODEs only on intervals $[t_0, t_0 + T]$ where $t_0 + T < t^*$.

Whether or not the problem is reducible to a sequence of IVPs for ODEs is reflected in the different hypotheses required for the existence and uniqueness of the solution.

Two significant cases of equations involving a strongly vanishing delay are the classical Volterra equation

$$y'(t) = f\left(t, y(t), \int_0^t k(t, t-s, y(s))\, ds\right), \quad t \geq 0,$$

and the *pantograph equation*

$$y'(t) = f(t, y(t), y(qt)), \quad t \geq 0, \quad 0 < q < 1.$$

The former corresponds to equation (2.3), with $\tau_1(t) = t$ and $\tau_2(t) = 0$ at any point t; the latter corresponds to (2.1), with $\tau_1(t) = (1-q)t$, which equals zero only at $t = 0$.

4. Existence and uniqueness

Now, for the different types of RFDEs we give conditions under which the IP (1.2) has a *unique local solution*, that is, there exists $T > 0$ such that the IP has a unique solution on $(-\infty, t_0 + T]$. Most of the results of this and the next section are taken from Maset (2009). To this end, we let L_T denote the linear space $L^\infty([0,T], \mathbb{R}^d)$, i.e., the linear space of the equivalence classes of the Lebesgue-measurable and essentially bounded functions $[0,T] \to \mathbb{R}^d$, and let $L_T^\diamond$ be the linear space of the Lebesgue-measurable and bounded functions $[0,T] \to \mathbb{R}^d$. Note that $L_T^\diamond$ can be embedded in L_T by identifying any function with its equivalence class.

Moreover, we assume the data set X satisfies the following assumptions.

(DS1) For any $\varphi \in X$ and $s \in (-\infty, 0]$, we have $\varphi_s \in X$.

(DS2) For any $\varphi \in X$, $T > 0$ and $z \in L_T$, we have $v(\varphi, z)_T \in X$, where $v(\varphi, z) : (-\infty, T] \to \mathbb{R}^d$ is the continuous function given by

$$v(\varphi, z)(t) = \begin{cases} \varphi(0) + \int_0^t z(s)\, ds & \text{if } t \in [0, T], \\ \varphi(t) & \text{if } t \in (-\infty, 0]. \end{cases} \quad (4.1)$$

Since assumption (DS1) holds, we also have

$$v(\varphi, z)_t = \bigl(v(\varphi, z)_T\bigr)_{t-T} \in X, \quad t \leq T, \quad (4.2)$$

in (DS2).

It is clear that both data sets $\mathcal{C}$ and $\mathcal{LC}$ satisfy (DS1) and (DS2).

Finally, as well as assumptions (DS1) and (DS2) on the data set, we assume that the RFDE (1.1) satisfies the following *Boundedness Assumption*.

(BA) For any $\sigma \in \mathbb{R}$, $\varphi \in X$, $T > 0$ and $z \in L_T$, the function
$$t \mapsto F(\sigma + t, v(\varphi, z)_t), \quad t \in [0, T],$$
belongs to $L_T^\diamond$.

For the various types of RFDEs presented above, the assumption (BA) holds under minimal conditions. This is stated in the following two propositions.

Proposition 4.1. An NSDDDE (an NDDE) satisfies property (BA) if:

- the function f is measurable and locally bounded;
- the delays τ_j, $j = 1, \ldots, s$, and τ_j^*, $j = 1, \ldots, s^*$, are measurable and locally bounded.

An analogous proposition holds for SDDDEs and DDEs (which are not particular NSDDDEs or NDDEs, since the data set for the former is larger).

Proposition 4.2. An NSDDIDE (an NDIDE) satisfies property (BA) if:

- the function f is measurable and locally bounded;
- the delays τ_j, $j = 1, 2$, are measurable and locally bounded;
- the kernel k satisfies assumption (NK).

A similar proposition holds for SDDIDEs and DIDEs.

Since the Boundedness Assumption holds, for given $\sigma \in \mathbb{R}$, $\varphi \in X$ and $T > 0$, we can introduce the map
$$Q_T(\sigma, \varphi) : L_T \to L_T^\diamond$$
defined by
$$[Q_T(\sigma, \varphi)(z)](t) = F(\sigma + t, v(\varphi, z)_t), \quad t \in [0, T] \text{ and } z \in L_T. \quad (4.3)$$
For the analysis of the numerical methods introduced in the next section, it is useful to introduce the map
$$Q_T^\diamond(\sigma, \varphi) = Q_T(\sigma, \varphi)|_{C_T} : C_T \to L_T^\diamond, \quad (4.4)$$
where C_T is the subspace of L_T (and of $L_T^\diamond$) of the continuous functions.

In the following, we consider $Q_T(\sigma, \varphi)$ as a map $L_T \to L_T$ by embedding $L_T^\diamond$ in L_T.

There is a link between the map $Q_T(t_0, \phi)$ and a solution of (1.2) on $(-\infty, t_0 + T]$, which is given in the following basic theorem.

Theorem 4.1. Let $y : (-\infty, t_0 + T] \to \mathbb{R}^d$ and let x be the shift function given by
$$x(t) = y(t_0 + t), \quad t \in (-\infty, T]. \quad (4.5)$$

The function y is a solution of the IP (1.2) on $(-\infty, t_0 + T]$ if and only if
$$x = v(\phi, z^*)$$
for some fixed point z^* of the map $Q_T(t_0, \phi)$.

Proof. Let $y : (-\infty, t_0 + T] \to \mathbb{R}^d$ be a solution of the IP (1.2). Then the shift function x is continuous, $x_t \in X$ for all $t \in [0, T]$, the function
$$z^*(t) = F(t_0 + t, x_t), \quad t \in [0, T],$$
belongs to L_T° (and, then, to L_T) and, for $t \in (-\infty, T]$, we have
$$x(t) = \begin{cases} \phi(0) + \int_0^t F(t_0 + s, x_s)\, ds & \text{if } t \in [0, T], \\ \phi(t) & \text{if } t \in (-\infty, 0] \end{cases}$$
$$= v(\phi, z^*)(t).$$

Since
$$[Q_T(t_0, \phi)(z^*)](t) = F(t_0 + t, v(\phi, z^*)_t)$$
$$= F(t_0 + t, x_t) = z^*(t), \quad t \in [0, T],$$
z^* turns out to be a fixed point of the map $Q_T(t_0, \phi)$.

Vice versa, let $z^* \in L_T$ be a fixed point of the map $Q_T(t_0, \phi)$, and let
$$x(t) = v(\phi, z^*)(t), \quad t \in (-\infty, T].$$
The function x is continuous, $x_t = v(\phi, z^*)_t \in X$ for all $t \in [0, T]$ (recall (4.2)), the function
$$t \mapsto F(t_0 + t, x_t) = F(t_0 + t, v(\phi, z^*)_t), \quad t \in [0, T],$$
belongs to L_T° by the Boundedness Assumption, and, for $t \in (-\infty, T]$, we have
$$x(t) = v(\phi, z^*)(t)$$
$$= v(\phi, Q_T(t_0, \phi)(z^*))(t)$$
$$= \begin{cases} \phi(0) + \int_0^t [Q_T(t_0, \phi)(z^*)](s)\, ds & \text{if } t \in [0, T], \\ \phi(t) & \text{if } t \in (-\infty, 0] \end{cases}$$
$$= \begin{cases} \phi(0) + \int_0^t F(t_0 + s, v(\phi, z^*)_t)(s)\, ds & \text{if } t \in [0, T], \\ \phi(t) & \text{if } t \in (-\infty, 0] \end{cases}$$
$$= \begin{cases} \phi(0) + \int_0^t F(t_0 + s, x_s)\, ds & \text{if } t \in [0, T], \\ \phi(t) & \text{if } t \in (-\infty, 0]. \end{cases}$$

Hence,
$$y(t) = x(t - t_0), \quad t \in (-\infty, t_0 + T],$$
is a solution of (1.2) on $(-\infty, t_0 + T]$. $\square$

By the previous result, it is clear that there exists a unique solution of (1.2) on $(-\infty, t_0 + T]$ if and only if the map $Q_T(t_0, \phi)$ has a unique fixed point.

The reduction of the problem of existence and uniqueness of a solution on $(-\infty, t_0 + T]$ to the existence and uniqueness of fixed points of the map $Q_T(t_0, \phi)$ allows us to prove the following theorems.

Theorem 4.2. Consider a DDE (2.1). If:

- the function $f(t, y_0, y_1, \ldots, y_s)$ is measurable, locally bounded, locally Lipschitz-continuous with respect to the argument y_0 and locally Lipschitz-continuous with respect to those arguments y_j, $j = 1, \ldots, s$, such that the delay τ_j is strongly vanishing;
- the delays τ_j, $j = 1, \ldots, s$, are measurable and locally bounded;

then any IP (1.2) for the DDE has a unique local solution $\forall \phi \in \mathcal{C}$.

Theorem 4.3. Consider an SDDDE (2.5). If:

- the function $f(t, y_0, y_1, \ldots, y_s)$ is measurable, locally bounded and locally Lipschitz-continuous with respect to any argument y_j, $j = 0, 1, \ldots, s$;
- the delays $\tau_j(t, y)$, $j = 1, \ldots, s$, are measurable, locally bounded and locally Lipschitz-continuous with respect to the argument y;

then any IP (1.2) for the SDDDE has a unique local solution $\forall \phi \in \mathcal{LC}$.

Theorem 4.4. Consider an NDDE (2.9). If:

- the function $f(t, y_0, y_1, \ldots, y_s, z_1, \ldots, z_{s^*})$ is measurable, locally bounded, locally Lipschitz-continuous with respect to the argument y_0, locally Lipschitz-continuous with respect to those arguments y_j, $j = 1, \ldots, s$, such that the delay τ_j is strongly vanishing, and globally Lipschitz-continuous of constant l_j with respect to those arguments z_j, $j = 1, \ldots, s^*$, such that the delay τ_j^* is strongly vanishing;

-
$$\sum_{\substack{j=1,\ldots,s^* \\ \tau_j^* \text{ is strongly vanishing}}} l_j < 1;$$

- the delays τ_j, $j = 1, \ldots, s$, are measurable and locally bounded;
- the delays τ_j^*, $j = 1, \ldots, s^*$, are measurable, locally bounded and each strongly vanishing delay τ_j^* is such that:

  - for each $\sigma \in \mathbb{R}$, there exists $T^* > 0$ such that, for any subset A of $(0, T^*]$ of zero measure, the set
  $$\{t \in [0, T^*] \mid t - \tau_j^*(\sigma + t) \in A\}$$
  has measure zero;

then any IP (1.2) for the NDDE has a unique local solution $\forall \phi \in \mathcal{LC}$.

It is remarkable that, in order to obtain an existence theorem for the more restricted class of NSDDDEs, it is not sufficient to impose simultaneously the conditions required in the previous theorems for the DDEs with state-dependent delay and of neutral type. In fact, the following theorem, besides considering more restrictive conditions on the functional F, requires additional specific conditions on the initial function depending on the functional itself.

Theorem 4.5. Consider an NSDDDE (2.11). If:

- the function $f(t, y_0, y_1, \ldots, y_s, z_1, \ldots, z_{s^*})$ is measurable, locally bounded and locally Lipschitz-continuous with respect to the arguments $y_j, j = 0, 1, \ldots, s$ and $z_j, j = 1, \ldots, s^*$;
- the delays $\tau_j(t, y)$, $j = 1, \ldots, s$, are measurable, locally bounded and locally Lipschitz-continuous with respect to the argument y;
- the delays $\tau_j^*(t, y)$, $j = 1, \ldots, s^*$, are weakly non-vanishing, measurable, locally bounded and locally Lipschitz-continuous with respect to the argument y;

then any IP (1.2) for the NSDDDE has a unique local solution $\forall \phi \in \mathcal{LC}^1$, $\mathcal{LC}^1$ being the set of continuously differentiable functions with locally Lipschitz-continuous derivative, provided ϕ satisfies the splicing condition

$$\varphi'(0) = f\Big(t_0, \varphi(0), \varphi(-\tau_1(t_0)), \ldots, \varphi(-\tau_s(t_0)),$$
$$\varphi'(-\tau_1^*(t_0)), \ldots, \varphi'(-\tau_{s^*}^*(t_0))\Big). \tag{4.6}$$

Now we consider integro-differential equations.

Theorem 4.6. Consider an NDIDE (2.3). If:

- the function $f(t, y_0, y_1)$ is measurable, locally bounded, locally Lipschitz-continuous with respect to the argument y_0 and, if τ_1 or τ_2 is strongly vanishing, also locally Lipschitz-continuous with respect to the arguments y_1;
- the delays τ_1 and τ_2 are measurable and locally bounded;
- the kernel k satisfies assumption (NK) and, if τ_1 or τ_2 is strongly vanishing, also the assumption:

(NK1) There exists $\widehat{T} > 0$ such that, for any bounded subset B of $\mathbb{R} \times \mathbb{R}^d \times \mathbb{R}^d$, the Lipschitz constants

$$\theta \mapsto \sup_{\substack{(t,y_1,z) \in B \\ (t,y_2,z) \in B \\ y_1 \neq y_2}} \frac{|k(t, -\theta, y_1, z) - k(t, -\theta, y_2, z)|}{|y_1 - y_2|}, \quad \theta \in [-\widehat{T}, 0),$$

and

$$\theta \mapsto \sup_{\substack{(t,y,z_1)\in B \\ (t,y,z_2)\in B \\ z_1 \neq z_2}} \frac{|k(t,-\theta,y,z_1) - k(t,-\theta,y,z_2)|}{|z_1 - z_2|}, \quad \theta \in [-\widehat{T},0),$$

are integrable;

then any IP (1.2) for the NDIDE has a unique local solution $\forall \phi \in \mathcal{LC}$.

Theorem 4.7. Consider an NSDDIDE (2.12). If:

- the function $f(t,y_0,y_1)$ is measurable, locally bounded, locally Lipschitz-continuous with respect to the arguments y_0 and y_1;
- the delays $\tau_1(t,y)$ and $\tau_2(t,y)$ are measurable, locally bounded and locally Lipschitz-continuous with respect to the argument y;
- the kernel $k(t,\theta,y,z)$ satisfies conditions (NK), (NK1) and:

  (NK2) For any bounded subset B of $\mathbb{R} \times \mathbb{R}^d \times \mathbb{R}^d$, the function M_B defined in assumption (NK) is essentially locally bounded on the images $\tau_j(\mathbb{R} \times \mathbb{R}^d)$, $j = 1,2$;

then any IP (1.2) for the NSDDIDE has a unique local solution $\forall \phi \in \mathcal{LC}$.

Analogous theorems are valid for DIDEs and SDDIDEs by replacing assumption (NK) with (K) and by modifying assumptions (NK1) and (NK2) in an obvious way. Note that for RFDEs of integral type the existence and uniqueness of the solution is achieved in their own data set.

4.1. The solution map

In the foregoing theorems we have seen that existence and uniqueness of the local solutions is not always guaranteed for any initial data in the data set. In particular, when the delay is state-dependent, the existence and uniqueness is guaranteed on a restriction of the data set, namely $\mathcal{LC}$ for SDDDEs and the subset of $\mathcal{LC}^1$ satisfying the splicing conditions (4.6) for NSDDDEs.

Therefore, it is worth introducing the concept of *state set*, as the subset of the data set for which the local solution uniquely exists and can be prolonged to a maximal solution.

Definition 4.1. A subset Y of the data set X is called a *state set* if, for any $t_0 \in \mathbb{R}$ and $\phi \in Y$, the IP (1.2) has a unique local solution y, defined on $(-\infty, t_0 + T]$, and $y_t \in Y$ holds for all $t \in [t_0, t_0 + T]$.

If Y is a state set for the RFDE (1.1), then, for any $\sigma \in \mathbb{R}$ and $\varphi \in Y$, the IP (1.2) with $t_0 = \sigma$ and $\phi = \varphi$ has a unique local solution. Such a solution can be prolonged to a unique maximal solution $y(\sigma, \varphi) : (-\infty, \sigma + T_{\max}) \to \mathbb{R}^d$, where $T_{\max} = T_{\max}(\sigma, \varphi) \in (0, +\infty]$.

Summarizing, under the conditions stated in the previous theorems, we have the following state sets for the various classes of RFDEs:

- the set $\mathcal{C}$ is a state set for DDEs, DIDEs and SDDIDEs,
- the set $\mathcal{LC}$ is a state set for SDDDEs, NDDEs, NDIDEs, and NSDDIDEs,
- the subset of $\mathcal{LC}^1$ satisfying the splicing condition (4.6) is the state set for NSDDDEs.

The *solution map* for the RFDE is the map V associating the state

$$V(\sigma, \varphi, T) = y(\sigma, \varphi)_{\sigma+T} \in Y$$

at the point $\sigma + T$ to the triple (σ, φ, T), where $\sigma \in \mathbb{R}$, $\varphi \in Y$ and $T \in [0, T_{\max}(\sigma, \varphi))$. This state determines the future states $y(\sigma, \varphi)_{\sigma+T+\Delta}$, where $\Delta \in [0, T_{\max}(\sigma, \varphi) - T)$, by

$$V(\sigma, \varphi, T + \Delta) = V(\sigma + T, V(\sigma, \varphi, T), \Delta).$$

We can conclude this section by the following basic proposition based on Theorem 4.1.

Proposition 4.3. *The solution map can be expressed as*

$$V(\sigma, \varphi, T) = v(\varphi, z^*(\sigma, \varphi, T))_T, \qquad (4.7)$$

where $v(\cdot, \cdot)$ is defined in (4.1) and $z^*(\sigma, \varphi, T)$ is the unique fixed point of the map $Q_T(\sigma, \varphi)$ defined in (4.3).

5. Numerical methods for RFDEs

A numerical method for a RFDE with state set Y provides a map $\widetilde{V}$ approximating the solution map V. More precisely, the map $\widetilde{V}$ associates data

$$\widetilde{V}(\sigma, \varphi, h) \in Y$$

approximating the state $V(\sigma, \varphi, h)$ to the triple (σ, φ, h), where $\sigma \in \mathbb{R}$, $\varphi \in Y$ and $h \in [0, H_{\max}(\sigma, \varphi))$. Here, the third argument of $\widetilde{V}$ is denoted by h and not by T, since it has the meaning of *step-size* in the integration process described below. Moreover, note that its definition domain is $[0, H_{\max}(\sigma, \varphi))$, which can be different from the definition domain of the third argument T of V.

We deal with the problem of the computation of the solution $y = y(t_0, \phi)$ of the IP (1.2), where $\phi \in Y$, on the integration window $[t_0, t_0 + T]$, where $T \in (0, T_{\max}(t_0, \phi))$. We refer to it as the *integration problem* (t_0, ϕ, T).

When a numerical method providing a map $\widetilde{V}$ is applied to the integration problem (t_0, ϕ, T) with the mesh

$$\Delta = \{t_n \mid n = 0, 1, 2, \ldots, N_\Delta\}, \quad t_0 < t_1 < t_2 < \cdots < t_{N_\Delta} = t_0 + T, \quad (5.1)$$

it yields the finite sequence of states

$$\{\phi_n\}_{n=0,1,2,\ldots,N_\Delta},$$

where $\phi_n \in Y$ is an approximation of the exact state y_{t_n} at the mesh point t_n, given by the recursion

$$\phi_{n+1} = \widetilde{V}(t_n, \phi_n, h_{n+1}), \quad n = 0, 1, \ldots, N_\Delta - 1,$$
$$\phi_0 = \phi. \quad (5.2)$$

Here, $h_{n+1} = t_{n+1} - t_n$ is the $(n+1)$th step-size and satisfies $h_{n+1} \in [0, H_{\max}(t_n, \phi_n))$. Note that the sequence

$$\{y_{t_n}\}_{n=0,1,2,\ldots,N_\Delta}$$

of the exact states satisfies the recursion

$$y_{t_{n+1}} = V(t_n, y_{t_n}, h_{n+1}), \quad n = 0, 1, \ldots, N_\Delta - 1,$$
$$y_{t_0} = \phi.$$

In this paper we consider numerical methods such that

$$\widetilde{V}(\sigma, \varphi, h) = v\big(\varphi, \widetilde{z}^*(\sigma, \varphi, h)\big)_h \quad (5.3)$$

(recall Proposition 4.3), where $\widetilde{z}^*(\sigma, \varphi, h) \in L_h^\circ$ is a suitable function approximating the fixed point $z^*(\sigma, \varphi, h)$ of the map $Q_h(\sigma, \varphi)$. We define the *local error functions* $e(\sigma, \varphi, h)$ and $E(\sigma, \varphi, h)$ at (σ, φ) with step-size h by

$$e(\sigma, \varphi, h) = \widetilde{z}^*(\sigma, \varphi, h) - z^*(\sigma, \varphi, h) \quad (5.4)$$

and

$$E(\sigma, \varphi, h) = \big(v(\varphi, \widetilde{z}^*(\sigma, \varphi, h)) - v(\varphi, z^*(\sigma, \varphi, h))\big)|_{[0,h]}$$
$$= v(0, e(\sigma, \varphi, h))|_{[0,h]}$$
$$= \int e(\sigma, \varphi, h), \quad (5.5)$$

where, for $z \in L_h$, $\int z$ denotes the primitive of z, i.e.,

$$\left(\int z\right)(t) = \int_0^t z(s)\, ds, \quad t \in [0, h].$$

5.1. Convergence analysis

Now, we tackle the study of the error in the numerical solution of integration problems by a numerical method providing approximations of type (5.3).

To this end, we consider:

- a norm $|\cdot|$ on $\mathbb{R}^d$;
- a norm on the spaces L_h° and $C_h \subseteq L_h^\circ$,
$$\|z\| = \sup_{t \in [0,T]} |z(t)|, \quad z \in L_h^\circ.$$

We consider an integration problem (t_0, ϕ, T) and the numerical computation of the solution $y = y(t_0, \phi)$ on the integration window $[t_0, t_0 + T]$ given by the process (5.2). Since we need to measure the error between the states y_{t_n} and the approximated states ϕ_n, $n = 1, 2, \ldots, N_\Delta$, we introduce the distance
$$d(\varphi, \psi) = \sup_{\theta \in (-\infty, 0]} |\varphi(\theta) - \psi(\theta)| \in [0, +\infty]$$
between two data $\varphi, \psi \in C$.

Note that, for $n = 0, 1, \ldots, N_\Delta - 1$, we have
$$\phi_n = (\phi_{n+1})_{-h_{n+1}} \quad \text{and} \quad y_{t_n} = (y_{t_{n+1}})_{-h_{n+1}}$$
and then
$$d(\phi_n, y_{t_n}) \leq d(\phi_{n+1}, y_{t_{n+1}}).$$

Hence, we define
$$\mathbf{E}(\Delta) = d(\phi_{N_\Delta}, y_{t_0+T})$$
the *global error* in the numerical solution of the integration problem (t_0, ϕ, T) with mesh Δ. As we will see below, the global error is linked to the local errors at the mesh points
$$\begin{aligned} \mathbf{E}_{n+1}(\Delta) &= \|E(t_n, y_{t_n}, h_{n+1})\|, \\ \mathbf{E}_{n+1}^0(\Delta) &= |E(t_n, y_{t_n}, h_{n+1})(h_{n+1})|, \\ \mathbf{e}_{n+1}(\Delta) &= \|e(t_n, y_{t_n}, h_{n+1})\|, \end{aligned} \tag{5.6}$$
where $n = 0, 1, \ldots, N_\Delta - 1$.

Our aim is to study the infinitesimal order of $\mathbf{E}(\Delta)$ as $h_\Delta \to 0$, where
$$h_\Delta = \max_{n=1,2,\ldots,N_\Delta} h_n$$
is the maximum step-size in the mesh Δ.

In the next definition, we introduce the concept of global order of the given numerical method for RFDEs.

Definition 5.1. Let r be a positive integer and let $\mathcal{F}$ be a family of integration problems (t_0, ϕ, T) such that $y(t_0, \phi)|_{[t_0, t_0+T]}$ is piecewise smooth (i.e., piecewise C^m for some positive integer m). The method has *global order* r on $\mathcal{F}$ if, for any integration problem $(t_0, \phi, T) \in \mathcal{F}$, we have
$$\mathbf{E}(\Delta_k) = \mathcal{O}(h_{\Delta(k)}^r), \quad k \to +\infty,$$

for any sequence $\{\Delta^{(k)}\}$ of meshes on $[t_0, t_0 + T]$ such that

$$h_{\Delta^{(k)}} \to 0, \quad k \to +\infty,$$

and, for any k, $\Delta^{(k)}$ includes all the breaking points of $y(t_0, \phi)|_{[t_0, t_0+T]}$.

Now, we distinguish between RFDEs with data set $\mathcal{C}$ and RFDEs with data set $\mathcal{LC}$.

RFDEs with data set $\mathcal{C}$

For RFDEs with data set $\mathcal{C}$, we introduce the concepts of uniform order and discrete order of the given method.

Definition 5.2. Let q be a positive integer and let $\mathcal{F}$ be a family of integration problems (t_0, ϕ, T) for which $y(t_0, \phi)|_{[t_0, t_0+T]}$ is piecewise smooth. The method has *uniform order* q on $\mathcal{F}$ if, for any integration problem $(t_0, \phi, T) \in \mathcal{F}$, there exist constants $H > 0$ and $C > 0$ such that

$$\|E(t, y_t, h)\| \leq Ch^{q+1},$$

for any $t \in [t_0, t_0 + T)$ and $h \in [0, H_{\max}(t, y_t))$ such that $h \leq T - t$, and the interval $(t, t+h)$ does not contain breaking points of $y(t_0, \phi)$ and $h < H$.

Definition 5.3. Let p be a positive integer and let $\mathcal{F}$ be a family of integration problems (t_0, ϕ, T) for which $y(t_0, \phi)|_{[t_0, t_0+T]}$ is piecewise smooth. The method has *discrete order* p on $\mathcal{F}$ if, for any integration problem $(t_0, \phi, T) \in \mathcal{F}$, there exist constants $H > 0$ and $C > 0$ such that

$$|E(t, y_t, h)(h)| \leq Ch^{p+1},$$

for any $t \in [t_0, t_0 + T)$ and $h \in [0, H_{\max}(t, y_t))$ such that $h \leq T - t$, and the interval $(t, t+h)$ does not contain breaking points of $y(t_0, \phi)$ and $h < H$.

In order to link the uniform and discrete orders to the global order, we introduce the concept of stability of the method.

Definition 5.4. Let $\mathcal{F}$ be a family of integration problems (t_0, ϕ, T). The method is *stable* on $\mathcal{F}$ if, for any integration problem (t_0, ϕ, T), there exist $\delta > 0$, $\overline{H} > 0$ and $L \geq 0$ such that

$$\overline{H} \leq H_{\max}(t, \varphi), \quad t \in [t_0, T_0 + T),$$

and

$$\|\widetilde{z}(t, \varphi, h) - \widetilde{z}(t, y_t, h)\| \leq L \cdot d(\varphi, y_t), \quad h \in [0, \overline{H}).$$

for any $t \in [t_0, t_0 + T)$ and $\varphi \in Y$ such that $d(\varphi, y_t) \leq \delta$.

Here is the convergence theorem for RFDEs with data set $\mathcal{C}$.

Theorem 5.1. (Convergence) Consider the numerical solution of integration problems given by the recursive process (5.2) and assume that the numerical method is of type (5.3). Let $\mathcal{F}$ be a family of integration problems. If the method has uniform order q, discrete order p and it is stable on $\mathcal{F}$, then it has global order $q' = \min\{q+1, p\}$ on $\mathcal{F}$.

Proof. Consider the numerical solution of an integration problem (t_0, ϕ, T) belonging to the family $\mathcal{F}$. Let $\{\Delta^{(k)}\}$ be a sequence of meshes (with mesh points $t_n^{(k)}$, $n = 0, 1, \ldots, N_{\Delta^{(k)}}$, and step-sizes $h_{n+1}^{(k)}$, $n = 0, 1, \ldots, N_{\Delta^{(k)}} - 1$) such that $h_{\Delta^{(k)}} \to 0$, $k \to \infty$.

Let K_0 be such that
$$h_{\Delta^{(k)}} \leq \overline{H}, \quad k \geq K_0,$$
where $\overline{H}$ is given in Definition 5.4. Hence, for $k \geq K_0$, we have
$$h_{n+1}^{(k)} \leq \overline{H} \leq H_{\max}\big(t_n^{(k)}, y_{t_n^{(k)}}\big), \quad n = 0, 1, \ldots, N_{\Delta^{(k)}} - 1,$$
and so the local errors in (5.6) are defined.

Now, we prove the following relations for the errors.

(i) If
$$\max_{n=1,\ldots,N_{\Delta^{(k)}}} \mathbf{E}_n(\Delta^{(k)}) \to 0, \quad k \to \infty,$$
and
$$\max_{n=1,\ldots,N_{\Delta^{(k)}}-1} \frac{\mathbf{E}_n^0(\Delta^{(k)})}{h_n^{(k)}} \to 0, \quad k \to \infty,$$
then there exists K_1, $K_1 \geq K_0$, such that, for $k \geq K_1$, the sequence $\{\phi_n^{(k)}\}$ is defined, i.e.,
$$h_{n+1}^{(k)} \leq H_{\max}\big(t_n^{(k)}, \phi_n^{(k)}\big), \quad n = 0, 1, \ldots, N_{\Delta^{(k)}} - 1,$$
and
$$\mathbf{E}(\Delta^{(k)}) = \mathcal{O}\Big(\max_{n=1,\ldots,N_{\Delta^{(k)}}} \mathbf{E}_n(\Delta^{(k)})\Big)$$
$$+ \mathcal{O}\Big(\max_{n=1,\ldots,N_{\Delta^{(k)}}-1} \frac{\mathbf{E}_n^0(\Delta^{(k)})}{h_n^{(k)}}\Big), \quad k \to \infty.$$

In order to prove (i), let K_2 be such that $K_2 \geq K_1$ and
$$h_{\Delta^{(k)}} < \frac{1}{L}, \quad k \geq K_2 \tag{5.7}$$

and
$$\frac{e^{\frac{L}{1-h_\Delta^{(k)} L}T}}{1 - h_\Delta^{(k)} L} \cdot \max_{n=1,\ldots,N_{\Delta^{(k)}}} \mathbf{E}_n(\Delta^{(k)})$$
$$+ \frac{e^{\frac{L}{1-h_\Delta^{(k)} L}T} - 1}{L} \cdot \max_{n=1,\ldots,N_{\Delta^{(k)}}-1} \frac{\mathbf{E}_n^0(\Delta^{(k)})}{h_n^{(k)}} \leq \delta. \tag{5.8}$$

Now, we fix an index k such that $k \geq K_2$ (it will be dropped in the notation). We define, for $n = 1, \ldots, N_\Delta$,
$$\overline{\mathbf{E}}_n = \max_{i=1,\ldots,n} \mathbf{E}_i(\Delta),$$
$$\overline{\mathbf{E}}_n^0 = \max_{i=1,\ldots,n} \frac{\mathbf{E}_i^0(\Delta)}{h_i}.$$
Moreover, if ϕ_n is defined, $n = 0, \ldots, N_\Delta$, we set
$$\mathbf{d}_n = d(\phi_n, y_{t_n}),$$
$$\mathbf{d}_n^0 = |\phi_n(0) - y_{t_n}(0)|,$$
$$\overline{\mathbf{d}}_n^0 = \max_{i=0,\ldots,n} \mathbf{d}_i^0,$$
noting that
$$\overline{\mathbf{d}}_n = \max_{i=0,\ldots,n} \mathbf{d}_i.$$

For a given $N \in \{1, \ldots, N_\Delta - 1\}$, we assume that the sequence $\{\phi_n\}_{n=0}^N$ is defined and satisfies

$$\mathbf{d}_N \leq \frac{e^{\frac{L}{1-h_\Delta L}(t_{N-1}-t_0)}}{1 - h_\Delta L} \cdot \overline{\mathbf{E}}_N + \frac{e^{\frac{L}{1-h_\Delta L}(t_{N-1}-t_0)} - 1}{L} \cdot \overline{\mathbf{E}}_{N-1}^0, \quad (5.9)$$

setting $\overline{\mathbf{E}}_{N-1}^0 = 0$ for $N = 1$. To complete the induction, we prove that ϕ_{N+1} is defined and satisfies

$$\mathbf{d}_{N+1} \leq \frac{e^{\frac{L}{1-h_\Delta L}(t_N-t_0)}}{1 - h_\Delta L} \cdot \overline{\mathbf{E}}_{N+1} + \frac{e^{\frac{L}{1-h_\Delta L}(t_N-t_0)} - 1}{L} \cdot \overline{\mathbf{E}}_N^0. \quad (5.10)$$

Since (5.8) and (5.9) imply $\mathbf{d}_N = d(\phi_N, y_{t_N}) \leq \delta$, we have
$$h_{N+1} \leq h_\Delta < \overline{H} \leq H_{\max}(t_N, \phi_N),$$
recalling Definition 5.4. Hence, ϕ_{N+1} is defined. Moreover, we have
$$\|\widetilde{z}(t_n, \phi_n, h_{n+1}) - \widetilde{z}(t_N, y_{t_N}, h_{N+1})\| \leq L\mathbf{e}_N. \quad (5.11)$$

Let $n = 0, 1, \ldots, N$. By (5.11), we obtain
$$\mathbf{d}_{n+1}^0 = |\phi_{n+1}(0) - y_{t_{n+1}}(0)|$$
$$\leq |v(\phi_n, \widetilde{z}^*(t_n, \phi_n, h_{n+1}))(h_{n+1})$$
$$\quad - v(y_{t_n}, \widetilde{z}^*(t_n, y_{t_n}, h_{n+1}))(h_{n+1})|$$
$$\quad + |v(y_{t_n}, \widetilde{z}^*(t_n, y_{t_n}, h_{n+1}))(h_{n+1})$$
$$\quad - v(y_{t_n}, z^*(t_n, y_{t_n}, h_{n+1}))(h_{n+1})|$$
$$\leq \mathbf{d}_n^0 + h_{n+1} L \mathbf{d}_n + \mathbf{E}_{n+1}^0(\Delta).$$

Moreover, for $\theta \leq -h_{n+1}$, we have
$$\begin{aligned}|\phi_{n+1}(\theta) - y_{t_{n+1}}(\theta)| &= |v(\phi_n, \tilde{z}^*(t_n, \phi_n, h_n))(h_{n+1} + \theta) \\ &\quad - v(y_{t_n}, z^*(t_n, y_{t_n}, h_{n+1}))(h_{n+1} + \theta)| \\ &\leq |\phi_n(h_{n+1} + \theta) - y_{t_n}(h_{n+1} + \theta)| \\ &\leq d_n\end{aligned}$$

and, for $\theta \in [-h_{n+1}, 0]$, again by (5.11), we have
$$\begin{aligned}|\phi_{n+1}(\theta) - y_{t_{n+1}}(\theta)| &= |v(\phi_n, \tilde{z}^*(t_n, \phi_n, h_{n+1}))(h_{n+1} + \theta) \\ &\quad - v(y_{t_n}, z^*(t_n, y_{t_n}, h_{n+1}))(h_{n+1} + \theta)| \\ &\leq |v(\phi_n, \tilde{z}^*(t_n, \phi_n, h_{n+1}))(h_{n+1} + \theta) \\ &\quad - v(y_{t_n}, \tilde{z}^*(t_n, y_{t_n}, h_{n+1}))(h_{n+1} + \theta)| \\ &\quad + |v(y_{t_n}, \tilde{z}^*(t_n, y_{t_n}, h_{n+1}))(h_{n+1} + \theta) \\ &\quad - v(y_{t_n}, z^*(t_n, y_{t_n}, h_{n+1}))(h_{n+1} + \theta)| \\ &\leq d_n^0 + h_{n+1} L d_n + E_{n+1}(\Delta).\end{aligned}$$

Thus, we have
$$d_{n+1}^0 \leq d_n^0 + h_{n+1} L d_n + E_{n+1}^0(\Delta), \quad n = 0, 1, \ldots, N, \tag{5.12}$$

and
$$d_{n+1} \leq \max\{d_n, d_n^0 + h_{n+1} L d_n + E_{n+1}(\Delta)\}, \quad n = 0, 1, \ldots, N. \tag{5.13}$$

Now, for $n = 0, 1, \ldots, N$, we have
$$\overline{d}_{n+1}^0 \leq \overline{d}_n^0 + h_{n+1} L d_n + E_{n+1}^0(\Delta) \tag{5.14}$$

since $\overline{d}_{n+1}^0 = \max\{d_{n+1}^0, \overline{d}_n^0\}$ and
$$d_{n+1}^0 \leq \overline{d}_n^0 + h_{n+1} L d_n + E_{n+1}^0(\Delta)$$

by (5.12). Moreover, for $i = 0, 1, \ldots, N$, (5.13) yields
$$d_{i+1} \leq d_{i-k}^0 + h_{i+1-k} L d_{i-k} + E_{i+1-k}(\Delta)$$

for some $k = 0, 1, 2, \ldots$. Hence,
$$d_{i+1} \leq \frac{1}{1 - h_\Delta L} (\overline{d}_i^0 + \overline{E}_{i+1}), \quad i = 0, 1, \ldots, N. \tag{5.15}$$

By inserting (5.15) with $i + 1 = n$ in (5.14), we obtain
$$\overline{d}_{n+1}^0 \leq \left(1 + h_{n+1} \frac{L}{1 - h_\Delta L}\right) \overline{d}_n^0$$
$$+ h_{n+1} \left(\frac{L}{1 - h_\Delta M} \overline{E}_{n+1} + \overline{E}_{n+1}^0\right), \quad n = 0, 1, \ldots, N. \tag{5.16}$$

The recursion (5.16) yields

$$\overline{d}_N^0 \leq \left(e^{\frac{L}{1-h_\Delta L}(t_N-t_0)} - 1\right)\overline{E}_N + \frac{e^{\frac{L}{1-h_\Delta L}(t_N-t_0)} - 1}{\frac{L}{1-h_\Delta L}} \cdot \overline{E}_N^0.$$

Thus, by (5.15) with $i+1 = N+1$ we obtain (5.10).

Since ϕ_N is defined and (5.9) holds with $N=1$, we obtain

$$E(\Delta) = d_{N_\Delta} \leq \frac{e^{\frac{L}{1-h_\Delta L}(T-h_{N_\Delta})}}{1-h_\Delta L} \cdot \overline{E}_{N_\Delta} + \frac{e^{\frac{L}{1-h_\Delta L}(T-h_{N_\Delta})} - 1}{L} \cdot \overline{E}_{N_\Delta-1}^0,$$

and then (i) is proved.

Now, the theorem follows in a straightforward way from (i). □

RFDEs with data set $\mathcal{LC}$

For RFDEs (1.1) with data set $\mathcal{LC}$, we introduce the concept of order of the method.

Definition 5.5. Let q be a positive integer and let $\mathcal{F}$ be a family of integration problems (t_0, ϕ, T) such that $y(t_0, \phi)|_{[t_0, t_0+T]}$ is piecewise smooth. The method has *order* q on $\mathcal{F}$ if, for any integration problem $(t_0, \phi, T) \in \mathcal{F}$, there exist constants $H > 0$ and $C > 0$ such that

$$\|e(t, y_t, h)\| \leq Ch^q$$

for any $t \in [t_0, t_0 + T)$ and $h \in [0, H_{\max}(t, y_t))$ such that $h \leq T - t$, the interval $(t, t+h)$ does not contain breaking points of $y(t_0, \phi)$ and $h < H$.

As for RFDEs with data set $\mathcal{C}$, the order is linked to the global order by the concept of stability.

Definition 5.6. Let $\mathcal{F}$ be a family of integration problems (t_0, ϕ, T). The method is *stable* on $\mathcal{F}$ if, for any integration problem (t_0, ϕ, T), there exist $\delta > 0$, $\overline{H} > 0$, $L \geq 0$, $M \geq 0$ and $P \in [0, 1)$ such that

$$\overline{H} \leq H_{\max}(t, y_t), \quad t \in [t_0, t_0+T),$$

and

$$\|\tilde{z}(t, \varphi, h) - \tilde{z}(t, y_t, h)\| \leq L \cdot d(\varphi, y_t) + M \cdot d(\varphi'_{-\tau(t)}, y'_{t-\tau(t)})$$
$$+ P \cdot d(\varphi', y'_t), \quad h \in [0, \overline{H}],$$

for any $t \in [t_0, t_0+T)$ and $\varphi \in Y$ such that $d(\varphi, y_t) \leq \delta$. Here, τ is a function $[t_0, t_0+T] \to [0, +\infty)$ for which there exist $\xi_0, \xi_1, \ldots, \xi_{K-1}, \xi_K$ such that

$$t_0 = \xi_0 < \xi_1 < \cdots < \xi_{K-1} < \xi_K = t_0+T$$

and

$$t - \tau(t) \leq \xi_k, \quad t \in [\xi_k, \xi_{k+1}] \text{ and } k = 0, 1, \ldots, K-1.$$

Here is the convergence theorem for RFDEs with data set $\mathcal{LC}$.

Theorem 5.2. (Convergence) Consider the numerical solution of integration problems given by the recursive process (5.2) and assume that the numerical method is of type (5.3). Let $\mathcal{F}$ be a family of integration problems. If the method has order q and it is stable on $\mathcal{F}$, then it has global order q on $\mathcal{F}$.

Proof. Consider the numerical solution of an integration problem (t_0, ϕ, T) belonging to the the family $\mathcal{F}$. Let $\{\Delta^{(k)}\}$ be a sequence of meshes such that $h_{\Delta^{(k)}} \to 0$, $k \to \infty$.

Let K_0 be such that
$$h_{\Delta^{(k)}} \leq \overline{H}, \quad k \geq K_0,$$
where $\overline{H}$ is given in Definition 5.6. Hence, for $k \geq K_0$, the local errors in (5.6) are defined. As in the proof of Theorem 5.1, we now address the errors.

(i) If
$$\max_{n=1,\ldots,N_{\Delta^{(k)}}} \mathbf{e}_n(\Delta^{(k)}) \to 0, \quad k \to \infty,$$
then there exists K_1, $K_1 \geq K_0$, such that, for $k \geq K_1$, the sequence $\{\phi_n^{(k)}\}$ is defined and
$$\mathbf{E}(\Delta^{(k)}) = \mathcal{O}\left(\max_{n=1,\ldots,N_{\Delta^{(k)}}} \mathbf{e}_n(\Delta^{(k)})\right).$$

In order to prove (i), let K_2 be such that $K_2 \geq K_1$ and (5.7) hold, i.e.,
$$h_{\Delta^{(k)}} < \frac{1}{L}, \quad k \geq K_2,$$

and let
$$\frac{e^{\frac{L}{1-h_\Delta^{(k)}L}T}}{1-h_\Delta^{(k)}L} \cdot \max_{n=1,\ldots,N_{\Delta^{(k)}}} \mathbf{E}_n(\Delta^{(k)}) \leq \delta. \tag{5.17}$$

Now, we fix an index k such that $k \geq K_2$ (it will be dropped in the notation). We define, for $n = 1, \ldots, N_\Delta$,
$$\overline{\mathbf{E}}_n = \max_{i=1,\ldots,n} \mathbf{E}_i(\Delta),$$
$$\overline{\mathbf{e}}_n = \max_{i=1,\ldots,n} \mathbf{e}_i(\Delta).$$

Moreover, if ϕ_n is defined, $n = 0, \ldots, N_\Delta$, we set
$$\mathbf{d}_n = d(\phi_n, y_{t_n}),$$
$$\mathbf{d}'_n = d(\phi'_n, y'_{t_n}),$$

noting that
$$\mathbf{d}_n = \max_{i=0,\ldots,n} \mathbf{d}_i \quad \text{and} \quad \mathbf{d}'_n = \max_{i=0,\ldots,n} \mathbf{d}'_i.$$

We define the sets of indices
$$I_0 = \{0\},$$
$$I_k = \{n \in \{0,\ldots,N_\Delta\} \mid t_n \in [\xi_{k-1}, \xi_k]\}, \quad k = 1,\ldots,K,$$
and set, for $k = 0,\ldots,K$,
$$\widehat{\mathbf{d}}_k = \mathbf{d}_{\max I_k}, \quad \widehat{\mathbf{d}}'_k = \mathbf{d}'_{\max I_k} \quad \text{and} \quad \widehat{\mathbf{e}}_k = \bar{\mathbf{e}}_{\max I_k}.$$

For a given $N \in \{1,\ldots,N_\Delta - 1\}$, we assume that the sequence $\{\phi_n\}_{n=0}^N$ is defined and satisfies
$$\mathbf{d}_N \le \frac{e^{LC_1(t_N - t_0)} - 1}{LC_1} \cdot \bar{\mathbf{e}}_N, \tag{5.18}$$

where
$$C_1 = 1 + \frac{M+P}{1-P}\left(1 + \frac{MC_0}{1-P}\right) \tag{5.19}$$

and
$$C_0 = \sum_{i=1}^{k} \left(\frac{M}{1-P}\right)^{k-i}. \tag{5.20}$$

We complete the inductive proof by showing that ϕ_{N+1} is defined and satisfies
$$\mathbf{d}_{N+1} \le \frac{e^{LC_1(t_{N+1} - t_0)} - 1}{LC_1} \cdot \bar{\mathbf{e}}_{N+1}. \tag{5.21}$$

Since $\mathbf{d}_N = d(\phi_N, y_{t_N}) \le \delta$ holds by (5.17) and (5.18), we obtain that ϕ_{N+1} is defined. Moreover, recalling Definition 5.6, we have
$$\|\tilde{z}(t_n, \phi_n, h_{n+1}) - \tilde{z}(t_N, y_{t_N}, h_{N+1})\|$$
$$\le L\mathbf{d}_N + Md\big((\phi'_n)_{-\tau(t_n)}, y'_{t_n - \tau(t_n)}\big) + P\mathbf{d}'_n. \tag{5.22}$$

Let $n = 0, 1, 2, \ldots, N$. We look for a bound for $\mathbf{d}'_{n+1}$.
For $\theta \le -h_{n+1}$, we have
$$|\phi'_{n+1}(\theta) - y'_{t_{n+1}}(\theta)| = \big|v'\big(\phi_n, \tilde{z}^*(t_n, \phi_n, h_{n+1})\big)(h_{n+1} + \theta)$$
$$- v'\big(y_{t_n}, z^*(t_n, y_{t_n}, h_{n+1})\big)(h_{n+1} + \theta)\big|$$
$$= |\phi'_n(h_{n+1} + \theta) - y'_{t_n}(h_{n+1} + \theta)|$$
$$\le \mathbf{d}'_n.$$

For $\theta \in [-h_{n+1}, 0]$, by (5.22), we have

$$\begin{aligned}|\phi'_{n+1}(\theta) - y'_{t_{n+1}}(\theta)| &= |v'(\phi_n, \tilde{z}^*(t_n, \phi_n, h_{n+1}))(h_{n+1} + \theta) \\ &\quad - v'(y_{t_n}, z^*(t_n, y_{t_n}, h_{n+1}))(h_{n+1} + \theta)| \\ &\leq |\tilde{z}^*(t_n, \phi_n, h_{n+1})(h_{n+1} + \theta) \\ &\quad - \tilde{z}^*(t_n, y_{t_n}, h_{n+1})(h_{n+1} + \theta)| \\ &\leq |\tilde{z}^*(t_n, \phi_n, h_{n+1})(h_{n+1} + \theta) \\ &\quad - \tilde{z}^*(t_n, y_{t_n}, h_{n+1})(h_{n+1} + \theta)| \\ &\quad + |\tilde{z}^*(t_n, y_{t_n}, h_{n+1})(h_{n+1} + \theta) \\ &\quad - z^*(t_n, y_{t_n}, h_{n+1})(h_{n+1} + \theta)| \\ &\leq Ld_n + Md\big((\phi'_n)_{-\tau(t_n)}, (y'_{t_n})_{-\tau(t_n)}\big) \\ &\quad + Pd'_n + e_{n+1}(\Delta).\end{aligned}$$

Now, if $t_n \in [\xi_{k-1}, \xi_k]$, $k = 1, \ldots, K$, then $t_n - \tau(t_n) \leq \xi_{k-1}$ and so

$$d\big((\phi'_n)_{-\tau(t_n)}, y'_{t_n - \tau(t_n)}\big) \leq d(\phi'_m, y'_{t_m}) = \widehat{d}'_k,$$

where m is some index in the set I_{k-1}. Thus, if $t_n \in [\xi_{k-1}, \xi_k]$, we have

$$d'_{n+1} \leq \max\{d'_n, Ld_n + M\widehat{d}'_{k-1} + Pd'_n + e_{n+1}(\Delta)\}.$$

As a consequence, we obtain

$$d'_{n+1} \leq Ld_n + M\widehat{d}'_{k-1} + Pd'_n + \bar{e}_{n+1}$$

and then

$$d'_{n+1} \leq \frac{L}{1-P}d_n + \frac{M}{1-P}\widehat{d}'_{k-1} + \frac{1}{1-P}\bar{e}_{n+1}. \tag{5.23}$$

Inequality (5.23) yields

$$\widehat{d}'_k \leq \frac{M}{1-P}\widehat{d}'_{k-1} + \frac{L}{1-P}\widehat{d}_k + \frac{1}{1-P}\widehat{\bar{e}}_k. \tag{5.24}$$

and a recursive use of (5.24) gives

$$\widehat{d}'_k \leq \frac{C_0 L}{1-P}\widehat{d}_k + \frac{C_0}{1-P}\widehat{\bar{e}}_k,$$

where C_0 is given by (5.20). Then, by (5.23),

$$d'_{n+1} \leq \frac{L}{1-P}\left(1 + \frac{MC_0}{1-P}\right)d_n + \frac{1}{1-P}\left(1 + \frac{MC_0}{1-P}\right)\bar{e}_{n+1}. \tag{5.25}$$

Now, we look for a bound for $\mathbf{d}_{n+1}$. For $\theta \leq -h_{n+1}$, we have

$$|\phi_{n+1}(\theta) - y_{t_{n+1}}(\theta)| = |v(\phi_n, \tilde{z}^*(t_n, \phi_n, h_n))(h_{n+1} + \theta)$$
$$- v(y_{t_n}, z^*(t_n, y_{t_n}, h_{n+1}))(h_{n+1} + \theta)|$$
$$\leq |\phi_n(h_{n+1} + \theta) - y_{t_n}(h_{n+1} + \theta)|$$
$$\leq \mathbf{d}_n$$

and, for $\theta \in [-h_{n+1}, 0]$, by (5.22), we have

$$|\phi_{n+1}(\theta) - y_{t_{n+1}}(\theta)| = |v(\phi_n, \tilde{z}^*(t_n, \phi_n, h_{n+1}))(h_{n+1} + \theta)$$
$$- v(y_{t_n}, z^*(t_n, y_{t_n}, h_{n+1}))(h_{n+1} + \theta)|$$
$$\leq |v(\phi_n, \tilde{z}^*(t_n, \phi_n, h_{n+1}))(h_{n+1} + \theta)$$
$$- v(y_{t_n}, \tilde{z}^*(t_n, y_{t_n}, h_{n+1}))(h_{n+1} + \theta)|$$
$$+ |v(y_{t_n}, \tilde{z}^*(t_n, y_{t_n}, h_{n+1}))(h_{n+1} + \theta)$$
$$- v(y_{t_n}, z^*(t_n, y_{t_n}, h_{n+1}))(h_{n+1} + \theta)|$$
$$\leq \mathbf{d}_n + h_{n+1} L \mathbf{d}_n$$
$$+ h_{n+1} M d\big((\phi'_n)_{-\tau(t_n)}, (y'_{t_n})_{-\tau(t_n)}\big)$$
$$+ P \mathbf{d}'_n + \mathbf{E}_{n+1}(\Delta).$$

Thus,

$$\mathbf{d}_{n+1} \leq (1 + h_{n+1}L)\mathbf{d}_n + h_{n+1}(M + P)\mathbf{d}'_n + \mathbf{E}_{n+1}(\Delta). \tag{5.26}$$

By summarizing, we have proved the inequalities (5.25) and (5.26) for $n = 0, \ldots, N$. Hence, we obtain

$$\mathbf{d}_{n+1} \leq (1 + h_{n+1}LC_1)\mathbf{d}_n + h_{n+1}C_1\bar{\mathbf{e}}_{n+1}, \quad n = 0, \ldots, N, \tag{5.27}$$

where C_1 is given by (5.19). The recursion (5.27) yields (5.21). Moreover, since ϕ_N is defined and (5.18) holds with $N = 1$, we obtain

$$\mathbf{E}(\Delta) = \mathbf{d}_{N_\Delta} \leq \frac{e^{LC_1T} - 1}{LC_1} \cdot \bar{\mathbf{e}}_{N_\Delta},$$

and then (i) and the theorem follow. □

6. Functional continuous Runge–Kutta methods

The development of methods based on suitable modifications of RK methods for the numerical integration of IPs for RFDEs began in the late 1960s/early 1970s, and was essentially due to the pioneering papers of Feldstein (1964), Tavernini (1971) and Cryer and Tavernini (1972). In particular, Cryer and Tavernini (1972) considered the following generalization of the Euler and

Heun method, and proved that they are convergent of order one and two respectively. Specifically, we have the following methods.

Euler method:
$$\eta(t_n + \theta h_{n+1}) = \eta(t_n) + h_{n+1}\theta F(t_n, \eta_{t_n}), \quad \theta \in [0,1];$$

Heun method:
$$\eta(t_n + \theta h_{n+1}) = \eta(t_n) + h_{n+1}\left[\left(\theta - \frac{1}{2}\theta^2\right)F(t_n, \eta_{t_n}) + \frac{1}{2}\theta^2 F(t_{n+1}, Y_{t_{n+1}})\right],$$
$$\text{for } \theta \in [0,1],$$

where
$$Y(t_n + \theta h) = \eta(t_n) + h_{n+1}\theta F(t_n, \eta(t_n)), \quad \theta \in [0,1],$$
$$Y(t) = \eta(t), \quad t \in (-\infty, t_n].$$

Tavernini (1971) also proposed higher-order explicit methods obtained by a predictor–corrector implementation of polynomial collocation. First, he introduced a sequence $\{\eta^{(s)}\}_{s=2,3,\ldots,\bar{s}}$ of implicit methods given by

$$\eta^{(s)}(t_n + \theta h_{n+1}) = \eta(t_n) + h_{n+1}\sum_{i=1}^{s} b_i^{(s)}(\theta) F\left(t_n + c_i^{(s)} h_{n+1}, \eta_{t_n + c_i^{(s)} h_{n+1}}^{(s)}\right),$$
$$\text{for } \theta \in [0,1],$$

where, for $i = 1, \ldots, s$, $c_i^{(s)} \in [0,1]$ are distinct points, and $b_i^{(s)}(\cdot) : [0,1] \to \mathbb{R}$, $i = 1, \ldots, s$, are polynomials of degree s defined by the collocation conditions

$$(\eta^{(s)})'(t_n + c_i^{(s)} h_{n+1}) = F\left(t_n + c_i^{(s)} h_{n+1}, \eta_{t_n + c_i^{(s)} h_{n+1}}^{(s)}\right), \quad i = 1, \ldots, s.$$

In particular, he proposed the three equispaced nodes collocation methods, given in Table 6.1.

Then Tavernini considered the explicit method given by the recurrence relation

$$\eta^{(1)}(t_n + \theta h_{n+1}) = \eta(t_n) + h_{n+1}\theta F(t_n, \eta_{t_n}), \quad \theta \in [0,1],$$
$$\eta^{(1)}(t) = \eta(t), \quad t \leq t_n,$$

$$\eta^{(s)}(t_n + \theta h_{n+1}) = \eta(t_n) + h_{n+1}\sum_{i=1}^{s} b_i^{(s)}(\theta) F\left(t_n + c_i^{(s)} h_{n+1}, \eta_{t_n + c_i^{(s)} h_{n+1}}^{(s-1)}\right),$$
$$\text{for } \theta \in [0,1],$$

$$\eta^{(s)}(t) = \eta(t), \quad t \leq t_n,$$

for $s = 2, \ldots, \bar{s}$, and finally,

$$\eta(t_n + \theta h_{n+1}) = \eta^{(\bar{s})}(t_n + \theta h_{n+1}), \quad \theta \in [0,1].$$

Table 6.1. Abscissae and continuous weights of the collocation methods proposed by Tavernini (1971).

s	$c^{(s)}$	$b^{(s)}$
2	$(0,1)$	$b^{(2)}(\theta) = \left(\theta - \frac{1}{2}\theta^2,\ \frac{1}{2}\theta^2\right)$
3	$(0, \frac{1}{2}, 1)$	$b^{(3)}(\theta) = \left(\theta - \frac{3}{2}\theta^2 + \frac{2}{3}\theta^3,\ 2\theta^2 - \frac{4}{3}\theta^3,\ -\frac{\theta^2}{2} + \frac{2}{3}\theta^3\right)$
4	$(0, \frac{1}{3}, \frac{2}{3}, 1)$	$b^{(4)}(\theta) = \left(\theta - \frac{11}{4}\theta^2 + 3\theta^3 - \frac{9}{8}\theta^4,\ \frac{9}{2}\theta^2 - \frac{15}{2}\theta^3 + \frac{27}{8}\theta^4,\right.$
		$\left.-\frac{9}{4}\theta^2 + 6\theta^3 - \frac{27}{8}\theta^4,\ \frac{1}{2}\theta^2 - \frac{3}{2}\theta^3 + \frac{9}{8}\theta^4\right)$

For $\bar{s} = 1$ and $\bar{s} = 2$ the previous Euler and Heun methods, respectively, are obtained. In this way, explicit methods of arbitrary global order r for functional equations were obtained at the cost of $1 + \frac{r(r-1)}{2}$ evaluations of the functional F. Indeed, Tavernini also found a particular explicit method of global order 4 by using only 6 evaluations of F, instead of 7 as required by the approach described above. Surprisingly, after Tavernini this approach was not further investigated. Only recently, a general class of RK methods for RFDEs, including all implicit and explicit methods considered by Tavernini as particular instances, was proposed and investigated by Maset, Torelli and Vermiglio (2005). These methods, denoted by $(A(\theta), b(\theta), c)$ and called *functional continuous Runge–Kutta* methods (FCRK), are considered in the following section.

6.1. The general form of FCRK methods

Definition 6.1. Let ν be a positive integer. A ν-stage functional continuous Runge–Kutta method is a triple $(A(\theta), b(\theta), c)$, where $A(\theta)$ is an $\mathbb{R}^{\nu \times \nu}$-valued polynomial function such that $A(0) = 0$, $b(\theta)$ is an $\mathbb{R}^\nu$-valued polynomial function such that $b(0) = 0$, and $c \in \mathbb{R}^\nu$ with $0 \leq c_i \leq 1$, $i = 1, \ldots, \nu$.

The FCRK method $(A(\theta), b(\theta), c)$ provides the approximation

$$\widetilde{V}(\sigma, \varphi, h) = \widetilde{v}_h, \qquad (6.1)$$

where the function $\widetilde{v} : (-\infty, h] \to \mathbb{R}^d$ is given by

$$\widetilde{v}(\theta h) = \varphi(0) + h \sum_{i=1}^{\nu} b_i(\theta) K_i, \quad \theta \in [0, 1],$$

$$\widetilde{v}(t) = \varphi(t), \quad t \in (-\infty, 0],$$

the *derivatives* $K_i \in \mathbb{R}^d$, $i = 1, \ldots, \nu$, are given by

$$K_i = F(\sigma + c_i h, Y^i_{c_i h})$$

and the *stage functions* $Y^i : (-\infty, h] \to \mathbb{R}^d$, $i = 1, \ldots, \nu$, are given by

$$Y^i(\theta h) = \varphi(0) + h \sum_{j=1}^{\nu} a_{ij}(\theta) K_j, \quad \theta \in [0, 1],$$

$$Y^i(t) = \varphi(t), \quad t \in (-\infty, 0].$$

Note that the conditions $A(0) = 0$ and $b(0) = 0$ guarantee the continuity of the functions $\tilde{v}$ and Y^i, $i = 1, \ldots, \nu$.

The FCRK method $(A(\theta), b(\theta), c)$ will be denoted by the tableau

$$\begin{array}{c|c} c & A(\theta) \\ \hline & b(\theta) \end{array}.$$

In particular, the Euler and Heun methods are given by

$$\begin{array}{c|c} 0 & 0 \\ \hline & \theta \end{array} \tag{6.2}$$

and

$$\begin{array}{c|cc} 0 & 0 & 0 \\ 1 & \theta & 0 \\ \hline & \theta - \frac{1}{2}\theta^2 & \frac{1}{2}\theta^2 \end{array}, \tag{6.3}$$

respectively.

Moreover, we partition the set $I = \{1, \ldots, \nu\}$ of indices in the subsets

$$I^+ = \{c_i > 0 \mid i = 1, \ldots, \nu\}$$

and

$$I^0 = \{c_i = 0 \mid i = 1, \ldots, \nu\}.$$

Note that, if $i \in I^0$, then $K_i = F(\sigma, \varphi)$.

To show that the FCRK method $(A(\theta), b(\theta), c)$ provides an approximation of the solution map in the form (5.3), we introduce the following.

- The *prolongation linear operators*

$$\pi : (\mathbb{R}^d)^\nu \to C_h = C([0, h], \mathbb{R}^d), \tag{6.4}$$
$$\Pi_i : (\mathbb{R}^d)^\nu \to C_h, \quad i \in I,$$
$$\Pi : (\mathbb{R}^d)^\nu \to (C_h)^\nu,$$

given by

$$(\pi U)(t) = \sum_{i=1}^{\nu} b'_i\left(\frac{t}{h}\right) U_i,$$

$$(\Pi_i U)(t) = \sum_{j=1}^{\nu} a'_{ij}\left(\frac{t}{h}\right) U_j,$$

$t \in [0, h]$ and $U = (U_1, \ldots, U_\nu) \in (\mathbb{R}^d)^\nu$,

where the derivatives $b'_i(\theta)$ and $a'_{ij}(\theta)$ of the polynomial functions $b_i(\theta)$ and $a_{ij}(\theta)$ appear, and

$$\Pi U = (\Pi_1 U, \ldots, \Pi_\nu U), \quad U \in (\mathbb{R}^d)^\nu.$$

- The *restriction linear operator*

$$R : (C_h)^\nu \to (\mathbb{R}^d)^\nu$$

defined by

$$RZ = (Z_1(c_1 h), \ldots, Z_\nu(c_\nu h)), \quad Z \in (C_h)^\nu.$$

- The map

$$\mathbf{Q}_h^\diamond(\sigma, \varphi) : (C_h)^\nu \to (L_h^\diamond)^\nu$$

defined by

$$\mathbf{Q}_h^\diamond(\sigma, \varphi) Z = \left(Q_h^\diamond(\sigma, \varphi)(Z_1), \ldots, Q_h^\diamond(\sigma, \varphi)(Z_\nu)\right), \quad Z \in (C_h)^\nu,$$

where $L_h^\diamond$ is the space of the measurable and bounded functions $[0, h] \to \mathbb{R}^d$ and $Q_h^\diamond(\sigma, \varphi)$ is given by (4.4).

Proposition 6.1. *The FCRK method* $(A(\theta), b(\theta), c)$ *yields an approximation of the type* (5.3), *where*

$$\widetilde{z}^*(\sigma, \varphi, h) = \pi K,$$

and $K = (K_1, \ldots, K_\nu) \in (\mathbb{R}^d)^\nu$ *is a fixed point of the map*

$$R \mathbf{Q}_h^\diamond(\sigma, \varphi) \Pi : (\mathbb{R}^d)^\nu \to (\mathbb{R}^d)^\nu.$$

Proof. For the function $\widetilde{v}$ in (6.1), we have

$$\widetilde{v}(t) = \varphi(0) + h \sum_{i=1}^{\nu} b_i\left(\frac{t}{h}\right) K_i$$

$$= \varphi(0) + \int_0^t \sum_{i=1}^{\nu} b'_i\left(\frac{s}{h}\right) K_i \, ds$$

$$= v(\varphi, \pi K), \quad t \in [0, h].$$

Analogously, for the stage functions we obtain

$$Y^i(t) = \varphi(0) + \int_0^t \sum_{j=1}^{\nu} a'_{ij}\left(\frac{s}{h}\right) K_j \, ds$$

$$= v(\varphi, \Pi_i K)(t), \quad t \in [0, h],$$

and then, for $i = 1, \ldots, \nu$,

$$K_i = F(\sigma + c_i h, v(\varphi, \Pi_i K)_{c_i h})$$
$$= [Q_h^\diamond(\sigma, \varphi)(\Pi_i K)](c_i h)$$
$$= R(Q_h^\diamond(\sigma, \varphi)(\Pi_1 K), \ldots, Q_h^\diamond(\sigma, \varphi)(\Pi_\nu K))_i$$
$$= [R Q_h^\diamond(\sigma, \varphi)\Pi](K)_i. \qquad \square$$

An FCRK method $(A(\theta), b(\theta), c)$ is called *explicit* if $c_1 = 0$ and, for $i = 2, \ldots, \nu$, $a_{ij}(\theta) = 0$ for $j = i, \ldots, \nu$. The method is called *implicit* if it is not explicit.

If the method is explicit, the derivatives K_i, $i = 1, \ldots, \nu$, can be explicitly obtained in a recursive way: we compute

$$K_1 = F(\sigma, \varphi)$$

and then, successively for $i = 2, \ldots, \nu$,

$$K_i = F(\sigma + c_i h, Y^i_{c_i h}),$$

where

$$Y^i(\theta h) = \varphi(0) + h \sum_{j=1}^{i-1} a_{ij}(\theta) K_j, \quad \theta \in [0, 1],$$

$$Y^i(t) = \varphi(t), \quad t \in (-\infty, 0].$$

If the method is implicit, the derivative vector $K = (K_1, \ldots, K_\nu)$ is obtained as a solution of the fixed point equation

$$K = [R Q_h^\diamond(\sigma, \varphi) \Pi](K)$$

on $(\mathbb{R}^d)^\nu$.

6.2. Well-posedness of FCRK methods

In this section, we study fixed points of the map $R Q_h^\diamond(\sigma, \varphi) \Pi$. In order to avoid unwieldy notation, we omit the ordered pair (σ, φ).

First of all, observe that we cannot hope to have a unique fixed point of the map $R Q_h^\diamond \Pi$, even for small h. In fact, consider the scalar IP for ODEs

$$y'(t) = \lambda y(t)^2, \quad t \geq 0,$$
$$y(0) = y_0 \in \mathbb{R}.$$

The derivative K for the implicit Euler method is an approximation of $y'(h)$ and satisfies the equation
$$K = \lambda(y_0 + hK)^2.$$
Such an equation has the two solutions,
$$K_+ = \frac{2\lambda y_0^2}{1 - 2h\lambda y_0 + \sqrt{1 - 4h\lambda y_0}} \to \lambda y_0^2, \quad h \to 0,$$
and
$$K_- = \frac{2\lambda y_0^2}{1 - 2h\lambda y_0 - \sqrt{1 - 4h\lambda y_0}} \to \infty, \quad h \to 0.$$
The solution K_+ is an approximation of $y'(h)$, whereas K_- is a spurious solution which diverges as $h \to 0$.

In order to study the fixed points of the map $RQ_h^\diamond \Pi$, we introduce the following.

- A norm $|\cdot|$ on $\mathbb{R}^d$.
- A norm on $(\mathbb{R}^d)^\nu$,
$$\|U\|_\infty = \max_{i=1,\ldots,\nu} |U_i|, \quad U \in (\mathbb{R}^d)^\nu.$$
- A norm on the spaces $L_h^\diamond$ and $C_h \subseteq L_h^\diamond$,
$$\|z\| = \sup_{t \in [0,T]} |z(t)|, \quad z \in L_h^\diamond.$$
- For any $\rho > 0$, the closed ball in $(\mathbb{R}^d)^\nu$,
$$B_\rho = \{U \in (\mathbb{R}^d)^\nu \mid \|U\|_\infty \leq \rho\},$$
the closed ball in C_h,
$$C_{h,\rho} = \{z \in C_h \mid \|z\| \leq \rho\},$$
and the Lipschitz constant,
$$k_{h,\rho}^\diamond = \sup_{\substack{z_1, z_2 \in C_{h,\rho} \\ z_1 \neq z_2}} \frac{\|Q_h^\diamond(z_1) - Q_h^\diamond(z_2)\|}{\|z_1 - z_2\|}$$
of the map $Q_h^\diamond$ on $C_{h,\rho}$.

The next theorem concerns the fixed points of $RQ_h^\diamond \Pi$ (see Maset (2009)).

Theorem 6.1. (Well-posedness) If:

(A) There exist a function $a : (0, h_0] \times [\rho_0, +\infty) \to [0, +\infty)$, where $h_0 > 0$ and $\rho_0 > 0$, and a constant $b \in [0, +\infty)$ such that:

(1) $k_{h,\rho}^\diamond \leq a(h, \rho) + b$, $(h, \rho) \in (0, h_0] \times [\rho_0, +\infty)$,
(2) $\lim_{h \downarrow 0} a(h, \rho) = 0$, $\rho \in [\rho_0, +\infty)$,

(3) $b\Lambda < 1$, where

$$\Lambda = \max_{i \in I^+} \max_{\theta \in [0,c_i]} \sum_{j=1}^{\nu} |a'_{ij}(\theta)|, \qquad (6.5)$$

then there exist $\bar{\rho}_0 > 0$ and $\bar{h} > 0$ such that, for $0 < h < \bar{h}$, the map $RQ_h^\circ \Pi$ has a unique fixed point in $B_{\bar{\rho}_0}$, which is contained in the interior of $B_{\bar{\rho}_0}$. Moreover, any other fixed point of $RQ_h^\circ \Pi$ diverges as $h \to 0$. Finally, for any $\rho \geq \bar{\rho}_0$, there exists $\hat{h} = \hat{h}(\rho) > 0$ such that

$$k_{h,\Lambda\rho}^\circ \Lambda < 1 \qquad (6.6)$$

for $0 < h < \hat{h}$.

The hypothesis (A) in Theorem 6.1 holds for any $(\sigma, \varphi) \in \mathbb{R} \times Y$ for DDEs, SDDDEs, DIDEs, SDDIDEs, NDIDEs and NSDDIDEs, whenever the equations satisfy the conditions stated in the existence theorems of Section 4. For NDDEs, we have to require, in addition, that all the delays $\tau_j^*, j = 1, \ldots, s^*$, are non-vanishing. In all these cases, (1) holds with $b = 0$ and then (3) is satisfied with no restrictions on Λ.

For an NDDE with some strongly vanishing delay τ_j^*, the hypothesis (A) holds for any $(\sigma, \varphi) \in \mathbb{R} \times Y$ under the restriction $\Lambda \leq 1$. In fact, (1) holds for some $b \in [0, 1)$ and then (3) is fulfilled if $\Lambda \leq 1$.

Henceforth, we assume that the hypothesis (A) of Theorem 6.1 is satisfied for any $(\sigma, \varphi) \in \mathbb{R} \times Y$. Under this assumption, the FCRK method $(A(\theta), b(\theta), c)$ provides the approximation

$$\widetilde{V}(\sigma, \varphi, h) = v(\varphi, \widetilde{z}^*(\sigma, \varphi, h)), \quad (\sigma, \varphi) \in \mathbb{R} \times Y \text{ and } h \in [0, H_{\max}(\sigma, \varphi)),$$

where $\widetilde{z}^*(\sigma, \varphi, h) = pK$, $K = K(\sigma, \varphi, h)$ is the unique fixed point of the map $RQ_h^\circ(\sigma, \varphi)\Pi$ on $B_{\bar{\rho}_0}$, and $\bar{\rho}_0 = \bar{\rho}_0(\sigma, \varphi)$ and $H_{\max}(\sigma, \varphi) = \bar{h} = \bar{h}(\sigma, \varphi)$ are defined in Theorem 6.1.

7. Order conditions for FCRK methods

In this section, which is also based on the results by Maset, Torelli and Vermiglio (2005), we give an expansion of the local error functions defined in Section 5 in terms of the step-size h, and then we develop conditions for obtaining a given uniform or discrete order for RFDEs with data set $\mathcal{C}$ and a given order for RFDEs with data set $\mathcal{LC}$.

7.1. Study of the local error functions

We analyse the local error functions $e(\sigma, \varphi, h)$ and $E(\sigma, \varphi, h)$ given by (5.4) and (5.5). Here, we have $(\sigma, \varphi) \in \mathbb{R} \times Y$ and

$$h \in [0, \min\{T_{\max}(\sigma, \varphi), H_{\max}(\sigma, \varphi)\}),$$

and so both the functions $z^*(\sigma,\varphi,h)$ and $\tilde{z}^*(\sigma,\varphi,h) = pK(\sigma,\varphi,h)$ are defined. In this study, we assume that $z^*(\sigma,\varphi,h)$ is at least continuous. In order to avoid cumbersome notation, we omit the dependence on (σ,φ,h).
Let
$$Z^* = (z^*,\ldots,z^*) \in (C_h)^\nu.$$
It is clear that Z^* is a fixed point of the map $Q_h^\diamond$.

The local error functions e and E can be written as
$$e = \gamma + \pi\Delta \tag{7.1}$$
and
$$E = \int \gamma + \int \pi\Delta, \tag{7.2}$$
where
$$\gamma = \pi R Z^* - z^* \in C_h$$
and
$$\Delta = K - RZ^* \in (\mathbb{R}^d)^\nu.$$

As for the error γ, we can give the following bound.

Proposition 7.1. If z^* is of class C^m, where m is non-negative integer, then
$$\max_{\theta\in[0,1]}\left|\gamma(\theta h) - \sum_{k=0}^{m-1}\gamma_k(\theta)\frac{h^k}{k!}(z^*)^{(k)}(0)\right|$$
$$\leq \frac{h^m}{m!}\max_{\theta\in[0,1]}\left(\sum_{i=1}^\nu |b_i'(\theta)|c_i^m + \theta^m\right)\max_{t\in[0,h]}|z^{(m)}(t)|,$$
where, for $k = 0, 1, \ldots, m-1$,
$$\gamma_k(\theta) = \sum_{i=1}^\nu b_i'(\theta)c_i^k - \theta^k, \quad \theta \in [0,1].$$

Proof. We have, for $\theta \in [0,1]$,
$$\gamma(\theta h) = \sum_{i=1}^\nu b_i'(\theta)z^*(c_i h) - z^*(\theta h)$$
$$= \sum_{i=1}^\nu b_i'(\theta)\left(\sum_{k=0}^{m-1}\frac{c_i^k h^k}{k!}(z^*)^{(k)}(0)\right.$$
$$\left. + \frac{1}{(m-1)!}\int_0^1 (1-s)^{m-1}(z^*)^{(m)}(sc_i h)c_i^m h^m\, ds\right)$$

$$-\left(\sum_{k=0}^{m-1}\frac{\theta^k h^k}{k!}(z^*)^{(k)}(0)\right.$$
$$\left.+\frac{1}{(m-1)!}\int_0^1(1-s)^{m-1}(z^*)^{(m)}(s\theta h)\theta^m h^m\,ds\right)$$
$$=\sum_{k=0}^{m-1}\frac{h^k}{k!}\left(\sum_{i=1}^{\nu}b_i'(\theta)c_i^k-\theta^k\right)(z^*)^{(k)}(0)+\frac{h^m}{(m-1)!}$$
$$\cdot\int_0^1(1-s)^{m-1}\left(\sum_{i=1}^{\nu}b_i'(\theta)c_i^m(z^*)^{(m)}(sc_ih)-\theta^m(z^*)^{(m)}(s\theta h)\right)ds,$$

whenever z^* is of class C^{m+1}. □

As for the other error Δ in (7.1), we have
$$\Delta=R(\mathbf{Q}_h^\circ(Z^*+\Pi\Delta+\Gamma)-\mathbf{Q}_h^\circ(z^*)),\tag{7.3}$$

where
$$\Gamma=\Pi RZ^*-Z^*\in(C_h)^\nu.$$

Note that, for $i\in I^0$, $\Delta_i=0$.

In the next proposition, we establish that
$$\|\Delta\|_\infty=\mathcal{O}(\|\Gamma\|_\infty^+),\quad \|\Gamma\|_\infty^+\to 0,\tag{7.4}$$

where
$$\|\Gamma\|_\infty^+=\max_{i\in I^+}\|\Gamma_i|_{[0,c_ih]}\|.$$

Proposition 7.2. *If $h<\widehat{h}(\Lambda\bar{\rho}_1)$, where $\widehat{h}(\cdot)$ is defined in Theorem 6.1, Λ is defined in (6.5), $\bar{\rho}_1=\max\{\bar{\rho}_0,\frac{1}{\Lambda}\|z^*\|^\circ\}$ and $\bar{\rho}_0$ is defined in Theorem 6.1, then*
$$\|\Delta\|_\infty\leq\frac{k^\circ_{h,\Lambda\bar{\rho}_1}}{1-k^\circ_{h,\Lambda\bar{\rho}_1}\Lambda}\cdot\|\Gamma\|_\infty^+.\tag{7.5}$$

Proof. Since $K\in B_{\bar{\rho}_0}$ and
$$\Pi K=Z^*+\Pi\Delta+\Gamma,$$
we have
$$z^*+\Pi_i\Delta+\Gamma_i=\Pi_i K\in C_{h,\Lambda\bar{\rho}_0},\quad i\in I^+.$$

Thus, we have
$$z^*+\Pi_i\Delta+\Gamma_i\in C_{h,\Lambda\bar{\rho}_1},\quad i\in I^+,$$
and
$$z^*\in C_{h,\Lambda\bar{\rho}_1}.$$

Let $h < \widehat{h}(\Lambda \bar{\rho}_1)$. For any $i \in I^+$, we have
$$\begin{aligned}|\Delta_i| &= |Q_h^\diamond(z^* + \Pi_i \Delta + \Gamma_i)(c_i h) - Q_h^\diamond(z^*)(c_i h)| \\ &= |Q_{c_ih}^\diamond((z^* + \Pi_i \Delta + \Gamma_i)|_{[0,c_ih]})(c_i h) - Q_{c_ih}^\diamond(z^*|_{[0,c_ih]})(c_i h)| \\ &\leq \|Q_{c_ih}^\diamond((z^* + \Pi_i \Delta + \Gamma_i)|_{[0,c_ih]}) - Q_{c_ih}^\diamond(z^*|_{[0,c_ih]})\|^\diamond \\ &\leq k_{c_ih,\Lambda\bar{\rho}_1}\|(\Pi_i\Delta + \Gamma_i)|_{[0,c_ih]}\| \\ &\leq k_{c_ih,\Lambda\bar{\rho}_1}\left(\max_{\theta \in [0,c_i]} \sum_{j=1}^\nu |a'_{ij}(\theta)| \|\Delta\|_\infty + \|\Gamma_i|_{[0,c_ih]}\|\right) \\ &\leq k_{h,\Lambda\bar{\rho}_1}(\Lambda\|\Delta\|_\infty + \|\Gamma\|_\infty^+).\end{aligned}$$
Thus, (7.5) follows since $k_{h,\Lambda\bar{\rho}_1}\Lambda < 1$ holds by (6.6). □

As for the errors $\Gamma_i \in C_h$, $i \in I^+$, the following proposition holds.

Proposition 7.3. *If z^* is of class C^m, where m is a non-negative integer, then, for $i \in I^+$,*
$$\max_{\theta \in [0,c_i]}\left|\Gamma_i(\theta h) - \sum_{k=0}^{m-1} \Gamma_{ik}(\theta)\frac{h^k}{k!}(z^*)^{(k)}(0)\right|$$
$$\leq \frac{h^m}{m!}\max_{\theta \in [0,c_i]}\left(\sum_{i=1}^\nu |a'_{ij}(\theta)|c_i^m + \theta^m\right)\max_{t \in [0,c_ih]}|z^{(m)}(t)|,$$
where, for $k = 0, 1, \ldots, m-1$,
$$\Gamma_{ik}(\theta) = \sum_{i=1}^\nu a'_{ij}(\theta)c_j^k - \theta^k, \quad \theta \in [0, c_i].$$

Proof. The proof is analogous to the proof of Proposition 7.1. □

Now, we look for an expansion of the error Δ in terms of Γ. For our purposes, it is sufficient to consider the first-order expansion given in the next proposition.

Proposition 7.4. *Let us assume that the map $Q_h^\diamond$ is of class C^2. Then:*

(i) *the map $\mathbf{Q}_h^\diamond$ is of class C^2;*

and, under the hypothesis in Proposition 7.2,

(ii) *the linear map*
$$R(\mathbf{Q}_h^\diamond)'(z^*)\Pi : (\mathbb{R}^d)^\nu \to (\mathbb{R}^d)^\nu$$
has norm less than 1;

(iii)
$$\Delta = L_h\Gamma + \mathbf{R}(\Delta, \Gamma), \tag{7.6}$$

where L_h is the linear operator $(C_h)^\nu \to (\mathbb{R}^d)^\nu$ given by
$$L_h Z = \left(I_{(\mathbb{R}^d)^\nu} - R(\mathbf{Q}_h^\diamond)'(z^*)\Pi\right)^{-1} R(\mathbf{Q}_h^\diamond)'(Z^*)Z, \quad Z \in (C_h)^\nu,$$
and
$$\mathbf{R}(\Delta, \Gamma) = \frac{1}{2}\left(I_{(\mathbb{R}^d)^\nu} - R(\mathbf{Q}_h^\diamond)'(z^*)\Pi\right)^{-1}$$
$$\cdot R \int_0^1 (1-s)(\mathbf{Q}_h^\diamond)''\left(Z^* + s(\Pi\Delta + \Gamma)\right)(\Pi\Delta + \Gamma, \Pi\Delta + \Gamma)\,ds.$$

Proof. Point (i) is obvious. Note that
$$(\mathbf{Q}_h^\diamond)'(z)Y = \left((Q_h^\diamond)'(Z_1)Y_1, \ldots, (Q_h^\diamond)'(Z_\nu)Y_\nu\right),$$
$$\text{for } Z, Y \in (C_h)^\nu,$$
and
$$(\mathbf{Q}_h^\diamond)''(z)(Y, X) = \left((Q_h^\diamond)''(Z_1)(Y_1, X_1), \ldots, (Q_h^\diamond)''(Z_\nu)(Y_\nu, X_\nu)\right),$$
$$\text{for } Z, Y, X \in (C_h)^\nu.$$

Now, we prove (i) and (ii). Under the hypothesis of Proposition 7.2, *i.e.*, $h < \widehat{h}(\Lambda\bar{\rho}_1)$, we have
$$\|z^*\| < \Lambda\bar{\rho}_1$$
and
$$k_{h,\Lambda\bar{\rho}_1}\Lambda < 1.$$
The map $Q_h^\diamond$ is differentiable at z^*. Note that
$$[(Q_h^\diamond)'(z^*)u](0) = 0, \quad u \in C_h.$$
Moreover, for $0 < h_1 \leq h$, the map $Q_{h_1}^\diamond$ is differentiable at $z^*|_{[0,h_1]}$ and
$$(Q_{h_1}^\diamond)'(z^*|_{[0,h_1]})u|_{[0,h_1]} = ((Q_{h_1}^\diamond)'(z^*)u)|_{[0,h_1]}, \quad u \in C_h,$$
and
$$\|(Q_{h_1}^\diamond)'(z^*|_{[0,h_1]})\| \leq k_{h_1,\Lambda\bar{\rho}_1}.$$
Hence, for $U \in (\mathbb{R}^d)^\nu$, we have
$$|(R(\mathbf{Q}_h^\diamond)'(Z^*)\Pi U)_i| = |[(Q_h^\diamond)'(z^*)\Pi_i U](c_i h)|$$
$$= |[(Q_{c_ih}^\diamond)'(z^*|_{[0,c_ih]})\Pi_i U|_{[0,c_ih]}](c_i h)|$$
$$\leq \|(Q_{c_ih}^\diamond)'(z^*|_{[0,c_ih]})\Pi_i U|_{[0,c_ih]}\|$$
$$\leq k_{c_ih,\Lambda\bar{\rho}_1}\|\Pi U|_{[0,c_ih]}\|$$
$$\leq k_{c_ih,\Lambda\bar{\rho}_1} \max_{\theta \in [0,c_i]} \sum_{j=1}^\nu |a_{ij}(\theta)|\|U\|_\infty$$
$$\leq k_{h,\Lambda\bar{\rho}_1}\Lambda\|U\|_\infty$$

if $i \in I^+$ and
$$\left|(R(\mathbf{Q}_h^\diamond)'(Z^*)\Pi U)_i\right| = \left|[(Q_h^\diamond)'(z^*)\Pi_i U](0)\right| = 0$$
if $i \in I^0$. Thus
$$\|R(\mathbf{Q}_h^\diamond)'(Z^*)\Pi U\| \le k_{h,\Lambda\bar{p}_1}\Lambda < 1.$$

Finally, we prove (iii). Since the map $\mathbf{Q}_h^\diamond$ is of class C^2, we obtain
$$\mathbf{Q}_h^\diamond(Z^* + \Pi\Delta + \Gamma) - \mathbf{Q}_h^\diamond(z^*) = (\mathbf{Q}_h^\diamond)'(z^*)(\Pi\Delta + \Gamma)$$
$$+ \int_0^1 (1-s)(\mathbf{Q}_h^\diamond)''(Z^* + s\Pi\Delta + \Gamma)(\Pi\Delta + \Gamma, \Pi\Delta + \Gamma)\,ds,$$
and so, by (7.3),
$$\left(I_{(\mathbb{R}^d)^\nu} - R(\mathbf{Q}_h^\diamond)'(z^*)\Pi\right)\Delta = R(\mathbf{Q}_h^\diamond)'(z^*)\Gamma$$
$$+ R\int_0^1 (1-s)(\mathbf{Q}_h^\diamond)''(Z^* + s\Pi\Delta + \Gamma)(\Pi\Delta + \Gamma, \Pi\Delta + \Gamma)\,ds.$$
Since $\|R(\mathbf{Q}_h^\diamond)'(Z^*)\Pi\| < 1$, we obtain (7.6). □

Under the assumption that the map $(Q_h^\diamond)''$ is bounded in a neighbourhood of z^*, we obtain
$$\Delta = L_h\Gamma + \mathcal{O}((\|\Gamma\|_\infty^+)^2), \quad \|\Gamma\|_\infty^+ \to 0.$$
Moreover, we can write
$$L_h\Gamma = R(\mathbf{Q}_h^\diamond)'(Z^*)\Gamma + \sum_{k=1}^\infty (R(\mathbf{Q}_h^\diamond)'(z^*)\Pi)^k R(\mathbf{Q}_h^\diamond)'(Z^*)\Gamma.$$
In general, the first term $R(\mathbf{Q}_h^\diamond)'(z^*)\Gamma$ on the right-hand side does not have infinitesimal order, with respect to h, lower than the other terms. However, there are important cases where this happens. This is the subject of the next subsection, where we consider RFDEs (1.1) satisfying the following *Regularity Condition*.

(RC) For any $(\sigma, \varphi) \in \mathbb{R} \times X$, there exists a map $\overline{Q}_h^\diamond(\sigma, \varphi) : C_h \to L_h^\diamond$ such that
$$Q_h^\diamond(\sigma,\varphi)(z) = \overline{Q}_h^\diamond(\sigma,\varphi)\left(\int z\right), \quad z \in C_h.$$

Clearly, RFDEs with data set $\mathcal{C}$ satisfy the Regularity Condition. As for RFDEs with data set $\mathcal{LC}$, the Regularity Condition is satisfied for NDDEs or NSDDIDEs whenever all the delays are non-vanishing.

On local error functions for equations satisfying the Regularity Condition
For equations fulfilling the Regularity Condition, we introduce the map
$$\overline{\mathbf{Q}}_h^\diamond : (C_h)^\nu \to (L_h)^\nu$$

defined by
$$\overline{Q}_h^\circ Z = (\overline{Q}_h^\circ(Z_1), \ldots, \overline{Q}_h^\circ(Z_\nu)), \quad Z \in (C_h)^\nu.$$

Moreover, note that
$$(Q_h^\circ)'(z)v = (\overline{Q}_h^\circ)'\left(\int z\right)\int v, \quad z, v \in C_h,$$
$$(Q_h^\circ)''(z)v = (\overline{Q}_h^\circ)''\left(\int z\right)\left(\int v, \int w\right), \quad z, v, w \in C_h.$$

When the Regularity Condition holds, Proposition 7.4 can be restated as follows.

Proposition 7.5. Let us assume that the map $\overline{Q}_h^\circ$ is of class C^2. Under the hypothesis in Proposition 7.2, we have
$$\Delta = L_h \Gamma + \mathbf{R}(\Delta, \Gamma),$$
where
$$L_h \Gamma = \left(I - R(\overline{Q}_h^\circ)'\left(\int Z^*\right)\int \Pi\right)^{-1} R(\overline{Q}_h^\circ)'\left(\int Z^*\right)\int \Gamma$$
and
$$\mathbf{R}(\Delta, \Gamma) = \frac{1}{2}\left(I - R(\overline{Q}_h^\circ)'\left(\int Z^*\right)\int \Pi\right)^{-1}$$
$$\cdot R \int_0^1 (1-s)(\overline{Q}_h^\circ)''\left(\int Z^* + s\int (\Pi\Delta + \Gamma)\right)\left(\int (\Pi\Delta + \Gamma), \int (\Pi\Delta + \Gamma)\right) ds$$

with
$$\int Z^* = \left(\int z^*, \ldots, \int z^*\right) \in (C_h)^\nu$$
and
$$\int \Pi : (\mathbb{R}^d)^\nu \to (C_h)^\nu, \quad \left(\int \Pi\right) U = \left(\int \Pi_1 U_1, \ldots, \int \Pi_\nu U_\nu\right).$$

As a consequence of the previous proposition, we obtain
$$\Delta = L_h \Gamma + h^2 \cdot \mathcal{O}((\|\Gamma\|_\infty^+)^2), \quad \|\Gamma\|_\infty^+ \to 0, \tag{7.7}$$

under the assumption that the map $(\overline{Q}_h^\circ)''$ is bounded in a neighbourhood of $\int z^*$.

Now, we refine equation (7.7) by giving an expansion of the components Δ_i, where $i \in I^+$, in terms of powers of h. To this end, we let $c_1^*, \ldots, c_{\nu^*}^*$

denote the distinct positive abscissae. Moreover, we introduce:

- the continuous functions $g_k : [0, h] \to \mathbb{R}$, $k = 0, 1, 2, \ldots$, given by

$$g_k(\theta h) = \int_0^\theta \gamma_k(\beta) \, d\beta = \sum_{i=1}^\nu b_i(\theta) c_i^k - \frac{\theta^{k+1}}{k+1}, \quad \theta \in [0,1]; \qquad (7.8)$$

- for any $i \in I^+$, the continuous functions $G_{ik} : [0, h] \to \mathbb{R}$, $k = 0, 1, 2, \ldots$, given by

$$G_{ik}(\theta h) = \int_0^\theta \Gamma_{ik}(\beta) \, d\beta = \sum_{i=1}^\nu a_{ij}(\theta) c_j^k - \frac{\theta^{k+1}}{k+1}, \quad \theta \in [0,1]. \qquad (7.9)$$

Finally, we write $(\overline{Q}_h^\diamond)'$ and $(\overline{\mathbf{Q}}_h^\diamond)'$ instead of $(\overline{Q}_h^\diamond)'(\int z^*)$ and $(\overline{\mathbf{Q}}_h^\diamond)'(\int Z^*)$, respectively.

Proposition 7.6. Let us assume that:

- the map $Q_h^\diamond$ is of class C^2 and $(\overline{Q}_h^\diamond)''$ is bounded in a neighbourhood of $\int z^*$;

- the fixed point z^* of the map Q_h is of class C^3;

-
$$\sum_{j=1}^\nu a_{ij}(\theta) = \theta, \quad \theta \in [0, c_i] \text{ and } i \in I^+. \qquad (7.10)$$

Then, for any $i \in I^+$, we have

$$\Delta_i = h^2 [(\overline{Q}_h^\diamond)' \, G_{i1} \cdot (z^*)'(0)] (c_i h)$$
$$+ \frac{h^3}{2} [(\overline{Q}_h^\diamond)' \, G_{i2} \cdot (z^*)''(0)] (c_i h) + h^3 [(\overline{Q}_h^\diamond)' u_i] (c_i h)$$
$$+ \mathcal{O}(h^4), \qquad (7.11)$$

where

$$u_i(t) = \sum_{k=1}^{\nu^*} \left[(\overline{Q}_h^\diamond)' \sum_{\substack{j=1 \\ c_j = c_k^*}}^\nu a_{ij}\left(\frac{t}{h}\right) G_{j1} \cdot (z^*)'(0) \right] (c_k^* h), \quad t \in [0, h].$$

Proof. Since (7.10) holds, we have, for any $i \in I^+$,

$$\Gamma_{i0}(\theta) = 0, \quad \theta \in [0, c_i],$$

in Proposition 7.3. Hence $\|\Gamma\|_\infty^+ = \mathcal{O}(h)$ and so

$$\Delta = L_h \Gamma + \mathcal{O}(h^4).$$

Now, we write

$$L_h\Gamma = \left(I - R(\overline{\mathbf{Q}}_h^\diamond)' \int \Pi\right)^{-1} R(\overline{\mathbf{Q}}_h^\diamond)' \int \Gamma$$

$$= R(\overline{\mathbf{Q}}_h^\diamond)' \int \Gamma + \left[R(\overline{\mathbf{Q}}_h^\diamond)' \int \Pi\right] R(\overline{\mathbf{Q}}_h^\diamond)' \int \Gamma$$

$$+ \sum_{k=2}^\infty \left[R(\overline{\mathbf{Q}}_h^\diamond)' \int \Pi\right]^k R(\overline{\mathbf{Q}}_h^\diamond)' \int \Gamma.$$

Since $\|\int \Gamma\|_\infty^+ = \mathcal{O}(h^2)$ and $\|\int \Pi\| = \mathcal{O}(h)$, we obtain

$$\sum_{k=2}^\infty \left[R(\overline{\mathbf{Q}}_h^\diamond)' \int \Pi\right]^k R(\overline{\mathbf{Q}}_h^\diamond)' \int \Gamma = \mathcal{O}(h^4).$$

Moreover, for any $i \in I^+$, we have

$$\left(R(\overline{\mathbf{Q}}_h^\diamond)' \int \Gamma\right)_i = \left[(\overline{\mathbf{Q}}_h^\diamond)' \int \Gamma_i\right](c_i h),$$

where, by Proposition 7.3,

$$(\overline{\mathbf{Q}}_h^\diamond)' \int \Gamma_i = (\overline{\mathbf{Q}}_h^\diamond)' \int \left(\Gamma_{i1}\left(\frac{\cdot}{h}\right) h(z^*)'(0) + \Gamma_{i2}\left(\frac{\cdot}{h}\right) \frac{h^2}{2}(z^*)''(0) + \mathcal{O}(h^3)\right)$$

$$= (\overline{\mathbf{Q}}_h^\diamond)' \int_0^{\dot h} \left(\Gamma_{i1}(\theta) h^2 (z^*)'(0) + \Gamma_{i2}(\theta) \frac{h^3}{2}(z^*)''(0) + \mathcal{O}(h^4)\right) d\theta$$

$$= h^2 (\overline{\mathbf{Q}}_h^\diamond)' G_{i1} \cdot (z^*)'(0)$$

$$+ \frac{h^3}{2} (\overline{\mathbf{Q}}_h^\diamond)' G_{i2} \cdot (z^*)''(0)$$

$$+ \mathcal{O}(h^4).$$

Finally, for any $i \in I^+$,

$$\left(\left[R(\overline{\mathbf{Q}}_h^\diamond)' \int \Pi\right] R(\overline{\mathbf{Q}}_h^\diamond)' \int \Gamma\right)_i = \left[(\overline{\mathbf{Q}}_h^\diamond)' \int \Pi_i R(\overline{\mathbf{Q}}_h^\diamond)' \left(\int z^*\right) \int \Gamma\right](c_i h),$$

where

$$(\overline{\mathbf{Q}}_h^\diamond)' \int \Pi_i R(\mathbf{Q}_h^\diamond)' \int \Gamma$$

$$= (\overline{\mathbf{Q}}_h^\diamond)' \int \left(\sum_{j=1}^\nu a'_{ij}\left(\frac{\cdot}{h}\right) \left[(\overline{\mathbf{Q}}_h^\diamond)' \int \Gamma_j\right](c_j h)\right)$$

$$= h(\overline{\mathbf{Q}}_h^\diamond)' \int_0^{\dot h} \sum_{j=1}^\nu a'_{ij}(\theta) \left[(\overline{\mathbf{Q}}_h^\diamond)' \int \Gamma_j\right](c_j h) \, d\theta.$$

Since

$$(\overline{Q_h^\circ})' \int \Gamma_j = h^2 (\overline{Q_h^\circ})' \int_0^{\dot{h}} \Gamma_{j1}(\beta) \, d\beta (z^*)'(0) + \mathcal{O}(h^3)$$
$$= h^2 (\overline{Q_h^\circ})' G_{j1} \cdot (z^*)'(0) + \mathcal{O}(h^3),$$

we conclude that

$$\int_0^{\dot{h}} \sum_{j=1}^{\nu} a'_{ij}(\theta) \left[(\overline{Q_h^\circ})' \int \Gamma_j\right] (c_j h) \, d\theta$$

$$= h^2 \int_0^{\dot{h}} \sum_{j=1}^{\nu} a'_{ij}(\theta) [(\overline{Q_h^\circ})' G_{j1} \cdot (z^*)'(0)] (c_j h) \, d\theta + \mathcal{O}(h^3)$$

$$= h^2 \int_0^{\dot{h}} \sum_{\substack{k=1 \\ }}^{\nu^*} \sum_{\substack{j=1 \\ c_j = c_k^*}}^{\nu} a'_{ij}(\theta) [(\overline{Q_h^\circ})' G_{j1} \cdot (z^*)'(0)] (c_k^* h) \, d\theta + \mathcal{O}(h^3)$$

$$= h^2 \int_0^{\dot{h}} \sum_{k=1}^{\nu^*} \left[(\overline{Q_h^\circ})' \sum_{\substack{j=1 \\ c_j = c_k^*}}^{\nu} a'_{ij}(\theta) G_{j1} \cdot (z^*)'(0)\right] (c_k^* h) \, d\theta + \mathcal{O}(h^3),$$

and then

$$(\overline{Q_h^\circ})' \int \Pi_i \, R(\mathbf{Q}_h^\circ)' \int \Gamma$$

$$= h^3 (\overline{Q_h^\circ})' \int_0^{\dot{h}} \sum_{k=1}^{\nu^*} \left[(\overline{Q_h^\circ})' \sum_{\substack{j=1 \\ c_j = c_k^*}}^{\nu} a'_{ij}(\theta) G_{j1} \cdot (z^*)'(0)\right] (c_k^* h) \, d\theta$$
$$+ \mathcal{O}(h^4)$$

$$= h^3 (\overline{Q_h^\circ})' \sum_{k=1}^{\nu^*} \left[(\overline{Q_h^\circ})' \sum_{\substack{j=1 \\ c_j = c_k^*}}^{\nu} \int_0^{\dot{h}} a'_{ij}(\theta) \, d\theta \, G_{j1} \cdot (z^*)'(0)\right] (c_k^* h)$$
$$+ \mathcal{O}(h^4).$$

Thus, the expansion (7.11) follows. □

Now, we are able to give the following expansions for the local error functions e and E in terms of powers of h.

Theorem 7.1. Let us assume that:

- the map $\overline{Q_h^\circ}$ is of class C^2 and $(\overline{Q_h^\circ})''$ is bounded in a neighbourhood of z^*;
- the fixed point z^* of the map Q_h is of class C^4;

$$\sum_{j=1}^{\nu} a_{ij}(\theta) = 0, \quad \theta \in [0, c_i] \text{ and } i \in I^+. \tag{7.12}$$

Then, for $\theta \in [0,1]$,

$$e(\theta h) = \gamma_0(\theta) \cdot z(0)$$
$$+ h\gamma_1(\theta) \cdot (z^*)'(0)$$
$$+ \frac{h^2}{2}\gamma_2(\theta) \cdot (z^*)''(0)$$
$$+ h^2 \sum_{l=1}^{\nu^*} \left[(\overline{Q}_h^\circ)' \sum_{\substack{i=1 \\ c_i = c_l^*}}^{\nu} b_i'(\theta) G_{i1} \cdot (z^*)'(0) \right] (c_l^* h)$$
$$+ \frac{h^3}{6}\gamma_3(\theta) \cdot (z^*)'''(0)$$
$$+ \frac{h^3}{2} \sum_{l=1}^{\nu^*} \left[(\overline{Q}_h^\circ)' \sum_{\substack{i=1 \\ c_i = c_l^*}}^{\nu} b_i'(\theta) G_{i2} \cdot (z^*)''(0) \right] (c_l^* h)$$
$$+ h^3 \sum_{l=1}^{\nu^*} [(\overline{Q}_h^\circ)' w_l(\theta)](c_l^* h)$$
$$+ \mathcal{O}(h^4), \tag{7.13}$$

where

$$w_l(\theta)(t) = \sum_{k=1}^{\nu^*} \left[(\overline{Q}_h^\circ)' \sum_{\substack{i=1 \\ c_i = c_l^*}}^{\nu} \sum_{\substack{j=1 \\ c_j = c_k^*}}^{\nu} b_i'(\theta) a_{ij}\left(\frac{t}{h}\right) G_{j1} \cdot (z^*)'(0) \right] (c_k^* h),$$

for $t \in [0, h]$,

and

$$E(\theta h) = h g_0(\theta h) \cdot z^*(0)$$
$$+ h^2 g_1(\theta h) \cdot (z^*)'(0)$$
$$+ \frac{h^3}{2} g_2(\theta h) \cdot (z^*)''(0)$$
$$+ h^3 \sum_{l=1}^{\nu^*} \left[(\overline{Q}_h^\circ)' \sum_{\substack{i=1 \\ c_i = c_l^*}}^{\nu} b_i(\theta) G_{i1} \cdot (z^*)'(0) \right] (c_l^* h)$$
$$+ \frac{h^4}{6} g_3(\theta h) \cdot (z^*)'''(0)$$

$$+ \frac{h^4}{2} \sum_{l=1}^{\nu^*} \left[(\overline{Q}_h^\diamond)' \sum_{\substack{i=1 \\ c_i = c_l^*}}^{\nu} b_i(\theta) G_{i2} \cdot (z^*)''(0) \right] (c_l^* h)$$

$$+ h^4 \sum_{l=1}^{\nu^*} [(\overline{Q}_h^\diamond)' \, W_l(\theta)](c_l^* h)$$

$$+ \mathcal{O}(h^5), \tag{7.14}$$

where

$$W_l(\theta)(t) = \sum_{k=1}^{\nu^*} \left[(\overline{Q}_h^\diamond)' \sum_{\substack{i=1 \\ c_i = c_l^*}}^{\nu} \sum_{\substack{j=1 \\ c_j = c_k^*}}^{\nu} b_i(\theta) a_{ij} \left(\frac{t}{h}\right) G_{j1} \cdot (z^*)'(0) \right] (c_k^* h),$$

for $t \in [0, h]$.

Proof. The expansion (7.13) follows by (7.1) and Propositions 7.1 and 7.6. The expansion (7.14) follows by integrating (7.13). □

7.2. Order conditions

In this section, we establish conditions on FCRK methods to obtain, for RFDEs with data set $\mathcal{C}$, a prescribed uniform or discrete order and, for RFDEs with data set $\mathcal{LC}$, a prescribed order. Moreover, by using such order conditions, we construct explicit methods attaining a given global order.

RFDEs with data set $\mathcal{C}$

Since RFDEs with data set $\mathcal{C}$ satisfy the Regularity Assumption, we can use the expansion (7.14) in Theorem 7.1 for the local error $E(\sigma, \varphi, h)$.

We consider an FCRK method satisfying the condition (7.12) and a family $\mathcal{F}$ of integration problems (t_0, ϕ, T) such that:

(i) for all $t \in [t_0, t_0 + T)$, the map $\overline{Q}_h^\diamond(t, y_t)$ is of class C^2;

(ii) there exists $\varepsilon > 0$, $\overline{H} > 0$ and $M \geq 0$ such that, for all $t \in [t_0, t_0 + T)$ and $h \in (0, T - t]$,

$$\|(\overline{Q}_h^\diamond(t, y_t))''(z)\| \leq M$$

whenever $z \in C_h$ is such that $\|z - z^*(t, y_t, h)\| \leq \varepsilon$;

(iii) $y|_{[t_0, t_0 + T]}$ is piecewise C^5.

Note that conditions (i), (ii) and (iii) permit us to use expansion (7.14) and to assume that the term $\mathcal{O}(h^5)$ is uniformly bounded with respect to $t \in [t_0, t_0 + T)$. Moreover, we remark that a family of integration problems of DDEs, DIDEs, SDDDEs or SDDIDEs fulfils conditions (i), (ii) and (iii) if the function f, the delays and the kernel are sufficiently smooth (with respect to the variable t only piecewise smoothness is required).

Table 7.1. Uniform order conditions.

Order	Conditions
1	$\sum_{i=1}^{\nu} b_i(\theta) = \theta, \quad \theta \in [0,1]$
2	$\sum_{i=1}^{\nu} b_i(\theta) c_i = \frac{\theta^2}{2}, \quad \theta \in [0,1]$
3	$\sum_{i=1}^{\nu} b_i(\theta) c_i^2 = \frac{\theta^3}{3}, \quad \theta \in [0,1]$ For any $k = 1, \ldots, \nu^*$, $\sum_{\substack{i=1 \\ c_i = c_k^*}}^{\nu} b_i(\theta) \left(\sum_{j=1}^{\nu} a_{ij}(\beta) c_j - \frac{\beta^2}{2} \right) = 0, \quad \theta \in [0,1] \text{ and } \beta \in [0, c_k^*]$
4	$\sum_{i=1}^{\nu} b_i(\theta) c_i^3 = \frac{\theta^4}{4}, \quad \theta \in [0,1]$ For any $k = 1, \ldots, \nu^*$, $\sum_{\substack{i=1 \\ c_i = c_k^*}}^{\nu} b_i(\theta) \left(\sum_{j=1}^{\nu} a_{ij}(\beta) c_j^2 - \frac{\beta^3}{3} \right) = 0, \quad \theta \in [0,1] \text{ and } \beta \in [0, c_k^*]$ For any $l, k = 1, \ldots, \nu^*$, $\sum_{\substack{i=1 \\ c_i = c_l^*}}^{\nu} \sum_{\substack{j=1 \\ c_j = c_k^*}}^{\nu} b_i(\theta) a_{ij}(\beta) \left(\sum_{k=1}^{\nu} a_{jk}(\gamma) c_k - \frac{\gamma^2}{2} \right) = 0,$ for $\theta \in [0,1], \ \beta \in [0, c_l^*] \text{ and } \gamma \in [0, c_k^*]$

In Table 7.1, we give the conditions for getting uniform order two, three and four on the family $\mathcal{F}$.

The order conditions are obtained by expansion (7.14) by recalling (7.8) and (7.9). Such conditions are not only sufficient for getting the prescribed order, but they are also necessary.

As for the discrete order, the conditions for getting discrete order two, three and four are obtained by replacing $b_i(\theta)$ with $b_i = b_i(1)$ in Table 6.1.

The convergence Theorem 5.1 guarantees that a global order r on the family $\mathcal{F}$ is attained if the method has uniform order $r - 1$, has discrete order r and is stable on $\mathcal{F}$. We remark that, exactly as in the case of the above conditions (i), (ii) and (iii), the method is stable on a family of integration problems of DDEs, DIDEs, SDDEs and SDDIDEs if the function f, the delays and the kernel are sufficiently smooth.

It is clear that, for any $r \in \{1, 2, 3, 4\}$, by taking

$$a_{ij}(\theta) = b_j(\theta), \quad \theta \in [0,1], \quad i,j = 1, \ldots, \nu, \tag{7.15}$$

we obtain global order r if

$$\sum_{i=1}^{\nu} b_i(\theta) c_i^{k-1} = \frac{\theta^{k-1}}{k}, \quad \theta \in [0,1] \text{ and } k = 1, \ldots, r-1, \tag{7.16}$$

and

$$\sum_{i=1}^{\nu} b_i c_i^{r-1} = \frac{1}{r}. \tag{7.17}$$

Indeed, for methods of type (7.15), these conditions guarantee global order r even if $r > 4$.

As an example, the conditions for $r = 2$ show that global order two is attained by the one-stage method

$$\begin{array}{c|c} \frac{1}{2} & \theta \\ \hline & \theta \end{array},$$

which can be called the *functional midpoint* method. We also note that a one-stage method,

$$\begin{array}{c|c} c_1 & \theta \\ \hline & \theta \end{array}, \tag{7.18}$$

has uniform order one and global order one. The global order two is attained only if $c_1 = \frac{1}{2}$.

By using conditions (7.15) and (7.16), only implicit methods can be constructed. On the other hand, by considering methods not satisfying (7.15), we can obtain explicit methods or semi-implicit methods (*i.e.*, methods such that $a_{ij}(\theta) = 0$ for $j > i$). Here, we will consider only the case of explicit methods.

Now, for each $r \in \{1, 2, 3, 4\}$, we construct explicit methods (satisfying the condition (7.12) and attaining global order r.

$r = 1$

Global order one is obtained by one-stage explicit methods,

$$\begin{array}{c|c} 0 & 0 \\ \hline & b_1(\theta) \end{array},$$

with discrete order one. The condition for the discrete order one is $b_1 = 1$. Hence, the functional explicit Euler method has global order one.

$r = 2$
Global order two is obtained by two-stage explicit methods,

$$\begin{array}{c|cc} 0 & 0 & 0 \\ c_2 & \theta & 0 \\ \hline & b_1(\theta) & b_2(\theta) \end{array},$$

with discrete order two and uniform order one. The conditions for discrete order two and uniform order one are

$$b_1(\theta) + b_2(\theta) = \theta, \quad \theta \in [0, 1],$$

$$b_2 = \frac{1}{2c_2}.$$

In particular, the family of FCRK methods

$$\begin{array}{c|cc} 0 & 0 & 0 \\ c_2 & \theta & 0 \\ \hline & \theta - \frac{\theta^2}{2c_2} & \frac{\theta^2}{2c_2} \end{array}, \qquad (7.19)$$

with uniform order two, which satisfies

$$b_1(\theta)c_1 + b_2(\theta)c_2 = \frac{\theta^2}{2}, \quad \theta \in [0, 1],$$

has global order two. Particular elements of this family are the functional Heun method (6.3) (obtained for $c = 1$) and the method

$$\begin{array}{c|cc} 0 & 0 & 0 \\ \frac{1}{2} & \theta & 0 \\ \hline & \theta - \theta^2 & \theta^2 \end{array}$$

(obtained for $c = \frac{1}{2}$), which reduces to the Runge method for an ODE.

$r = 3$
Global order three is obtained by three-stage explicit methods,

$$\begin{array}{c|ccc} 0 & 0 & 0 & 0 \\ c_2 & \theta & 0 & 0 \\ c_3 & \theta - a_{32}(\theta) & a_{32}(\theta) & 0 \\ \hline & b_1(\theta) & b_2(\theta) & b_3(\theta) \end{array},$$

with discrete order three and uniform order two. The conditions for obtaining such orders are

$$b_1(\theta) = \theta - b_2(\theta) - b_3(\theta), \quad \theta \in [0,1],$$

$$b_2(\theta)c_2 + b_3(\theta)c_3 = \frac{\theta^2}{2}, \quad \theta \in [0,1],$$

$$b_2 c_2^2 + b_3 c_3^2 = \frac{1}{3},$$

and

$$b_2 = 0,$$

$$b_3 \left(a_{32}(\beta) c_2 - \frac{\beta^2}{2} \right) = 0, \quad \beta \in [0, c_3],$$

if $c_2 \neq c_3$ and

$$b_2 \left(\beta - \frac{\beta^2}{2} \right) + b_3 \left(a_{32}(\beta) c_2 - \frac{\beta^2}{2} \right) = 0, \quad \beta \in [0, c_3],$$

if $c_2 = c_3$.

By choosing $b_2(\theta) = 0$, the previous conditions select the family of methods

$$\begin{array}{c|cc}
0 & 0 & \\
c_2 & \theta & 0 & 0 \\
\frac{2}{3} & \theta - \frac{\theta^2}{2c_2} & \frac{\theta^2}{2c_2} & 0 \\
\hline
& \theta - \frac{3}{4}\theta^2 & 0 & \frac{3}{4}\theta^2
\end{array}.$$

On the other hand, by taking $c_2 = c_3$, the previous conditions reduce to

$$b_1(\theta) = \theta - b_2(\theta) - b_3(\theta), \quad \theta \in [0,1],$$

$$b_2(\theta) + b_3(\theta) = \frac{\theta^2}{2c_2}, \quad \theta \in [0,1],$$

$$b_2 + b_3 = \frac{1}{3c_2^2},$$

$$b_3 \neq 0,$$

$$a_{32}(\beta) = \frac{\beta^2}{2c_2}\left(1 + \frac{b_2}{b_3}\right) - \frac{b_2}{c_2 b_3}, \quad \beta \in [0,1],$$

which are equivalent to

$$c_2 = \frac{2}{3},$$

$$b_1(\theta) = \theta - b_2(\theta) - b_3(\theta), \quad \theta \in [0,1],$$

$$b_2(\theta) = \frac{3\theta^2}{4} - b_3(\theta), \quad \theta \in [0,1],$$

$$a_{32}(\beta) = \frac{9\beta^2}{16b_3}, \quad \beta \in [0, c_3].$$

$r = 4$

Global order four is obtained by explicit methods of discrete order four and uniform order three.

It is known that explicit continuous RK methods attain uniform order three if they have at least four stages (see Table 8.3 in Section 8). Hence, at least four stages have to be used for an FCRK method of uniform order three. One can prove that there do not exist explicit five-stage FCRK methods of discrete order four. Hence, we consider six-stage explicit methods:

$$
\begin{array}{c|cccccc}
0 & 0 & 0 & 0 & 0 & 0 & 0 \\
c_2 & \theta & 0 & 0 & 0 & 0 & 0 \\
c_3 & \theta - a_{32}(\theta) & a_{32}(\theta) & 0 & 0 & 0 & 0 \\
c_4 & \theta - \sum_{j=2}^{3} a_{4j}(\theta) & a_{42}(\theta) & a_{43}(\theta) & 0 & 0 & 0 \\
c_5 & \theta - \sum_{j=2}^{4} a_{5j}(\theta) & a_{52}(\theta) & a_{53}(\theta) & a_{54}(\theta) & 0 & 0 \\
c_6 & \theta - \sum_{j=2}^{5} a_{6j}(\theta) & a_{62}(\theta) & a_{63}(\theta) & a_{64}(\theta) & a_{65}(\theta) & 0 \\
\hline
 & b_1(\theta) & b_2(\theta) & b_3(\theta) & b_4(\theta) & b_5(\theta) & b_6(\theta)
\end{array}
$$

(7.20)

Proposition 7.7. An explicit six-stage RK method (7.20) is of uniform order three and discrete order four if (and only if for distinct abscissae)

$$\frac{c_5 + c_6}{3} - \frac{c_5 c_6}{2} = \frac{1}{4},$$

$$b_1(\theta) = \theta - b_3(\theta) - b_4(\theta) - b_5(\theta) - b_6(\theta), \quad \theta \in [0,1],$$

$$b_2(\theta) = 0, \quad \theta \in [0,1],$$

$$b_3 = b_4 = 0,$$

$$b_3(\theta)c_3 + b_4(\theta)c_4 + b_5(\theta)c_5 + b_6(\theta)c_6 = \frac{\theta^2}{2}, \quad \theta \in [0,1],$$

$$b_3(\theta)c_3^2 + b_4(\theta)c_4^2 + b_5(\theta)c_5^2 + b_6(\theta)c_6^2 = \frac{\theta^3}{3}, \quad \theta \in [0,1],$$

$$a_{32}(\beta) = \frac{\beta^2}{2c_2}, \quad \beta \in [0, c_2],$$

$$a_{42}(\beta)c_2 + a_{43}(\beta)c_3 = \frac{\beta^2}{2}, \quad \beta \in [0, c_4],$$

$$a_{52}(\beta) = 0, \quad \beta \in [0, c_5],$$

$$a_{53}(\beta) = \frac{\beta^2 c_4}{2c_3(c_4 - c_3)} - \frac{\beta^3}{3c_3(c_4 - c_3)}, \quad \beta \in [0, c_5],$$

$$a_{54}(\beta) = -\frac{\beta^2 c_3}{2c_4(c_4 - c_3)} + \frac{\beta^3}{3c_4(c_4 - c_3)}, \quad \beta \in [0, c_5],$$

$$a_{62}(\beta) = 0, \quad \beta \in [0, c_6],$$

$$a_{63}(\beta)c_3 + a_{64}(\beta)c_4 + a_{65}(\beta)c_5 = \frac{\beta^2}{2}, \quad \beta \in [0, c_6],$$

$$a_{63}(\beta)c_3^2 + a_{64}(\beta)c_4^2 + a_{65}(\beta)c_4^2 = \frac{\beta^3}{3}, \quad \beta \in [0, c_6]. \tag{7.21}$$

Proof. Sufficient conditions (and also necessary in the case of distinct abscissae) for uniform order three and discrete order four are as follows, divided into three blocks (see Table 7.1).

(1) $\quad b_1(\theta) + b_2(\theta) + b_3(\theta) + b_4(\theta) + b_5(\theta) + b_6(\theta) = \theta, \quad \theta \in [0, 1],$

$$b_2(\theta)c_2 + b_3(\theta)c_3 + b_4(\theta)c_4 + b_5(\theta)c_5 + b_6(\theta)c_6 = \frac{\theta^2}{2}, \quad \theta \in [0, 1],$$

$$b_2(\theta)c_2^2 + b_3(\theta)c_3^2 + b_4(\theta)c_4^2 + b_5(\theta)c_5^2 + b_6(\theta)c_6^2 = \frac{\theta^3}{3}, \quad \theta \in [0, 1],$$

$$b_2 c_2^3 + b_3 c_3^3 + b_4 c_4^3 + b_5 c_5^3 + b_6 c_6^3 = \frac{1}{4}.$$

(2) $\quad b_2(\theta)\left(-\frac{\beta^2}{2}\right) = 0, \quad \theta \in [0, 1], \; \beta \in [0, c_2],$

$$b_2 \cdot \left(-\frac{\beta^3}{3}\right) = 0, \quad \beta \in [0, c_2],$$

$$b_3(\theta)\left(a_{32}(\beta)c_2 - \frac{\beta^2}{2}\right) = 0, \quad \theta \in [0, 1], \; \beta \in [0, c_3],$$

$$b_3 \cdot \left(a_{32}(\beta)c_2^2 - \frac{\beta^3}{3}\right) = 0, \quad \beta \in [0, c_3],$$

$$b_4(\theta)\left(a_{42}(\beta)c_2 + a_{43}(\beta)c_3 - \frac{\beta^2}{2}\right) = 0, \quad \theta \in [0, 1], \; \beta \in [0, c_4],$$

$$b_4 \cdot \left(a_{42}(\beta)c_2^2 + a_{43}(\beta)c_3^2 - \frac{\beta^3}{3}\right) = 0, \quad \beta \in [0, c_4],$$

$$b_5(\theta)\left(a_{52}(\beta)c_2 + a_{53}(\beta)c_3 + a_{54}(\beta)c_4 - \frac{\beta^2}{2}\right) = 0,$$
$$\text{for } \theta \in [0,1], \ \beta \in [0, c_5],$$

$$b_5 \cdot \left(a_{52}(\beta)c_2^2 + a_{53}(\beta)c_3^2 + a_{54}(\beta)c_4^2 - \frac{\beta^3}{3}\right) = 0, \quad \beta \in [0, c_5],$$

$$b_6(\theta)\left(a_{62}(\beta)c_2 + a_{63}(\beta)c_3 + a_{64}(\beta)c_4 + a_{65}(\beta)c_5 - \frac{\beta^2}{2}\right) = 0,$$
$$\text{for } \theta \in [0,1], \ \beta \in [0, c_6],$$

$$b_6 \cdot \left(a_{62}(\beta)c_2^2 + a_{63}(\beta)c_3^2 + a_{64}(\beta)c_4^2 + a_{65}(\beta)c_5^2 - \frac{\beta^3}{3}\right) = 0,$$
$$\text{for } \beta \in [0, c_6].$$

(3) $\quad b_3 a_{32}(\beta)\left(-\frac{\gamma^2}{2}\right) = 0, \quad \beta \in [0, c_3], \ \gamma \in [0, c_2],$

$$b_4 a_{42}(\beta)\left(-\frac{\gamma^2}{2}\right) = 0, \quad \beta \in [0, c_4], \ \gamma \in [0, c_2],$$

$$b_4 a_{43}(\beta)\left(a_{32}(\gamma)c_2 - \frac{\gamma^2}{2}\right) = 0, \quad \beta \in [0, c_4], \ \gamma \in [0, c_3],$$

$$b_5 a_{52}(\beta)\left(-\frac{\gamma^2}{2}\right) = 0, \quad \beta \in [0, c_5], \ \gamma \in [0, c_2],$$

$$b_5 a_{53}(\beta)\left(a_{32}(\gamma)c_2 - \frac{\gamma^2}{2}\right) = 0, \quad \beta \in [0, c_5], \ \gamma \in [0, c_3],$$

$$b_5 a_{54}(\beta)\left(a_{42}(\gamma)c_2 + a_{43}(\gamma)c_3 - \frac{\gamma^2}{2}\right) = 0, \quad \beta \in [0, c_5], \ \gamma \in [0, c_4],$$

$$b_6 a_{62}(\beta)\left(-\frac{\gamma^2}{2}\right) = 0, \quad \beta \in [0, c_6], \ \gamma \in [0, c_2],$$

$$b_6 a_{63}(\beta)\left(a_{32}(\gamma)c_2 - \frac{\gamma^2}{2}\right) = 0, \quad \beta \in [0, c_6], \ \gamma \in [0, c_3],$$

$$b_6 a_{64}(\beta)\left(a_{42}(\gamma)c_2 + a_{43}(\gamma)c_3 - \frac{\gamma^2}{2}\right) = 0, \quad \beta \in [0, c_6], \ \gamma \in [0, c_4],$$

$$b_6 a_{65}(\beta)\left(a_{52}(\gamma)c_2 + a_{53}(\gamma)c_3 + a_{54}(\gamma)c_4 - \frac{\gamma^2}{2}\right) = 0,$$
$$\text{for } \beta \in [0, c_6], \ \gamma \in [0, c_5].$$

The first condition in block (2) implies $b_2(\theta) = 0$, the third and fourth conditions in (2) imply $b_3 = 0$, and the fifth and sixth conditions, together with the second condition in block (3), imply $b_4 = 0$.

For a method with $b_2(\theta) = 0$ and $b_3 = b_4 = 0$, the conditions in block (1) are satisfied only if $b_5, b_6 \neq 0$ and are equivalent to

$$b_1(\theta) = \theta - b_3(\theta) - b_4(\theta) - b_5(\theta) - b_6(\theta), \quad \theta \in [0,1],$$

$$b_3(\theta)c_3 + b_4(\theta)c_4 + b_5(\theta)c_5 + b_6(\theta)c_6 = \frac{1}{2}, \quad \theta \in [0,1],$$

$$b_3(\theta)c_3^2 + b_4(\theta)c_4^2 + b_5(\theta)c_5^2 + b_6(\theta)c_6^2 = \frac{1}{3}, \quad \theta \in [0,1],$$

$$\frac{c_5 + c_6}{3} - \frac{c_5 c_6}{2} = \frac{1}{4}.$$

For a method with $b_2(\theta) = 0$, $b_3 = b_4 = 0$ and $b_5, b_6 \neq 0$, the conditions in block (2) are equivalent to

$$b_3(\theta)\left(a_{32}(\beta)c_2 - \frac{\beta^2}{2}\right) = 0, \quad \theta \in [0,1], \ \beta \in [0, c_3],$$

$$b_4(\theta)\left(a_{42}(\beta)c_2 + a_{43}(\beta)c_3 - \frac{\beta^2}{2}\right) = 0, \quad \theta \in [0,1], \ \beta \in [0, c_4],$$

$$a_{52}(\beta)c_2 + a_{53}(\beta)c_3 + a_{54}(\beta)c_4 = \frac{\beta^2}{2}, \quad \beta \in [0, c_5],$$

$$a_{52}(\beta)c_2^2 + a_{53}(\beta)c_3^2 + a_{54}(\beta)c_4^2 = \frac{\beta^3}{3}, \quad \beta \in [0, c_5],$$

$$a_{62}(\beta)c_2 + a_{63}(\beta)c_3 + a_{64}(\beta)c_4 + a_{65}(\beta)c_5 = \frac{\beta^2}{2}, \quad \beta \in [0, c_6],$$

$$a_{62}(\beta)c_2^2 + a_{63}(\beta)c_3^2 + a_{64}(\beta)c_4^2 + a_{65}(\beta)c_4^2 = \frac{\beta^3}{3}, \quad \beta \in [0, c_6],$$

and the conditions in block (3) are equivalent to

$$a_{52}(\beta) = 0, \quad \beta \in [0, c_5],$$

$$a_{53}(\beta)\left(a_{32}(\gamma)c_2 - \frac{\gamma^2}{2}\right) = 0, \quad \beta \in [0, c_5], \ \gamma \in [0, c_3],$$

$$a_{54}(\beta)\left(a_{42}(\gamma)c_2 + a_{43}(\gamma)c_3 - \frac{\gamma^2}{2}\right) = 0, \quad \beta \in [0, c_5], \ \gamma \in [0, c_4],$$

$$a_{62}(\beta) = 0, \quad \beta \in [0, c_6],$$

$$a_{63}(\beta)\left(a_{32}(\gamma)c_2 - \frac{\gamma^2}{2}\right) = 0, \quad \beta \in [0, c_6], \ \gamma \in [0, c_3],$$

$$a_{64}(\beta)\left(a_{42}(\gamma)c_2 + a_{43}(\gamma)c_3 - \frac{\gamma^2}{2}\right) = 0, \quad \beta \in [0, c_6], \ \gamma \in [0, c_4],$$

$$a_{65}(\beta)\left(a_{52}(\gamma)c_2 + a_{53}(\gamma)c_3 + a_{54}(\gamma)c_4 - \frac{\gamma^2}{2}\right) = 0,$$

$$\text{for } \beta \in [0, c_6], \ \gamma \in [0, c_5].$$

So, the conditions in blocks (2) and (3) are equivalent to

$$a_{32}(\beta)c_2 = \frac{\beta^2}{2}, \quad \beta \in [0, c_3],$$

$$a_{42}(\beta)c_2 + a_{43}(\beta)c_3 = \frac{\beta^2}{2}, \quad \beta \in [0, c_4],$$

$$a_{52}(\beta) = 0, \quad \beta \in [0, c_5],$$

$$a_{53}(\beta)c_3 + a_{54}(\beta)c_4 = \frac{\beta^2}{2}, \quad \beta \in [0, c_5],$$

$$a_{53}(\beta)c_3^2 + a_{54}(\beta)c_4^2 = \frac{\beta^3}{3}, \quad \beta \in [0, c_5],$$

$$a_{62}(\beta) = 0, \quad \beta \in [0, c_6],$$

$$a_{63}(\beta)c_3 + a_{64}(\beta)c_4 + a_{65}(\beta)c_5 = \frac{\beta^2}{2}, \quad \beta \in [0, c_6],$$

$$a_{63}(\beta)c_3^2 + a_{64}(\beta)c_4^2 + a_{65}(\beta)c_4^2 = \frac{\beta^3}{3}, \quad \beta \in [0, c_6].$$

Now, conditions (7.21) follow. □

The set $(c_5, c_6) \in [0, 1]^2$ satisfying the first of conditions (7.21) is shown in Figure 7.1. For example, by taking

$$c_2 = 1, \quad c_3 = \frac{1}{2}, \quad c_4 = 1, \quad c_5 = \frac{1}{2}, \quad c_6 = 1,$$

$$b_2(\theta) = 0, \quad b_3(\theta) = 0, \quad b_4(\theta) = 0,$$

$$a_{43}(\beta) = 0, \quad a_{65}(\beta) = 0,$$

we obtain the particular method of global order four which was proposed in Tavernini (1971).

We remark that there does not exist an explicit six-stage RK method (7.20) with distinct abscissae of uniform order four.

RFDEs with data set $\mathcal{LC}$

First, we consider equations satisfying the Regularity Assumption. For such equations, we can use expansion (7.13) in Theorem 7.1 for the local error function $e(\sigma, \varphi, h)$.

We consider an FCRK method satisfying condition (7.12) and a family $\mathcal{F}$ of integration problems (t_0, ϕ, T) such that conditions (i), (ii) and (iii) of

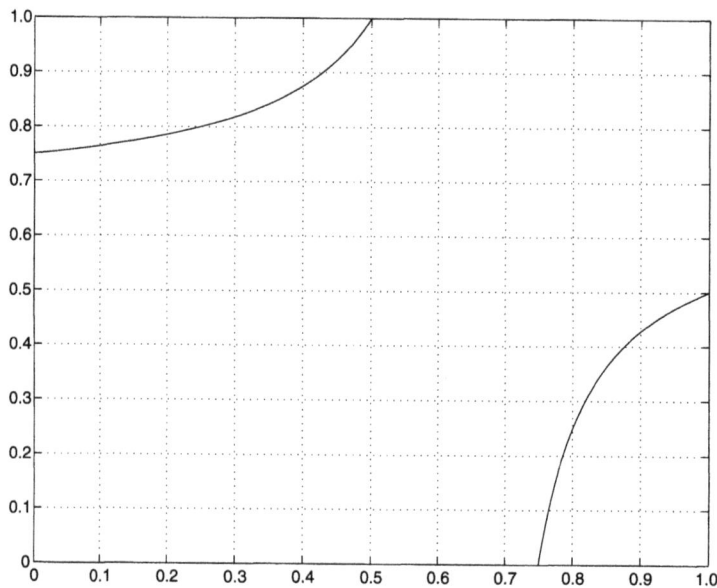

Figure 7.1. The curves are the set of couples $(c_5, c_6) \in [0,1]^2$ satisfying the first of conditions (7.21).

the previous subsection hold. A family of integration problems of NDDEs, NDIDEs and NSDDDEs with non-vanishing delays fulfils conditions (i), (ii) and (iii) if the function f, the delays and the kernel are sufficiently smooth.

In Table 7.2, we give the conditions for getting order one, two, three and four on the family $\mathcal{F}$.

It is clear that Table 7.2 reduces to Table 7.1 in the case of RFDEs with data set $\mathcal{C}$. Hence, the functional explicit Euler method has order one and the two-stage methods in the family (7.19) have order two.

Now, we consider equations not satisfying the Regularity Assumption and a family $\mathcal{F}$ of integration problems such that only condition (iii) of the previous subsection holds. By (7.1), (7.4) and Propositions 7.1 and 7.3, methods of type (7.15) attain order q if

$$\sum_{i=1}^{\nu} b'_i(\theta) c_i^{k-1} = \theta^{k-1}, \quad \theta \in [0,1] \text{ and } k = 1, \ldots, q,$$

or, equivalently,

$$\sum_{i=1}^{\nu} b_i(\theta) c_i^{k-1} = \frac{\theta^k}{k}, \quad \theta \in [0,1] \text{ and } k = 1, \ldots, q.$$

Table 7.2. Order conditions.

Order	Conditions
1	$\sum_{i=1}^{\nu} b_i'(\theta) = 1, \quad \theta \in [0,1]$
2	$\sum_{i=1}^{\nu} b_i'(\theta) c_i = \theta, \quad \theta \in [0,1]$
3	$\sum_{i=1}^{\nu} b_i'(\theta) c_i^2 = \theta^2, \quad \theta \in [0,1]$ For any $k = 1, \ldots, \nu^*$, $\sum_{\substack{i=1 \\ c_i = c_k^*}}^{\nu} b_i'(\theta) \left(\sum_{j=1}^{\nu} a_{ij}(\beta) c_j - \frac{\beta^2}{2} \right) = 0, \quad \theta \in [0,1], \ \beta \in [0, c_k^*]$
4	$\sum_{i=1}^{\nu} b_i'(\theta) c_i^3 = \theta^3, \quad \theta \in [0,1]$ For any $k = 1, \ldots, \nu^*$, $\sum_{\substack{i=1 \\ c_i = c_k^*}}^{\nu} b_i'(\theta) \left(\sum_{j=1}^{\nu} a_{ij}(\beta) c_j^2 - \frac{\beta^3}{3} \right) = 0, \quad \theta \in [0,1], \ \beta \in [0, c_k^*]$ For any $l, k = 1, \ldots, \nu^*$, $\sum_{\substack{i=1 \\ c_i = c_l^*}}^{\nu} \sum_{\substack{j=1 \\ c_j = c_k^*}}^{\nu} b_i'(\theta) a_{ij}(\beta) \left(\sum_{k=1}^{\nu} a_{jk}(\gamma) c_k - \frac{\gamma^2}{2} \right) = 0,$ for $\theta \in [0,1], \ \beta \in [0, c_l^*], \ \gamma \in [0, c_k^*]$

Hence, the one stage methods (7.18) (including the functional explicit Euler method) have order one and the two-stage semi-implicit methods

$$
\begin{array}{c|cc}
0 & 0 & 0 \\
c_2 & \theta - \frac{\theta^2}{2c_2} & \frac{\theta^2}{2c_2} \\
\hline
 & \theta - \frac{\theta^2}{2c_2} & \frac{\theta^2}{2c_2}
\end{array}
$$

have order two.

We conclude by remarking that, in both cases of equations satisfying and not satisfying the Regularity Assumption, the convergence Theorem 5.2 guarantees that a global order r on the family $\mathcal{F}$ is attained if the method has order r and is stable on $\mathcal{F}$. The method turns out to be stable on a family of integration problems of NDDEs, NDIDEs and NSDDIDEs if the function f, the delays and the kernel are sufficiently smooth.

8. The standard approach

In this section we outline the *standard approach* based on continuous RK methods, as described in the Introduction, applied to the specific classes of DDEs,

$$y'(t) = f(t, y(t), y(t - \tau(t, y(t)))), \quad t_0 \leq t \leq t_f, \qquad (8.1)$$
$$y(t) = \phi(t), \quad t \leq t_0,$$

and NDDEs

$$y'(t) = f(t, y(t), y(t - \tau(t, y(t))), y'(t - \tau(t, y(t)))), \quad t_0 \leq t \leq t_f, \qquad (8.2)$$
$$y(t) = \phi(t), \quad t \leq t_0.$$

In order to simplify the notation, we consider one single delay. Moreover, from now on, the end of the integration interval will be denoted by t_f instead of $t_0 + T$, which was used in the previous sections.

Given a mesh $\Delta = \{t_0, t_1, \ldots, t_n, \ldots, t_N = t_f\}$, the standard approach for (8.1) consists in solving step by step, by means of the chosen continuous RK method, the local problems

$$w'_{n+1}(t) = f(t, w_{n+1}(t), x(t - \tau(t, w_{n+1}(t)))), \quad t_n \leq t \leq t_{n+1}, \qquad (8.3)$$
$$w_{n+1}(t_n) = y_n,$$

where

$$x(s) = \begin{cases} \phi(s) & \text{for } s \leq t_0, \\ \eta(s) & \text{for } t_0 \leq s \leq t_n, \\ w_{n+1}(s) & \text{for } t_n \leq s \leq t_{n+1}, \end{cases}$$

and $\eta(s)$ is the continuous approximate solution computed by the method itself up to t_n.

Analogously, the standard approach for (8.2) consists in solving step by step the local problems

$$w'_{n+1}(t) = f(t, w_{n+1}(t), x(t - \tau(t, w_{n+1}(t))), z(t - \tau(t, w_{n+1}(t)))),$$
$$\text{for } t_n \leq t \leq t_{n+1},$$
$$w_{n+1}(t_n) = \eta(t_n),$$

where
$$x(s) = \begin{cases} \phi(s) & \text{for } s \leq t_0, \\ \eta(s) & \text{for } t_0 \leq s \leq t_n, \\ w_{n+1}(s) & \text{for } t_n \leq s \leq t_{n+1}, \end{cases}$$

$$z(s) = \begin{cases} \phi'(s) & \text{for } s \leq t_0, \\ \lambda(s) & \text{for } t_0 \leq s \leq t_n, \\ w'_{n+1}(s) & \text{for } t_n \leq s \leq t_{n+1}, \end{cases}$$

$\eta(t)$ is the continuous approximation of $y(t)$ and $\lambda(t)$ is an approximation of $y'(t)$ given by

$$\lambda(t) = \eta'(t) \tag{8.4}$$

or by

$$\lambda(t) = \mathcal{P}\Big(f(\cdot, \eta(\cdot), \eta(\cdot - \tau(\cdot, \eta(\cdot))), \lambda(\cdot - \tau(\cdot, \eta(\cdot))))\Big)(t), \tag{8.5}$$

where, in each mesh interval $[t_k, t_{k+1}]$, $\mathcal{P}$ is an interpolation operator in a suitable polynomial space of degree possibly other than $\deg(\eta')$ and nodes in $[t_k, t_{k+1}]$.

Here we report a condensed survey on continuous RK methods for ODEs as a basic tool for the implementation of the standard approach for RFDEs. Then we provide the main results on the error analysis of the resulting methods for equations (8.1) and (8.2), also in view of the particular issues treated in the forthcoming Sections 9, 10 and 11. These topics are covered in the book by Bellen and Zennaro (2003) and, hence, neither proofs nor bibliographic references are given here.

8.1. Continuous RK methods for ODEs

Given a mesh $\Delta = \{t_0, t_1, \ldots, t_n, \ldots, t_N = t_f\}$, a ν-stage RK method for the numerical solution of the ODE

$$\begin{aligned} y'(t) &= g(t, y(t)), \quad t_0 \leq t \leq t_f, \\ y(t_0) &= y_0, \end{aligned} \tag{8.6}$$

has the form (in the so-called Y notation)

$$Y_{n+1}^i = y_n + h_{n+1} \sum_{j=1}^{\nu} a_{ij} g(t_{n+1}^j, Y_{n+1}^j), \quad i = 1, \ldots, \nu, \tag{8.7}$$

$$y_{n+1} = y_n + h_{n+1} \sum_{i=1}^{\nu} b_i g(t_{n+1}^i, Y_{n+1}^i), \tag{8.8}$$

where $t_{n+1}^i = t_n + c_i h_{n+1}$, $c_i = \sum_{j=1}^{\nu} a_{ij}$, $i = 1, \ldots, \nu$, $h_{n+1} = t_{n+1} - t_n$ and ν is referred to as the number of *stages*. The b_is are called *weights* of the quadrature formula (8.8) and the c_is are called *abscissae* and, for most common methods, they belong to $[0,1]$. Since the RK method (8.7), (8.8) is characterized by the weights b_i and the matrix coefficients $A = (a_{ij})_{i,j=1}^{\nu}$, it will be denoted by (A, b, c). It is worth observing that in many papers and books the RK formulae are written in an equivalent different form, the so-called K *notation*. So the RK method (8.7), (8.8) takes the form

$$K_{n+1}^i = g\left(t_{n+1}^i, y_n + h_{n+1} \sum_{j=1}^{\nu} a_{ij} K_{n+1}^j\right), \quad i = 1, \ldots, \nu,$$

$$y_{n+1} = y_n + h_{n+1} \sum_{i=1}^{\nu} b_i K_{n+1}^i.$$

Note that K notation is obtained by setting

$$K_{n+1}^i = g(t_{n+1}^i, Y_{n+1}^i), \quad i = 1, \ldots, \nu,$$

in (8.7), (8.8).

Although in developing and implementing RK methods for ODEs the two notations are basically equivalent, in the application of RK methods to DDEs it will often be preferable to adopt the K notation.

The computational complexity of the method is mainly determined by the number of stages and by the form of the coefficient matrix A. It is well known that when the matrix A is lower triangular with zero diagonal elements, the method is called *explicit* and the computational cost is lower, whereas when the matrix A is full, the method is called *implicit* and the computational cost is higher.

The one-step interpolants of the RK method (8.7), (8.8) are constructed step by step by making use of information from the underlying mesh interval $[t_n, t_{n+1}]$ only, possibly by including some additional stages, that is, by some extra evaluations of the right-hand side function $g(t, y)$ in (8.6).

Interpolants constructed using no extra stages are called *interpolants of the first class* and the resulting continuous extension $\eta(t)$ is defined, in each subinterval of the mesh Δ, by a one-step continuous quadrature rule of the form

$$\eta(t_n + \theta h_{n+1}) = y_n + h_{n+1} \sum_{i=1}^{\nu} b_i(\theta) g(t_{n+1}^i, Y_{n+1}^i), \quad 0 \leq \theta \leq 1, \quad (8.9)$$

or, in K notation,

$$\eta(t_n + \theta h_{n+1}) = y_n + h_{n+1} \sum_{i=1}^{\nu} b_i(\theta) K_{n+1}^i, \quad 0 \leq \theta \leq 1,$$

where the $b_i(\theta)$s are polynomials of suitable degree $\leq \delta$ satisfying

$$b_i(0) = 0 \quad \text{and} \quad b_i(1) = b_i, \quad i = 1, \ldots, \nu, \quad (8.10)$$

so as to define a continuous piecewise polynomial function.

Interpolants constructed by means of additional stages are called *interpolants of the second class* and the continuous extension is given by

$$\eta(t_n + \theta h_{n+1}) = y_n + h_{n+1} \sum_{i=1}^{s} b_i(\theta) g(t_{n+1}^i, Y_{n+1}^i), \quad 0 \le \theta \le 1, \quad (8.11)$$

or, in K notation, by

$$\eta(t_n + \theta h_{n+1}) = y_n + h_{n+1} \sum_{i=1}^{s} b_i(\theta) K_{n+1}^i, \quad 0 \le \theta \le 1, \quad (8.12)$$

where the $b_i(\theta)$s are again polynomials of suitable degree $\le \delta$ satisfying the continuity conditions

$$\begin{aligned} b_i(0) &= 0, & i &= 1, \ldots, s, \\ b_i(1) &= b_i, & i &= 1, \ldots, \nu, \\ b_i(1) &= 0, & i &= \nu+1, \ldots, s. \end{aligned} \quad (8.13)$$

The additional $s - \nu$ stages are given by

$$Y_{n+1}^i = y_n + h_{n+1} \sum_{j=1}^{s} a_{ij} g(t_{n+1}^j, Y_{n+1}^j), \quad i = \nu+1, \ldots, s, \quad (8.14)$$

or, in K notation, by

$$K_{n+1}^i = g\left(t_{n+1}^i, y_n + h_{n+1} \sum_{j=1}^{s} a_{ij} K_{n+1}^j\right), \quad i = \nu+1, \ldots, s,$$

so that the original coefficient matrix $A = (a_{ij})_{i,j=1}^{\nu}$ is embedded into the block lower triangular matrix

$$A' = \begin{pmatrix} A & 0 \\ (a_{ij})_{i=\nu+1,j=1}^{s,\nu} & (a_{ij})_{i,j=\nu+1}^{s} \end{pmatrix}. \quad (8.15)$$

The overall *continuous Runge–Kutta* methods (8.7), (8.8), (8.9) and (8.7), (8.8), (8.14), (8.11), denoted by $(A, b(\theta), c)$ and $(A', b(\theta))$, respectively, are the continuous extensions of the RK method (A, b, c) and δ will be referred to as the *degree of the interpolant*. In contrast, the method (A, b, c) will be called the *underlying (discrete) RK method*.

It is worth remarking that, in general, $\eta(t_n + c_i h_{n+1}) \ne Y_{n+1}^i$. Nevertheless, equality holds for every right-hand side $g(t, y)$ whenever $b_i(c_j) = a_{ji}$, as appears evident by comparing (8.11) and (8.14). So we have

$$\{\eta(t_n + c_i h_{n+1}) = Y_{n+1}^i \;\; \forall i\} \iff \{b_i(c_j) = a_{ji} \;\; \forall i, j\}. \quad (8.16)$$

An interpolant, either of the first or second class, determines a matrix B, whose elements are $b_{ij} = b_j(c_i)$.

Definition 8.1. A continuous RK method is called natural if $A = B$ ($A' = B$).

As for the order of RK methods and their interpolants, we have the following definition.

Definition 8.2. We say that the RK method (8.7), (8.8) is consistent of order (or, equivalently, has order) p if $p \geq 1$ is the largest integer such that, for all C^p-continuous right-hand side functions $g(t,y)$ in (8.6) and for all mesh points, we have that

$$|z_{n+1}(t_{n+1}) - y_{n+1}| = \mathcal{O}(h_{n+1}^{p+1}),$$

uniformly with respect to y_n^* in any bounded subset of $\mathbb{R}^d$ and to $n = 0, \ldots, N-1$, where $z_{n+1}(t)$ is the *local solution* to the *local problem*

$$\begin{aligned} z'_{n+1}(t) &= g(t, z_{n+1}(t)), \quad t_n \leq t \leq t_{n+1}, \\ z_{n+1}(t_n) &= y_n^*. \end{aligned} \qquad (8.17)$$

We say that the interpolant (8.9) or (8.11) is consistent of uniform order (or, equivalently, has uniform order) q if $q \geq 1$ is the largest integer such that, for all C^q-continuous right-hand side functions $g(t,y)$ and for all mesh points, we have that

$$\max_{t_n \leq t \leq t_{n+1}} |z_{n+1}(t) - \eta(t)| = \mathcal{O}(h_{n+1}^{q+1}).$$

The convergence results are summarized by the following theorem.

Theorem 8.1. Let the RK method (8.7), (8.8) be consistent of *order p* and let the right-hand side function $g(t,y)$ in (8.6) be C^p-continuous. Then the method is convergent of order (or, equivalently, has global order) p on any bounded interval $[t_0, t_f]$, that is,

$$\max_{1 \leq n \leq N} |y(t_n) - y_n| = \mathcal{O}(h^p), \qquad (8.18)$$

where $h = \max_{1 \leq n \leq N} h_n$.

If the interpolant (8.9) or (8.11) has uniform order q, then the continuous RK method (8.7), (8.8), (8.9) or (8.7), (8.8), (8.14), (8.11) is uniformly convergent of order (or, equivalently, has uniform global order) $q' = \min\{p, q+1\}$, that is,

$$\max_{t_0 \leq t \leq t_f} |y(t) - \eta(t)| = \mathcal{O}(h^{q'}). \qquad (8.19)$$

We shall often refer to the order of consistency and to the order of convergence of the RK method (8.7), (8.8) as the *discrete order* and the *discrete global order* of the continuous RK method (8.7), (8.8), (8.9) or (8.7), (8.8), (8.14), (8.11).

The following theorem provides additional results on the derivatives of the continuous extension.

Theorem 8.2. If, in addition to the hypotheses of Theorem 8.1, the interpolant is a piecewise polynomial of degree $\delta \geq q$ and the right-hand side function $g(t, y)$ in (8.6) is $C^{\max\{\delta, p\}}$-continuous, then the following convergence, boundedness and unboundedness estimates hold for the derivatives of the global error function:

$$\max_{t_0 \leq t \leq t_f} |y^{(j)}(t) - \eta^{(j)}(t)| = \mathcal{O}(h^{q+1-j}), \quad j = 1, \ldots, \delta, \quad (8.20)$$

where the derivatives of $\eta(t)$ at the mesh points are taken in the left/right sense.

The estimates (8.19) and (8.20) show that the first derivative retains the global uniform order of the interpolant if and only if the interpolant has the maximum attainable uniform order p. It is also evident that, in order to get the uniform order q, the interpolant must be of degree $\delta \geq q$. On the other hand, polynomials of degree $\delta > q$ are unnecessary, as shown by the following theorem.

Theorem 8.3. Assume that the RK method (8.7), (8.8) has a continuous extension $\eta(t)$ of order q and degree $d > q$. Then there exists another continuous extension $\tilde{\eta}(t)$ of order q whose degree is also q.

Remark 8.1. By Theorems 8.2 and 8.3 we may observe that, not only is the employment of interpolants of degree higher than q unnecessary, but interpolants of degree $\delta > q + 1$ are even dangerous in that the derivatives of order k, with $q + 2 \leq k \leq \delta$, may diverge as $h \to 0$. For these reasons we shall assume that continuous extensions of order q will be always made by interpolants of degree $\delta = q$.

It is important to give an answer to the following two questions.

Question 1. What is the maximum uniform order an RK method of order p can achieve by means of an interpolant of the first class?

Question 2. What is the (minimum) number of stages necessary to construct a continuous RK method of uniform order $p - 1$ or even p?

So far, we can give the following upper bound to the uniform order of an interpolant (for both classes).

Theorem 8.4. Assume that the RK method (8.7), (8.8) has a continuous extension $\eta(t)$ given by (8.11). Then its uniform order q cannot exceed s^*, the number of distinct abscissae of the extended RK method represented by (8.15).

Table 8.1. Order conditions for continuous RK methods.

Order	Conditions
1	$\sum_{i=1}^{\nu} b_i(\theta) = \theta$
2	$\sum_{i=1}^{\nu} b_i(\theta) c_i = \frac{1}{2}\theta^2$
3	$\sum_{i=1}^{\nu} b_i(\theta) c_i^2 = \frac{1}{3}\theta^3$ $\sum_{i,j=1}^{\nu} b_i(\theta) a_{ij} c_j = \frac{1}{6}\theta^3$
4	$\sum_{i=1}^{\nu} b_i(\theta) c_i^3 = \frac{1}{4}\theta^4$ $\sum_{i,j=1}^{\nu} b_i(\theta) c_i a_{ij} c_j = \frac{1}{8}\theta^4$ $\sum_{i,j=1}^{\nu} b_i(\theta) a_{ij} c_j^2 = \frac{1}{12}\theta^4$ $\sum_{i,j,k=1}^{\nu} b_i(\theta) a_{ij} a_{jk} c_k = \frac{1}{24}\theta^4$

The above result is obvious after observing that formula (8.11) is a continuous quadrature rule based on exactly s^* distinct abscissae.

Since the construction of interpolants of the second class is a rather technical matter, here we confine ourselves to analysing only the interpolants of the first class and, consequently, we shall not give an answer to Question 2. However, we shall briefly consider the direct construction of continuous RK methods, without passing necessarily through interpolants of the second class of a given discrete RK formula.

A general analysis of the uniform order for the continuous extension (8.9) is based on the property that, for any $0 < \theta \leq 1$, it can be viewed as the discrete method $(\frac{A}{\theta}, \frac{b(\theta)}{\theta}, \frac{c}{\theta})$ with step-size θh_{n+1}. So, we immediately get the uniform order conditions for the polynomials $b_i(\theta)$ from the well-known order conditions of the RK methods. The conditions up to order $p = 4$ are shown in Table 8.1.

In order to answer Question 1, each method has to be analysed individually by checking the order conditions. In general, we can give only a partial answer by means of the following theorem, the proof of which does not directly involve the order conditions.

Theorem 8.5. Every RK method (8.7), (8.8) of order $p \geq 1$ has a continuous extension $\eta(t)$ of order (and degree) $q = 1, \ldots, \lfloor \frac{p+1}{2} \rfloor$.

Theorem 8.6. If an RK method (8.7), (8.8) has a continuous extension $\eta(t)$ of order (and degree) $q \geq 2$, then it also has another continuous extension $\tilde{\eta}(t)$ of order (and degree) $\tilde{q}$ for each $\tilde{q} \leq q - 1$.

In conclusion, we can answer Question 1 by saying that, in general, only interpolants up to order $\lfloor \frac{p+1}{2} \rfloor$ are ensured to exist. On the other hand, it might well be that the maximum uniform order reachable by means of interpolants of the first class is actually $> \lfloor \frac{p+1}{2} \rfloor$ and, possibly, even $= p$.

Definition 8.3. We say that an RK method (8.7), (8.8) of discrete order p is superconvergent if the maximum uniform order q reachable by means of interpolants of the first class is $\leq p - 1$.

In other words, *superconvergence* is attained at the end-point of the step-interval with respect to the maximum uniform accuracy order q. Of course, it might well be that the interpolant attains a higher order $p' > q$, not necessarily equal to the discrete order p, also at some additional points inside the step-interval. They will be called *inner superconvergence points*.

Collocation methods

A particular class of continuous RK methods that has been studied extensively are the *one-step collocation* methods. However, the interest in piecewise collocation is mostly due to the simplicity in determining the order of convergence and superconvergence via the non-linear variation-of-constants formula and to the optimal stability properties as discrete methods, rather than to its intrinsically continuous nature.

The one-step collocation method can be defined as follows. Choose ν distinct abscissae $c_1, \ldots, c_\nu \in [0, 1]$ and, in each mesh interval $[t_n, t_{n+1}]$, compute the polynomial $\eta(t)$ of degree $\leq \nu$ satisfying

$$\eta'(t_{n+1}^i) = g(t_{n+1}^i, \eta(t_{n+1}^i)), \quad i = 1, \ldots, \nu, \quad \eta(t_n) = y_n.$$

It is easy to check that such methods can be rewritten as a continuous implicit RK method (8.7), (8.9), where

$$a_{ij} = \int_0^{c_i} \ell_j(\xi)\, d\xi, \quad i, j = 1, \ldots, \nu,$$

$$b_i(\theta) = \int_0^\theta \ell_i(\xi)\, d\xi, \quad i = 1, \ldots, \nu,$$

$\ell_i(\xi)$ being the Lagrange polynomial coefficient

$$\prod_{k=1, k\neq i}^{\nu} \frac{\xi - c_k}{c_i - c_k}.$$

In particular, we have $b_i(c_j) = a_{ji}$ and, therefore, any collocation method is a natural continuous RK method.

For any choice of the abscissae $c_1, \ldots, c_\nu \in [0, 1]$, the collocation method has order $p \geq \nu$ and the uniform order of the interpolant (8.9) is $q = \nu$. Consequently, by Theorem 8.1, the collocation method is a continuous RK method of global uniform order $q' = \nu$ (if $p = \nu$) or $q' = \nu + 1$ (if $p > \nu$). In this sense the collocation method is optimal in that it achieves the maximum attainable uniform order for the given number of stages. In particular, if the abscissae are the shifted roots of the Legendre orthogonal polynomial of degree ν, then the method has order $p = 2\nu$. This is the most famous example of superconvergence.

Direct construction of continuous RK methods

So far we have considered continuous extensions of *a priori* given discrete RK methods. Now we consider the other philosophy of constructing directly a continuous RK method, without necessarily starting from a given discrete formula.

As already pointed out in this section, a general analysis of the uniform order for the continuous extensions (8.9) is based on the property that, for any $0 < \theta \leq 1$, it can be viewed as a discrete method $\left(\frac{A}{\theta}, \frac{b(\theta)}{\theta}, \frac{c}{\theta}\right)$ with scaled step-size θh_{n+1}. So we immediately get the uniform order conditions for the parameters c_i and a_{ij} and for the polynomials $b_i(\theta)$ from the well-known order conditions of the RK methods (see Table 8.1 for conditions up to order $p = 4$).

Let $N(p)$ and $CN(q)$ be the minimum number of stages for which there exist RK methods of (discrete) order p and continuous RK methods of uniform order q, respectively. Similarly, let $EN(p)$ and $CEN(q)$ be the same quantities restricted to the class of *explicit* RK methods and *continuous explicit* RK methods.

In the general case, it is well known that

$$N(p) = \left\lfloor \frac{p+1}{2} \right\rfloor \quad \text{and} \quad CN(q) = q$$

and that these optimal bounds are attained, for instance, by collocation methods.

For explicit RK and continuous explicit RK methods the results are often obtained by making somewhat sophisticated analyses of the continuous order conditions.

Table 8.2. The minimum number of stages necessary for an explicit RK method to attain the discrete order p.

p	1	2	3	4	5	6	7	8	≥ 9
$EN(p)$	1	2	3	4	6	7	9	11	$\geq p+3$

Table 8.3. The minimum number of stages necessary for a continuous explicit RK method to attain the uniform order q.

q	1	2	3	4	5	6	≥ 7
$CEN(q)$	1	2	4	6	8	11	$\geq 2q-2$

All the order barriers for explicit methods are summarized in Tables 8.2 and 8.3.

Now we concentrate on continuous explicit RK methods with a minimum number of stages $CEN(q)$. It easily turns out that, for given $q \geq 2$, a whole family of such methods exists, which depends on a certain number of parameters. So the parameters can be selected in order to guarantee some nice properties of the method, such as minimization of a suitable estimate of the local error constant and maximization of the absolute stability region of the underlying discrete method.

Another nice characteristic of some continuous explicit RK methods is the FSAL (*first same as last*) property. The FSAL property means that the last stage can be re-used as the first stage $K_{n+1}^1 = g(t_{n+1}, y_{n+1})$ of the next step. This implies that the actual cost of the method is reduced by one function evaluation per step. Of course, because the method is explicit, the re-usable stage can be involved only for computation of the interpolant $\eta(t_n + \theta h_{n+1})$ for $\theta \neq 1$ and not for computation of $y_{n+1} = \eta(t_{n+1})$.

8.2. RK methods for DDEs

Once the continuous RK method $(A, b(\theta), c)$ is chosen, the standard approach for the DDE (8.1) turns out to be

$$\eta(t_n + \theta h_{n+1}) \qquad (8.21)$$
$$= y_n + h_{n+1} \sum_{i=1}^{s} b_i(\theta) f\left(t_{n+1}^i, Y_{n+1}^i, \eta\left(t_{n+1}^i - \tau(t_{n+1}^i, Y_{n+1}^i)\right)\right),$$

for $0 \leq \theta \leq 1$, and

$$Y_{n+1}^i = y_n + h_{n+1} \sum_{j=1}^{s} a_{ij} f\left(t_{n+1}^j, Y_{n+1}^j, \eta(t_{n+1}^j - \tau(t_{n+1}^j, Y_{n+1}^j))\right), \quad (8.22)$$

for $i = 1, \ldots, s$.

In this section, the method will be called the *RK method for DDEs* or, in short, the *DDE method*, and $(A, b(\theta), c)$ will be referred to as the *underlying continuous RK method*.

Note that the use of RK methods with an abscissa $c_i > 1$ could lead to an advanced deviated argument $t_{n+1}^i - \tau(t_{n+1}^i, Y_{n+1}^i) > t_{n+1}$, where the continuous extension $x(s)$ should be computed in some subsequent step. Therefore, in order to avoid such a disappointing situation, we assume that the abscissae satisfy the constraint

$$0 \leq c_i \leq 1, \quad i = 1, \ldots, s. \quad (8.23)$$

However, even under condition (8.23), it may well be that, for some index i, the argument $t_{n+1}^i - \tau(t_{n+1}^i, Y_{n+1}^i)$ of $\eta(s)$ lies in the current interval $[t_n, t_{n+1}]$. We shall call this occurrence *overlapping*. It is convenient to define the *spurious stage*

$$\tilde{Y}_{n+1}^i = \eta\left(t_{n+1}^i - \tau(t_{n+1}^i, Y_{n+1}^i)\right)$$

which, in the case of overlapping, is given by formula (8.21) itself for

$$\theta = \theta_{n+1}^i = c_i - \frac{\tau(t_{n+1}^i, Y_{n+1}^i)}{h_{n+1}}.$$

It is worth remarking that the overall method becomes implicit even if the underlying continuous RK method is explicit. This makes a remarkable difference with respect to the explicit FCRK methods, described in Section 6, which preserve their explicitness even in the case of overlapping.

On the contrary, if overlapping does not occur, the spurious stage is simply given by the interpolant $\eta(t)$ as computed in the past.

In any case, in the mesh interval $[t_n, t_{n+1}]$ the method takes the form (in Y notation)

$$\eta(t_n + \theta h_{n+1}) = y_n + h_{n+1} \sum_{i=1}^{s} b_i(\theta) f\left(t_{n+1}^i, Y_{n+1}^i, \tilde{Y}_{n+1}^i\right), \quad 0 \leq \theta \leq 1,$$
$$(8.24)$$

$$Y_{n+1}^i = y_n + h_{n+1} \sum_{j=1}^{s} a_{ij} f\left(t_{n+1}^j, Y_{n+1}^j, \tilde{Y}_{n+1}^j\right), \quad i = 1, \ldots, s, \quad (8.25)$$

where the spurious stages $\tilde{Y}^i_{n+1}$ are implicitly given by

$$\tilde{Y}^i_{n+1} = y_n + h_{n+1}\sum_{j=1}^{s} b_j(\theta^i_{n+1}) f\left(t^j_{n+1}, Y^j_{n+1}, \tilde{Y}^j_{n+1}\right) \tag{8.26}$$

if the overlapping condition $t^i_{n+1} - \tau(t^i_{n+1}, Y^i_{n+1}) > t_n$ holds, and by the known value

$$\tilde{Y}^i_{n+1} = \eta\left(t^i_{n+1} - \tau(t^i_{n+1}, Y^i_{n+1})\right) \tag{8.27}$$

otherwise.

Note that, whereas the system (8.24), (8.25), (8.27) has to be solved only for the stage values Y^j_{n+1}, $j = 1, \ldots, s$, the system enlarged by (8.26) for some i has to be solved also for the relevant spurious stages $\tilde{Y}^i_{n+1}$.

Indeed, the dimension of the system is not increased. In fact, by using K notation

$$K^i_{n+1} = f\left(t^i_{n+1}, Y^i_{n+1}, \tilde{Y}^i_{n+1}\right),$$

we get the following system to be solved for K^i_{n+1}, $i = 1, \ldots, s$:

$$\eta(t_n + \theta h_{n+1}) = y_n + h_{n+1}\sum_{i=1}^{s} b_i(\theta) K^i_{n+1}, \quad 0 \le \theta \le 1, \tag{8.28}$$

$$K^i_{n+1} = f\left(t^i_{n+1}, y_n + h_{n+1}\sum_{j=1}^{s} a_{ij} K^j_{n+1}, \tilde{Y}^i_{n+1}\right), \quad i = 1, \ldots, s, \tag{8.29}$$

where

$$\tilde{Y}^i_{n+1} = y_n + h_{n+1}\sum_{j=1}^{s} b_j\left(c_i - \frac{\tau(t^i_{n+1}, y_n + h_{n+1}\sum_{k=1}^{s} a_{ik} K^k_{n+1})}{h_{n+1}}\right) K^j_{n+1} \tag{8.30}$$

if the overlapping condition

$$t^i_{n+1} - \tau\left(t^i_{n+1}, y_n + h_{n+1}\sum_{k=1}^{s} a_{ik} K^k_{n+1}\right) > t_n$$

holds, and

$$\tilde{Y}^i_{n+1} = \eta\left(t^i_{n+1} - \tau\left(t^i_{n+1}, y_n + h_{n+1}\sum_{k=1}^{s} a_{ik} K^k_{n+1}\right)\right) \tag{8.31}$$

otherwise.

Despite it being impossible to express all RK methods for DDEs in terms of the stage values Y^i_{n+1} only, there are particular classes, essentially collocation methods, that allow us to express the spurious stages $\tilde{Y}^i_{n+1}$ in the system (8.25) in terms of the Y^i_{n+1}. This is the case for any *natural* continuous RK method (see Definition 8.1) with s distinct abscissae $c_1, \ldots, c_s$ such that $c_i \neq 0$, $i = 1, \ldots, s$, and a continuous extension $\eta(t_n + \theta h_{n+1})$ of degree s. In fact, in this case the polynomial $\eta(t)$ may be written using the Lagrange interpolation formula through the $s+1$ values $y_n (= \eta(t_n))$ and $Y_{n+1}^i (= \eta(t_n + c_i h_{n+1}))$, $i = 1, \ldots, s$, that is,

$$\eta(t_n + \theta h_{n+1}) = \ell_0(\theta) y_n + \sum_{i=1}^{s} \ell_i(\theta) Y_{n+1}^i, \qquad (8.32)$$

where ℓ_j, $j = 0, \ldots, s$ are the Lagrange polynomial coefficients relevant to the nodes $c_0 = 0$ and c_i, $i = 1, \ldots, s$. Therefore $\tilde{Y}_{n+1}^i$, which is equal to $\eta(t_{n+1}^i - \tau(t_{n+1}^i, Y_{n+1}^i))$, may be written using (8.32) for

$$\theta = \theta_{n+1}^i = c_i - \frac{\tau(t_{n+1}^i, Y_{n+1}^i)}{h_{n+1}}.$$

The Gaussian collocation and Radau IIA methods satisfy the above condition and are natural choices for the construction of DDE methods.

For both Y and K notation, the method is well-posed for any sufficiently small h_{n+1}, as stated by the following theorem.

Theorem 8.7. (Well-posedness) Assume that the local problem (8.3) possesses a unique solution $w_{n+1}(t)$. Then, for sufficiently small step-size h_{n+1}, equations (8.21)–(8.22) admit a unique solution $\eta(t)$.

As for the convergence analysis of the DDE methods, we have the following result, assuming we are able to compute and include the discontinuity points in the mesh, even in the state-dependent delay case.

Theorem 8.8. (Convergence) Consider the DDE

$$y'(t) = f(t, y(t), y(t - \tau(t, y(t)))), \quad t_0 \leq t \leq t_f,$$
$$y(t) = \phi(t), \quad t \leq t_0,$$

where $f(t, y, x)$ is C^p-continuous in $[t_0, t_f] \times \mathbb{R}^d \times \mathbb{R}^d$, the initial function $\phi(t)$ is C^p-continuous and the delay $\tau(t, y)$ is C^p-continuous in $[t_0, t_f] \times \mathbb{R}^d$. Moreover, assume that the mesh $\Delta = \{t_0, t_1, \ldots, t_n, \ldots, t_N = t_f\}$ includes all the discontinuity points lying in $[t_0, t_f]$ where the solution $y(t)$ is not at least C^p-continuous. If the underlying continuous RK method has discrete order p and uniform order q, then the DDE method (8.24), (8.25), (8.26), (8.27) has discrete global order and uniform global order $q' = \min\{p, q+1\}$, that is,

$$\max_{1 \leq n \leq N} |y(t_n) - y_n| = \mathcal{O}(h^{q'})$$

and

$$\max_{t_0 \leq t \leq t_f} |y(t) - \eta(t)| = \mathcal{O}(h^{q'}),$$

where $h = \max_{1 \leq n \leq N} h_n$.

According to Theorem 8.8, if the underlying continuous RK method has discrete order p and uniform order q, then we can either be satisfied with a DDE method with, possibly lower, uniform global order $q' = \min\{p, q+1\}$, or increase the uniform order of the underlying interpolant up to at least $p-1$ in order to preserve the uniform global order p.

We can summarize the last option in the following corollary.

Corollary 8.1. Under the hypotheses of Theorem 8.8 with $q \geq p-1$, the continuous numerical solution $\eta(t)$ is such that

$$\max_{t_0 \leq t \leq t_f} |y(t) - \eta(t)| = \mathcal{O}(h^p).$$

Theorem 8.8 and Corollary 8.1 just guarantee that, by using an interpolant of order $p-1$, the global order p of the discrete method is preserved for any choice of the mesh. A sharper error estimate and convergence analysis of the standard approach reveals that, under some restrictions on the mesh, the condition $q = p-1$ is no longer necessary for the method to preserve the global order p. In other words, superconvergence is possible. On the other hand, an efficient DDE code ought to be implemented in a variable step-size mode by performing the error control. In this case, if we try to estimate the local error by a method of higher order $p+1$, uniform approximation of order $p-1$ for the deviated arguments $y(t-\tau)$ is not sufficient and must be raised to p. For a deep analysis of these aspects, we again refer the interested reader to Bellen and Zennaro (2003).

We remark that the DDE method with underlying continuous RK method $(A, b(\theta), c)$ provides an approximation of the solution map of form (5.3). In particular, we have

$$V(t_n, \eta_{t_n}, h_{n+1}) = \pi K_{n+1},$$

where π is the prolongation operator defined in (6.4) and

$$K_{n+1} = (K_{n+1}^1, \ldots, K_{n+1}^\nu) \in (\mathbb{R}^d)^\nu.$$

Consequently, Theorem 8.8 above may also be obtained as a corollary to the general convergence Theorem 5.1.

It is also worth remarking that a DDE method based on a natural continuous RK method $(A, b(\theta), c)$ (see Definition 8.1) provides the same approximation of the solution map as the one provided by the particular implicit FCRK method $(A(\theta), b(\theta), c)$, where

$$a_{ij}(\theta) = b_j(\theta), \quad i, j = 1, \ldots, \nu.$$

In fact, for such an FCRK method, all the stage functions Y^i, $i = 1, \ldots, \nu$, coincide with the function $\eta(t - t_n)$, $t \in (-\infty, h_{n+1}]$. On the contrary, if the underlying continuous RK method is not natural, then the DDE method does not fall into the class of FCRK methods introduced in Section 6.

Consequently, the well-posedness result expressed by Theorem 8.7 may also be obtained as a corollary to Theorem 6.1 only when the underlying continuous RK method is natural.

RK methods for NDDEs

With respect to the NDDE (8.2), for the choice (8.4), the DDE method (8.22) and (8.21) in Y notation modifies to the following *RK method for NDDEs*:

$$Y_{n+1}^i = y_n + h_{n+1}\sum_{j=1}^{s} a_{ij} f\left(t_{n+1}^j, Y_{n+1}^j, \tilde{Y}_{n+1}^j, \tilde{Z}_{n+1}^j\right), \qquad (8.33)$$

for $i = 1, \ldots, s$,

$$\eta(t_n + \theta h_{n+1}) = y_n + h_{n+1}\sum_{i=1}^{s} b_i(\theta) f\left(t_{n+1}^i, Y_{n+1}^i, \tilde{Y}_{n+1}^i, \tilde{Z}_{n+1}^i\right), \qquad (8.34)$$

for $0 \le \theta \le 1$, and

$$\lambda(t_n + \theta h_{n+1}) = \sum_{i=1}^{s} b_i'(\theta) f\left(t_{n+1}^i, Y_{n+1}^i, \tilde{Y}_{n+1}^i, \tilde{Z}_{n+1}^i\right), \qquad (8.35)$$

for $0 \le \theta \le 1$, where

$$\tilde{Y}_{n+1}^j = \eta\left(t_{n+1}^j - \tau(t_{n+1}^j, Y_{n+1}^j)\right) \quad \text{and} \quad \tilde{Z}_{n+1}^j = \lambda\left(t_{n+1}^j - \tau(t_{n+1}^j, Y_{n+1}^j)\right).$$

Note that, for the arguments $s_j = t_{n+1}^j - \tau(t_{n+1}^j, Y_{n+1}^j)$, the values $\eta(s_j)$ and $\lambda(s_j)$ may or may not be known. If overlapping occurs, that is if, for some index i, the argument $s_i > t_n$, then the *spurious stages* $\tilde{Y}_{n+1}^i$ and $\tilde{Z}_{n+1}^i$ are unknown, and are given by (8.34) and (8.35) for

$$\theta = \theta_{n+1}^i = c_i - \frac{\tau(t_{n+1}^i, Y_{n+1}^i)}{h_{n+1}},$$

that is,

$$\tilde{Y}_{n+1}^i = y_n + h_{n+1}\sum_{j=1}^{s} b_j(\theta_{n+1}^i) f\left(t_{n+1}^j, Y_{n+1}^j, \tilde{Y}_{n+1}^j, \tilde{Z}_{n+1}^j\right),$$

$$\tilde{Z}_{n+1}^i = \sum_{j=1}^{s} b_j'(\theta_{n+1}^i) f\left(t_{n+1}^j, Y_{n+1}^j, \tilde{Y}_{n+1}^j, \tilde{Z}_{n+1}^j\right).$$

On the contrary, if the arguments of $\eta(s)$ and $\lambda(s)$ lie outside the current interval $[t_n, t_{n+1}]$, then the values $\tilde{Y}_{n+1}^j$ and $\tilde{Z}_{n+1}^j$ are given by the

interpolants $\eta(s)$ and $\eta'(s)$ as computed at the past points
$$t^i_{n+1} - \tau(t^i_{n+1}, Y^i_{n+1}) = t_{n+1-m} + \theta h_{n+1-m}$$
for suitable values of m and θ.

As with DDEs with no neutral terms, the spurious stages $\tilde{Y}^i_{n+1}$ and $\tilde{Z}^i_{n+1}$, if any, only apparently increase the dimension of the system to be solved at each step. In fact, by using K notation
$$K^i_{n+1} = f(t^i_{n+1}, Y^i_{n+1}, \tilde{Y}^i_{n+1}, \tilde{Z}^i_{n+1}),$$
all the stages Y^i_{n+1}, $\tilde{Y}^i_{n+1}$ and $\tilde{Z}^i_{n+1}$, as well as the arguments θ^i_{n+1}, turn out to depend on K^i_{n+1} only.

Remark 8.2. As in the non-neutral case, for any natural continuous RK method with s distinct abscissae $c_1, \ldots, c_s$ such that $c_i \neq 0$, $i = 1, \ldots, s$, and continuous extension $\eta(t)$ of degree s, the system to be solved at each step may be stated in terms of the sole Y^i_{n+1}s. In fact, the polynomial $\eta(t_n + \theta h_{n+1})$ may be written using the Lagrange interpolation formula through the $s+1$ values $y_n(= \eta(t_n))$ and $Y^i_{n+1}(= \eta(t_n + c_i h_{n+1}))$, $i = 1, \ldots, s$, that is,
$$\eta(t_n + \theta h_{n+1}) = \ell_0(\theta) y_n + \sum_{i=1}^{s} \ell_i(\theta) Y^i_{n+1}, \tag{8.36}$$
where ℓ_j, $j = 0, \ldots, s$ are the Lagrange polynomial coefficients on the nodes $c_0 = 0$ and c_i, $i = 1, \ldots, s$. Therefore, $\tilde{Y}^i_{n+1} = \eta(t^i_{n+1} - \tau(t^i_{n+1}, Y^i_{n+1}))$ may be written by (8.36) for $\theta = \theta^i_{n+1} = c_i - \tau(t^i_{n+1}, Y^i_{n+1})/h_{n+1}$. Similarly, $\tilde{Z}^i_{n+1} = \lambda(t^i_{n+1} - \tau(t^i_{n+1}, Y^i_{n+1}))$ may be written by using the derivative of (8.36) for $\theta = \theta^i_{n+1}$.

For the choice (8.5), the RK method for NDDEs (in Y notation) is given by (8.33), (8.34) along with
$$\lambda(t_n + \theta h_{n+1}) = \sum_{i=0}^{s^*} \ell_i(\theta) f(\bar{t}^i_{n+1}, U^i_{n+1}, \tilde{U}^i_{n+1}, \tilde{V}^i_{n+1}), \quad 0 \leq \theta \leq 1, \tag{8.37}$$
where $\bar{t}^i_{n+1} = t_n + \bar{c}_i h_{n+1}$ and $\ell_i(\theta)$, $i = 0, \ldots, s^*$, are the nodes and the Lagrange polynomial coefficients of the interpolation operator $\mathcal{P}$. Here, besides the values
$$\tilde{Y}^j_{n+1} = \eta(t^j_{n+1} - \tau(t^j_{n+1}, Y^j_{n+1})) \quad \text{and} \quad \tilde{Z}^j_{n+1} = \lambda(t^j_{n+1} - \tau(t^j_{n+1}, Y^j_{n+1})),$$
there are additional values
$$U^j_{n+1} = \eta(\bar{t}^j_{n+1}),$$
$$\tilde{U}^j_{n+1} = \eta(\bar{t}^j_{n+1} - \tau(\bar{t}^j_{n+1}, U^j_{n+1})) \quad \text{and} \quad \tilde{V}^j_{n+1} = \lambda(\bar{t}^j_{n+1} - \tau(\bar{t}^j_{n+1}, U^j_{n+1})).$$

Note that, according to the argument $s_j = t^j_{n+1} - \tau(t^j_{n+1}, Y^j_{n+1})$, the values $\eta(s_j)$ and $\lambda(s_j)$ may or may not be known. If $s_j > t_n$, then $\tilde{Y}^j_{n+1}$ and $\tilde{Z}^j_{n+1}$ are unknown and must be computed by (8.34) and (8.37), respectively. In particular, for the application of (8.37) in the current interval, U^j_{n+1}, $\tilde{U}^j_{n+1}$ and $\tilde{V}^j_{n+1}$ need to be known. Here the U^j_{n+1}s are certainly unknown, whereas knowledge of $\tilde{U}^j_{n+1}$ and $\tilde{V}^j_{n+1}$ depends on the location of the further argument $\bar{t}^j_{n+1} - \tau(\bar{t}^j_{n+1}, U^j_{n+1})$.

Summarizing, if for some index j some of the arguments are $> t_n$, then the relevant *spurious stages* $\tilde{Y}^j_{n+1}$, $\tilde{Z}^j_{n+1}$, U^j_{n+1}, $\tilde{U}^j_{n+1}$ or $\tilde{V}^j_{n+1}$ are unknown and are given by (8.34) and (8.37) for suitable values of θ. More precisely,

$$\tilde{Y}^j_{n+1} = \eta(t_n + \theta^j_{n+1} h_{n+1}) \quad \text{and} \quad \tilde{Z}^j_{n+1} = \lambda(t_n + \theta^j_{n+1} h_{n+1})$$

with $\theta^j_{n+1} = c_j - \tau(t^j_{n+1}, Y^j_{n+1})/h_{n+1}$,

$$U^j_{n+1} = \eta(t_n + \bar{c}^j h_{n+1}),$$

$$\tilde{U}^j_{n+1} = \eta(t_n + \bar{\theta}^j_{n+1} h_{n+1}) \quad \text{and} \quad \tilde{V}^j_{n+1} = \lambda(t_n + \bar{\theta}^j_{n+1} h_{n+1})$$

with $\bar{\theta}^j_{n+1} = \bar{c}_j - \tau(\bar{t}^j_{n+1}, U^j_{n+1})/h_{n+1}$.

The dimension of the system may still be reduced by using K notation but, unlike option (8.4), as well as the K values

$$K^j_{n+1} = f(t^j_{n+1}, Y^j_{n+1}, \tilde{Y}^j_{n+1}, \tilde{Z}^j_{n+1}), \quad j = 1, \ldots, s,$$

we have the additional values

$$H^j_{n+1} = f(\bar{t}^j_{n+1}, U^j_{n+1}, \tilde{U}^j_{n+1}, \tilde{V}^j_{n+1}), \quad j = 0, \ldots, s^*.$$

Remark 8.3. As with option (8.5), the number of unknowns in the system to be solved at each step may be reduced. In fact, if the underlying continuous RK method is natural and if the interpolation formula (8.37) is based on the nodes $\bar{c}_i = c_i$, $i = 1, \ldots, s^* = s$, and on another node $\bar{c}_0 \neq c_i$, then, for $j = 1, \ldots, s$,

$$Y^j_{n+1} = U^j_{n+1},$$
$$\tilde{Y}^j_{n+1} = \tilde{U}^j_{n+1},$$
$$\tilde{Z}^j_{n+1} = \tilde{V}^j_{n+1},$$

and, therefore, also

$$H^j_{n+1} = K^j_{n+1}.$$

In this case the spurious stages reduce to only Y^j_{n+1}, $\tilde{Y}^j_{n+1}$ and $\tilde{Z}^j_{n+1}$ in Y notation, and to merely

$$K^j_{n+1} = f(t^j_{n+1}, Y^j_{n+1}, \tilde{Y}^j_{n+1}, \tilde{Z}^j_{n+1})$$

in the equivalent K notation. Note also that, for the new set of stage values

$$Z_{n+1}^j = \lambda(t_n + c_j h_{n+1}), \quad j = 1, \ldots, s,$$

by (8.37) we have

$$Z_{n+1}^j = K_{n+1}^j.$$

On the other hand, independently of the choice of the $\bar{c}_i$s, if $c_i \neq 0$, $i = 1, \ldots, s$, as in the non-neutral case, we can express each $\tilde{Y}_{n+1}^j$ in terms of the Y_{n+1}^js and, hence, the overall method is based on the stage values Y_{n+1}^j and $\tilde{Z}_{n+1}^j$. However, in no case can the RK method reduce to just the Y values.

The convergence result extending Theorem 8.8 may be stated as follows.

Theorem 8.9. Consider the state-dependent NDDE (8.2), where the right-hand side $f(t, y, x, w)$ is C^p-continuous in $[t_0, t_f] \times \mathbb{R}^d \times \mathbb{R}^d \times \mathbb{R}^d$, the delay $\tau(t, y)$ is C^p-continuous in $[t_0, t_f] \times \mathbb{R}^d$ and the initial function $\phi(t)$ is C^p-continuous. Moreover, assume that the mesh $\Delta = \{t_0, t_1, \ldots, t_n, \ldots, t_N = t_f\}$ includes all the discontinuity points lying in $[t_0, t_f]$ where the solution $y(t)$ is not at least C^p-continuous. If the underlying continuous RK method $(A, b(\theta), c)$ has discrete order p and uniform order q, and the approximation $\lambda(t)$ has uniform order r, then the resulting RK method for NDDEs has discrete global order and uniform global order $q' = \min\{p, q+1, r+1\}$, that is,

$$\max_{1 \leq n \leq N} |y(t_n) - y_n| = \mathcal{O}(h^{q'})$$

and

$$\max_{t_0 \leq t \leq t_f} |y(t) - \eta(t)| = \mathcal{O}(h^{q'}),$$

where $h = \max_{1 \leq n \leq N} h_n$. In particular, if $\lambda(t)$ is given by the option (8.4), then $r = q - 1$ and, hence, $q' = q$.

Note that, for the option (8.5), $\lambda(t)$ is given by (8.37) and the interpolation operator $\mathcal{P}$ has order $r = s^*$. Therefore, on the basis of Theorem 8.9, it is useless to take $s^* > q$. On the other hand, the choice $s^* = q$ preserves the optimal order $q' = \min\{p, q+1\}$ and makes the option (8.5), along with the conditions in Remark 8.3, preferable to (8.4).

Finally, it is worth remarking that, when the option (8.4) is adopted, the RK method for NDDEs yields an approximation of the solution of the form (5.3). On the contrary, this is not the case when the other option (8.5) is used.

Therefore, the above convergence result can also be obtained as a corollary to the general convergence Theorem 5.2 only if option (8.4) is adopted.

9. Implementation issues in the standard approach

We have seen in the previous section that two points are important for an efficient and accurate implementation of the standard approach, namely location of the breaking points, if any, as defined in Section 3 and detection of possible overlapping. From a practical point of view, it would be important to answer *a priori* the following two questions which are particularly difficult when the delay is state-dependent:

(Q_1) Whether or not, for the local problem (8.3), overlapping occurs and, in particular, whether or not the approximated delayed function $\eta(t-\tau)$ is known at a given point t of the current interval $[t_n, t_{n+1}]$, so as to make the right choice between (8.26) and (8.27) (or between (8.30) and (8.31)) for the computation of the $\tilde{Y}^i_{n+1}$s.

(Q_2) Whether or not, for the step-size h_{n+1}, the current interval $[t_n, t_{n+1}]$ includes some breaking point ξ and, more specifically, how to tune the step-size h_{n+1} in order to have $t_{n+1} = \xi$, as we would like to have in the proximity of ξ, as required by Theorems 8.8 and 8.9.

As far as question (Q_1) is concerned, overlapping can be avoided for a sufficiently small step-size by assuming that:

(H_1) There exists a constant $\tau_0 > 0$ such that $\tau(t, y(t)) \geq \tau_0$ for all $t \in [t_0, t_f]$.

Moreover, when the delay is actually state-dependent, it is often welcome to be able to assume the stronger condition (already considered in Section 3):

(H_1^*) There exists a constant $\tau_0 > 0$ such that $\tau(t, z) \geq \tau_0$ for all $t \in [t_0, t_f]$ and $z \in \mathbb{R}^d$.

In fact, (H_1^*) prevents the state-dependent delay from vanishing even if it is computed in a perturbation of the true local solution $w_{n+1}(t)$ of (8.3).

Under the hypothesis (H_1^*) on the delay, for sufficiently small step-size, namely $h_{n+1} = t_{n+1} - t_n \leq \tau_0$, the function $\eta(s)$ is known for every $s = t - \tau(t, z)$ with $t \in [t_n, t_{n+1}]$ and for all $z \in \mathbb{R}^d$. Therefore, overlapping is avoided for any approximation of the local solution $w_{n+1}(t)$.

On the contrary, if $h_{n+1} > \tau_0$, it might well be that

$$t - \tau(t, w_{n+1}(t)) > t_n$$

for some $t \in (t_n, t_{n+1}]$ and, consequently, overlapping could occur.

When the hypothesis (H_1), or (H_1^*), does not hold, the delay τ necessarily vanishes at some point ξ and, thus, overlapping inevitably occurs whenever the interval $(t_n, t_{n+1}]$ includes ξ. This occurrence leads to some complications in the case of state-dependent delays. In fact, in this case, we cannot choose between (8.26) or (8.27) ((8.30) or (8.31)) *a priori*, and the choice may vary during the computation of the $\tilde{Y}_{n+1}^i$s. This point will be deeply investigated in Section 11, as well as considering the neutral case.

As for question (Q_2), the location of breaking points has been discussed by various authors from both the theoretical and implementational perspectives. Two approaches have been pursued in the literature. They are based, respectively, on discontinuity tracking, that is, on a direct computation of discontinuities from the deviated arguments (see, *e.g.*, Willé and Baker (1992)) and on defect control (see, *e.g.*, Enright, Jackson, Nørsett and Thomsen (1988) and Shampine and Thompson (2000)).

The first, usually referred to as the *tracking of discontinuities*, is based on finding the discontinuities $\xi_{k,j}$ satisfying

$$\xi_{k,j} - \tau(\xi_{k,j}, w_{n+1}(\xi_{k,j})) = \xi_{k-1,i} \quad \text{for some } i$$

(see (3.6)) and to include them as mesh points. It is just worth mentioning that for state-dependent delays the task appears very hard to accomplish. How to do this, and how to achieve the accuracy necessary for preserving the order of the overall integration procedure, is presented later in this section. Although expensive, this strategy appears the most robust.

The second approach, relying on step-size control, gives up tracking the discontinuities, which are instead assumed to be automatically included in the mesh by suitable variable step-size strategies based on the estimation of the local error or on the computation of the defect. In general, the codes are simpler but undergo a larger number of rejected steps and may lead to a sequence of very small step-sizes in the neighbourhood of a low-order discontinuity point ξ.

9.1. Tracking the breaking points

An accurate tracking of breaking points is important in order to compute and automatically insert them into the mesh of integration. We shall discuss this topic here in the context of DDEs and, in Section 10, in the even more challenging context of NDDEs.

Although the approach can be extended to any DDE method, here we consider collocation methods with ν distinct abscissae $c_1, \ldots, c_\nu$ such that $c_i \neq 0$, $i = 1, \ldots, \nu$, whose continuous approximation at the nth step is expressed in the form

$$\eta(t_n + \theta h_{n+1}) = \sum_{i=0}^{\nu} \ell_i(\theta) Y_{n+1}^i, \quad \theta \in [0, 1],$$

where $Y_n^0 = y_n$ and the abscissa $c_0 = 0$ is added to the other abscissae of the method (see (8.32)). The algorithms we are going to describe here are extensively discussed in Guglielmi and Hairer (2008) and implemented in the code Radar5 by Guglielmi and Hairer (2001). They are mainly intended for state-dependent delays, but they may be used in general because they are designed with the aim of computing only those breaking points which are important in terms of the required accuracy.

To compute the set $\mathcal{B}$ of breaking points recursively, we start by initializing $\mathcal{B} = \{t_0\}$. Then the iteration step consists in finding the zeros of the function

$$d_\zeta^*(t) = \alpha(t, y(t)) - \zeta, \tag{9.1}$$

where $\zeta \in \mathcal{B}$ is a previous breaking point. However, since $y(t)$ cannot be found exactly, we have to consider a suitable approximation $\eta(t)$, e.g., the continuous extension of the collocation method, and solve the approximate equation

$$d_\zeta(t) = \alpha(t, \eta(t)) - \zeta. \tag{9.2}$$

The solution ξ of this equation is added to $\mathcal{B}$.

The novel method we present is split into two phases: a first one, where the presence of a breaking point is detected, and a second one, where the breaking point is actually computed.

Detection and accurate computation of breaking points

For a given step-size h, the first phase consists in checking the possible presence of a breaking point in the interval $[t_n, t_n + h]$. To this end we consider the continuous extension computed at the previous accepted step,

$$\hat{\eta}(t_{n-1} + \theta h_n) = \sum_{i=0}^{\nu} \ell_i(\theta) Y_n^i, \quad \theta \geq 1,$$

to be used for extrapolation in the current step.

After setting h as a predicted new step-size, we look for zeros of the functions

$$d_\zeta(\theta) = \alpha(t_{n-1} + \theta h_n, \hat{\eta}(t_{n-1} + \theta h_n)) - \zeta, \quad \theta \in [1, 1 + h/h_n],$$

for all previously computed breaking points $\zeta \in \mathcal{B}$. The presence of a new breaking point is guessed if $d_\zeta(t_n) \cdot d_\zeta(t_n + h) < 0$ for some $\zeta \in \mathcal{B}$. This idea is related to that used by Enright and Hayashi (1997) in their explicit solver.

Let $\xi^{[0]}$ be the *detected* breaking point, that is, the solution of the equation

$$\alpha(\xi^{[0]}, \hat{\eta}(\xi^{[0]})) - \zeta = 0. \tag{9.3}$$

In general, $\xi^{(0)}$ provides a poor approximation to the exact breaking point due to the fact that we are making use of an extrapolation of the collocation polynomial $\hat{\eta}$. Note that a better approximation of the solution in $[t_n, t_n+h]$ could reveal the absence of breaking points, i.e., of solutions for (9.3) in that interval.

In any case, once a breaking point is detected inside the interval $[t_n, t_n+h]$, we assume it actually exists and, hence, we try to compute it accurately in order to preserve the high order and accuracy of the numerical method. The heuristic, which we shall explain theoretically and illustrate experimentally, is that of coupling the RK equations and the equation for the breaking point (9.2).

Therefore, we consider the system of the RK equations (see (8.24)–(8.27)) coupled with (9.2), that is,

$$0 = \alpha\big(t_n + h_{n+1}, \eta(t_n + h_{n+1})\big) - \zeta, \tag{9.4}$$

$$Y_{n+1}^i = y_n + h_{n+1} \sum_{j=1}^{\nu} a_{ij} f\big(t_n + c_j h_{n+1}, Y_{n+1}^j, \tilde{Y}_{n+1}^j\big), \quad i = 1, \ldots, \nu, \tag{9.5}$$

which is solved with respect to both the stages $Y_{n+1}^1, \ldots, Y_{n+1}^\nu$ and the step-size h_{n+1}.

If equations (9.4)–(9.5) are solved successfully, the point $\xi = t_n + h_{n+1}$ is inserted into the set of computed breaking points $\mathcal{B}$.

Since we are considering collocation methods, we have to solve (9.4)–(9.5) by a Newton process, especially if the problem is stiff. However, instead of applying it to the whole system, it is convenient to split the problem in order to take advantage of its structure.

Solving (9.4)–(9.5) by an iterative scheme

For given h_{n+1} the system (9.5) is usually solved by the well-known simplified Newton iteration that exploits the structure of the Jacobian (see, e.g., Hairer and Wanner (1996)). In order to solve (9.4)–(9.5) efficiently with respect to the unknowns $\{Y_{n+1}^i\}$ and h_{n+1}, it would be important not to lose such a structure (see Guglielmi and Hairer (2001, 2008)).

Aiming to preserve the block-diagonal structure, it is possible to solve the system (9.4)–(9.5) in an iterative way. In particular, we denote by $Y_{n+1}^{j\,[k]}$, $j = 1, \ldots, \nu$, and $h_{n+1}^{[k]}$ the stage values and the step-size at the kth iteration of the iterative process (this means that $t_n + h_{n+1}^{[k]}$ gives the current approximation of the breaking point).

Starting with $h_{n+1}^{[0]} = \xi^{[0]} - t_n$, where $\xi^{[0]}$ is the approximation to the breaking point obtained solving (9.3) in the detection phase, and using some initial approximation to the stage values obtained, for example, by extrapolation from the previous step, we consider the following two-step iteration.

(I1) Solve equation (9.4) with respect to the unknown $h_{n+1}^{[k+1]}$, i.e.,

$$0 = \alpha\big(t_n + h_{n+1}^{[k+1]}, \eta^{[k]}(t_n + h_{n+1}^{[k+1]})\big) - \zeta, \tag{9.6}$$

with fixed stage values $\{Y_{n+1}^{j\,[k]}\}_{j=1}^{\nu}$, that is, with a fixed vector-valued polynomial $\eta^{[k]}$ given by

$$\eta^{[k]}(t_n + \theta h_{n+1}^{[k]}) = \sum_{i=0}^{\nu} \ell_i(\theta) Y_{n+1}^{i\,[k]}, \quad \theta \geq 0.$$

(I2) Solve the system (9.5) with respect to the unknowns $\{Y_{n+1}^{j\,[k+1]}\}_{j=1}^{\nu}$ with fixed step-size $h_{n+1}^{[k+1]}$ by means of a simplified Newton iteration, i.e.,

$$Y_{n+1}^{i\,[k+1]} = y_n + h_{n+1}^{[k+1]} \sum_{j=1}^{\nu} a_{ij} f\big(t_n + c_j h_{n+1}^{[k+1]}, Y_{n+1}^{j\,[k+1]}, \tilde{Y}_{n+1}^{j\,[k+1]}\big),$$

for $i = 1, \ldots, \nu$, (9.7)

where $\tilde{Y}_{n+1}^{j\,[k+1]} = \eta\big(\alpha(t_n + c_j h_{n+1}^{[k+1]}, Y_{n+1}^{j\,[k+1]})\big)$.

It is clear that, assuming that the iterative scheme converges, its efficiency depends on the speed of convergence. The following lemma shows that this iterative method converges (see Guglielmi and Hairer (2008) for the proof). Although linear, the convergence turns out to be fast, since the convergence ratio depends on the νth power of the step-size.

Lemma 9.1. Assume that the solution $h_{n+1}^{[k+1]}$ of (9.6) is simple for all k and that the simplified Newton iteration applied to (9.7) converges in a suitable neighbourhood of h_{n+1}. Then,

$$\big|h_{n+1}^{[k+1]} - h_{n+1}\big| \leq C \cdot \big(h_{n+1}^{[k]}\big)^{\nu} \cdot \big|h_{n+1}^{[k]} - h_{n+1}\big|,$$

where h_{n+1} is the exact solution of (9.4)–(9.5) and C a suitable constant.

In practice we have experienced that the required accuracy is achieved after very few iterations. This implies that the cost is only slightly greater than that of the standard step (for which no breaking point is detected), since it essentially consists of solving (9.5) a small number of times.

In fact, in the code Radar5, which implements the described procedure, the possible presence of a breaking point is not checked at every step but, instead, the function $d_\zeta(t)$ is monitored only in the following cases:

(i) if the Newton process does not converge,
(ii) if the estimated error is not under the given required tolerance,
(iii) if the estimated error increases with respect to the previous step of a factor larger than a prescribed value (the default value is 5).

Unlike most of the previous strategies, in this way one computes only those breaking points that are relevant to the required accuracy.

Other authors have considered techniques for approximating the breaking points (see, *e.g.*, Feldstein and Neves (1984) and Hauber (1997)). The algorithm of Feldstein and Neves (1984) checks at all steps whether

$$d_\zeta(t_n + \theta h_{n+1}) = \alpha\big(t_n + \theta h_{n+1}, \eta(t_n + \theta h_{n+1})\big) - \zeta \qquad (9.8)$$

changes sign for some $\zeta \in \mathcal{B}$. In such a case, the zero of d_ζ is computed and the new breaking point is inserted in $\mathcal{B}$. Note that the use of η to compute the new breaking point is not reliable in general. In fact, in the current integration interval the solution would not be smooth and consequently $\eta(t)$ would be a bad approximation of the solution. A modification of this idea was considered by Hauber (1997), who proposed extrapolating the continuous output of the preceding step, *i.e.*, to replace (9.8) by

$$d_\zeta(t_{n-1} + \theta h_n) = \alpha\big(t_{n-1} + \theta h_n, \hat{\eta}(t_{n-1} + \theta h_n)\big) - \zeta, \quad \theta > 1.$$

Although this idea allows us to overcome the problem due to the lack of smoothness of the solution, using the collocation polynomial computed in the interval $[t_{n-1}, t_n]$ for $t > t_n$ may also determine an inaccurate computation. The work of Enright and Hayashi (1997) also uses this kind of extrapolation to cross breaking points.

The basic idea presented here is related to the fact that, in the algorithm which computes the RK step, the step-size is not fixed but variable. This allows for a more accurate computation of breaking points and for an improvement of the convergence theorem, as illustrated below by Theorem 9.1.

9.2. *Convergence and accuracy of breaking points*

This section is devoted to illustrating the theoretical aspects associated with the solution of (9.4)–(9.5).

Accuracy of the computed breaking points
Concerning the coupling of the Runge–Kutta equations and the equation for the breaking point (9.4)–(9.5), the following error bound is obtained (see Guglielmi and Hairer (2008) for the proof).

Theorem 9.1. Let $y(t)$ be the solution of (8.1) and let ζ^* and ξ^* be exact breaking points of the problem such that $\alpha(\xi^*, y(\xi^*)) = \zeta^*$. Furthermore, let ζ be an approximation of ζ^*. If

$$\frac{d}{dt}(\alpha(t, y(t)))\big|_{t=\xi^*} \neq 0, \qquad (9.9)$$

then the computed breaking point $\xi = t_n + h_{n+1}$, obtained by solving (9.4)–(9.5), satisfies the error estimate

$$|\xi - \xi^*| \leq C\big(\|y_{n+1} - y(t_{n+1})\| + |\zeta - \zeta^*|\big)$$

for some constant $C > 0$.

This means that the breaking points are computed to the same order as the numerical solution at grid points.

Although the order is the same as that obtained using an extrapolation of the dense output computed in the previous interval (see (9.3)), the error constant is expected to be much smaller. Usually, in fact, stiffly accurate collocation methods such as Radau-IIA methods exhibit an error at mesh points which is much smaller than the uniform error in the integration interval. Therefore, making breaking points to coincide with mesh points should improve accuracy, as is actually confirmed by several numerical experiments.

Convergence

By means of Theorem 9.1 above, it is easy to refine Theorem 8.8 and to avoid the assumption that the mesh contains all the exact breaking points where the solution $y(t)$ is not at least C^p-continuous.

Theorem 9.2. (Convergence) Consider the DDE

$$y'(t) = f(t, y(t), y(\alpha(t, y(t)))), \quad t_0 \leq t \leq t_f,$$
$$y(t) = \phi(t), \quad t \leq t_0,$$

with simple breaking points (*i.e.*, (9.9) holds). Assume that the hypotheses of Theorem 8.8 hold except that, instead of the exact breaking points, those obtained by solving (9.4)–(9.5) are included in the mesh.

If the underlying collocation method has discrete order p (and uniform order $q = \nu$), then the DDE method (8.24)–(8.27) has discrete global order and uniform global order $q' = \min\{p, \nu + 1\}$.

Table 9.1. Numerical results for equation (9.10), where FE stands for the number of function evaluations, ERR for the error at the final point and ERRBP for an average error in the computation of the breaking points.

	Radar5: old version		Radar5: new version		
$-\log(\text{tol})$	FE	ERR	FE	ERR	ERRBP
2	94	$0.41\,10^{-1}$	97	$0.13\,10^{-3}$	$0.55\,10^{-4}$
4	146	$0.55\,10^{-3}$	147	$0.14\,10^{-5}$	$0.63\,10^{-6}$
6	247	$0.40\,10^{-3}$	198	$0.32\,10^{-7}$	$0.13\,10^{-7}$
8	443	$0.15\,10^{-5}$	276	$0.60\,10^{-9}$	$0.25\,10^{-9}$
10	733	$0.85\,10^{-7}$	490	$0.52\,10^{-10}$	$0.21\,10^{-10}$
12	1622	$0.85\,10^{-9}$	932	$0.46\,10^{-12}$	$0.20\,10^{-12}$

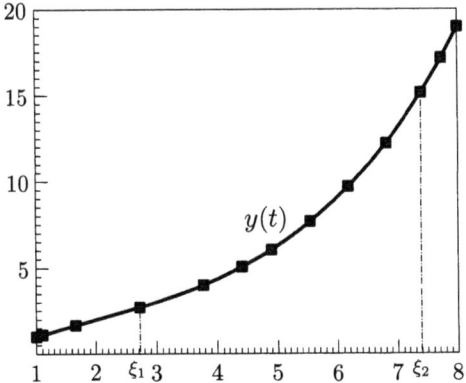

Figure 9.1. The solution of (9.10)
and its numerical approximation.

Example 9.1. Consider the equation (see Neves (1975) and Paul (1994))

$$y'(t) = \frac{y(t)\, y(\log(y(t)))}{t}, \quad 1 \leq t \leq 8, \qquad (9.10)$$

$$y(t) = 1, \quad t \leq 1.$$

The exact solution is

$$y(t) = \begin{cases} 1, & t < \xi_0^*, \\ t, & \xi_0^* \leq t < \xi_1^*, \\ e^{t/e}, & \xi_1^* \leq t < \xi_2^*, \\ \cdots & \cdots \end{cases}$$

and the breaking points are $\xi_0^* = 1$ (of order 0), $\xi_1^* = e$ (of order 1), $\xi_2^* = e^2$ (of order 2), etc.

Table 9.1 shows the numerical integration of (9.10) by the code Radar5. In version 1.1 of the code the breaking points were not computed explicitly, but only implicitly, through the error control, and the step-size was simply driven by error estimates. In the current version, 2.1, an explicit computation of the breaking points is implemented, according to the algorithm previously described, which solves (9.4)–(9.5).

In Figure 9.1 we show the exact solution and its numerical approximation, obtained with relative and absolute error tolerances per step, $R_{\text{tol}} = A_{\text{tol}} = 5 \cdot 10^{-5}$. The computed breaking points are

$$\xi_1 = 2.71828358623074\cdots,$$
$$\xi_2 = 7.38905155206748\cdots.$$

The first numerical breaking point is essentially exact because the solution is linear in the first interval. The second numerical breaking point has an error $|\xi_2 - \xi_2^*| = 4.5469 \cdot 10^{-6}$. The number of accepted steps is 13 and the number of rejected steps is 2. The relative error at the final point is ERR $\approx 1.001 \cdot 10^{-4}$.

10. Neutral problems with state-dependent delays

Now we consider neutral problems of the form (8.2). Without loss of generality, we focus our attention on systems in autonomous form

$$y'(t) = f\bigl(y(t),\, y(\alpha(y(t))),\, y'(\alpha(y(t)))\bigr), \quad t_0 \leq t \leq t_f,$$
$$y(t) = \phi(t), \quad t \leq t_0, \tag{10.1}$$

where, as usual, $\alpha(y(t))$ denotes the deviated argument.

In general, at the initial point t_0 the right-hand derivative

$$y'(t_0) = f\bigl(\phi(t_0),\, \phi(\alpha(\phi(t_0))),\, \phi'(\alpha(\phi(t_0)))\bigr)$$

is different from the left-hand derivative $\phi'(t_0)$, i.e., it does not satisfy the splicing condition (4.6) assumed in Theorem 4.5. This irregularity at t_0 is propagated by the deviated argument $\alpha(y(t))$ to further breaking points, where the first derivative of the solution is not continuous.

Due to such jump discontinuities in the first derivative of the solution, because of a breaking point, problem (10.1) has to be considered as a discontinuous differential equation (see, e.g., Filippov (1964, 1988)). Moreover, the delay being state-dependent, existence and uniqueness of a classical solution are no longer assured, independently of the regularity of f. Therefore, the solution might either terminate, or even bifurcate, in the presence of a breaking point. This leads us in a natural way to consider *weak* (or *generalized*) solutions, which may allow the integrator to prolong the solution beyond those breaking points where the classical solution ceases to exist. To this end, we consider some possible *regularizations*, and define weak solutions to be the limits of the solutions of the regularized problems as the regularization parameters tend to zero.

As in Baker and Paul (2006) and Bellen and Guglielmi (2009), we give the following definition where, with respect to Definition 1.1, the value of y' is assigned at any point of the integration interval. As before, we let $\mathcal{B}$ denote the set of breaking points.

Definition 10.1. We say that a function $y(t)$ is a solution to problem (10.1) in $[t_0, t_f]$ if:

(i) it is continuous on $[t_0, t_f]$;

(ii) it is continuously differentiable in $[t_0, t_f] \setminus \mathcal{B}$;

(iii) it satisfies (10.1) in $[t_0, t_f] \setminus \mathcal{B}$;

(iv) at those breaking points $\xi \in \mathcal{B}$ where (10.1) is not satisfied, we have

$$\lim_{t \searrow \xi} y'(t) = f(y(\xi), y(\alpha(y(\xi))), z),$$

where

$$z = \lim_{t \searrow \xi} y'(\alpha(y(t))).$$

Thus $y'(t)$ is the usual two-sided derivative for all $t \in [t_0, t_f]$, except for the breaking points ξ where (10.1) is not satisfied. At such points we take it to be the one-sided right derivative $\lim_{t \searrow \xi} y'(t)$.

In general, for neutral equations there is no smoothing effect. Therefore, breaking points of order zero may be propagated throughout the integration interval.

Since Theorem 4.5 is not applicable, in order to study the existence of a solution, we make the assumption that the set $\mathcal{B}$ is finite.

10.1. Neutral problems as discontinuous differential equations

Let $\xi > \zeta$ be a breaking point of order zero, that is,

$$\alpha(y(\xi)) = \zeta,$$

where ζ is a previous breaking point, the ancestor of ξ, where the derivative of the solution has a jump discontinuity.

Let

$$\begin{aligned} x^+(s) &= y(s) \quad \text{for } s \geq \zeta, \\ x^-(s) &= y(s) \quad \text{for } s < \zeta, \end{aligned} \tag{10.2}$$

and let $x'^+(s)$ and $x'^-(s)$ be the corresponding derivatives.

Since we assumed that the set $\mathcal{B}$ is finite, they are defined and smooth in a suitable neighbourhood of ζ. Then we can locally write problem (10.1) in the form

$$y'(t) = h(y(t)) = \begin{cases} h^+(y(t)) & \text{if } \alpha(y(t)) > \zeta, \\ h^-(y(t)) & \text{if } \alpha(y(t)) < \zeta, \\ h^+(y(\xi)) & \text{if } t = \xi \text{ and } \alpha(y(t)) \nearrow \zeta, \\ h^-(y(\xi)) & \text{if } t = \xi \text{ and } \alpha(y(t)) \searrow \zeta, \end{cases} \tag{10.3}$$

where

$$\begin{aligned} h^+(y(t)) &= f(y(t), x^+(\alpha(y(t))), x'^+(\alpha(y(t)))), \\ h^-(y(t)) &= f(y(t), x^-(\alpha(y(t))), x'^-(\alpha(y(t)))). \end{aligned} \tag{10.4}$$

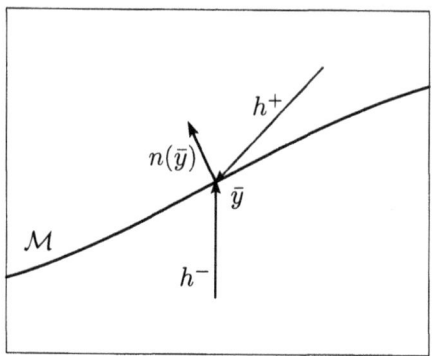

Figure 10.1. The vector fields h^+ and h^- and the normal n to the manifold oriented towards the region $\{y \mid g(y) > 0\}$.

Note that (10.3) is a differential equation with discontinuous right-hand side. In fact, the discontinuity occurs at at $t = \zeta$, where

$$x'^+\bigl(\alpha(y(\xi))\bigr) \neq x'^-\bigl(\alpha(y(\xi))\bigr).$$

On the contrary, the solution is continuous, i.e., $x^+\bigl(\alpha(y(\xi))\bigr) = x^-\bigl(\alpha(y(\xi))\bigr)$. Let us introduce the so-called *switching function*

$$g(y(t)) = \alpha(y(t)) - \zeta, \tag{10.5}$$

whose zeros identify the instants when the right-hand side of (10.4) switches from h^+ to h^- or *vice versa*.

If we introduce the manifold

$$\mathcal{M} = \{y \mid g(y) = 0\},$$

which separates the two regions where the vector field of the differential equation is smooth, we have the situation illustrated in Figure 10.1.

Consider $\bar{y} \in \mathcal{M}$, that is, $g(\bar{y}) = 0$, and assume that $\bar{t}$ is such that $y(\bar{t}) = \bar{y}$. Then let ∇ denote the gradient with respect to y and set

$$n(\bar{y}) = \frac{\nabla g(\bar{y})}{\|\nabla g(\bar{y})\|} \quad \text{if } \nabla g(\bar{y}) \neq 0.$$

Finally, consider the quantities

$$\langle n(\bar{y}), h^+(\bar{y})\rangle \quad \text{and} \quad \langle n(\bar{y}), h^-(\bar{y})\rangle. \tag{10.6}$$

If the conditions

$$\begin{aligned}\langle n(\bar{y}), h^+(\bar{y})\rangle &< 0, \\ \langle n(\bar{y}), h^-(\bar{y})\rangle &> 0,\end{aligned} \tag{10.7}$$

occur, then the vector fields h^+ and h^- have a normal direction with respect

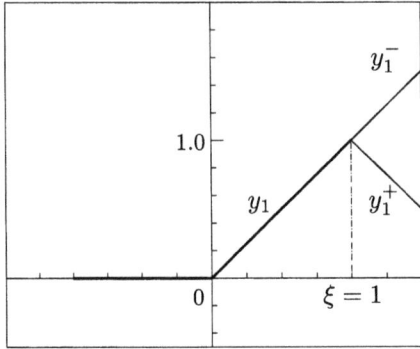
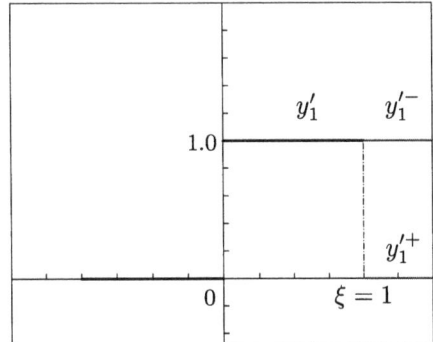

Figure 10.2. The solution of (10.9) terminates at $\xi = 1$.

to the manifold at $\bar{y}$ which is oriented towards the manifold itself. This means that the classical solution to (10.3) ceases to exist. This situation is illustrated in Figure 10.1.

On the contrary, in the two cases when

$$\langle n(\bar{y}), h^+(\bar{y})\rangle \cdot \langle n(\bar{y}), h^-(\bar{y})\rangle > 0, \tag{10.8}$$

a unique classical solution keeps on existing in a right neighbourhood of $\bar{t}$.

Finally, if $\langle n(\bar{y}), h^+(\bar{y})\rangle > 0$, $\langle n(\bar{y}), h^-(\bar{y})\rangle = 0$ and $\langle n(y), h^-(y)\rangle \leq 0$ in a neighbourhood of $\bar{y}$ (or in the specular case $\langle n(\bar{y}), h^-(\bar{y})\rangle < 0$, $\langle n(\bar{y}), h^+(\bar{y})\rangle = 0$ and $\langle n(y), h^+(y)\rangle \geq 0$ in a neighbourhood of $\bar{y}$) two solutions are admissible, so that uniqueness is lost.

Example 10.1. Let us consider the system

$$\begin{aligned} y_1'(t) &= 1 - 2\, y_1'(y_1(t) - 1), \\ y_2'(t) &= 2 - \frac{1}{2} y_2'(y_1(t) - 1), \end{aligned} \tag{10.9}$$

with initial data $y_1(t) = y_2(t) \equiv 0$ for $t \leq 0$.

The solution exists until $t = 1$ and is given by $y_1(t) = t, y_2(t) = 2t$. Then it terminates at $t = \xi = 1$.

We have

$$\bar{y} = \begin{pmatrix} 1 \\ 2 \end{pmatrix}, \quad \nabla g(\bar{y}) = \begin{pmatrix} 1 \\ 0 \end{pmatrix}, \quad h^+(\bar{y}) = \begin{pmatrix} -1 \\ 1 \end{pmatrix}, \quad h^-(\bar{y}) = \begin{pmatrix} 1 \\ 2 \end{pmatrix}.$$

We see that conditions (10.7) are satisfied. In fact,

$$\langle \nabla g(\bar{y}), h^+(\bar{y})\rangle = -1,$$
$$\langle \nabla g(\bar{y}), h^-(\bar{y})\rangle = 1.$$

This implies termination of the classical solution.

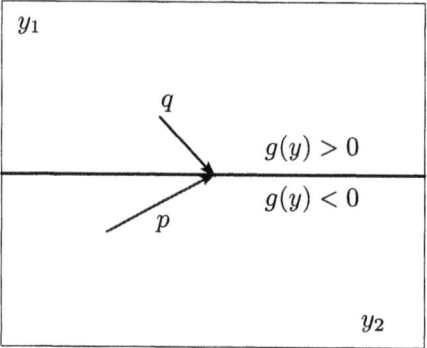

Figure 10.3. Problem (10.9) reformulated as a discontinuous differential equation.

In other words, as illustrated in Figure 10.2, we can explain termination by considering, for $t \in [1, 1+\delta]$ with δ sufficiently small, the pair of differential equations

$$y'^{+}(t) = h^{+}(y^{+}(t)),$$
$$y'^{-}(t) = h^{-}(y^{-}(t)), \quad (10.10)$$

where

$$h^{+}(y^{+}(t)) = q = \begin{pmatrix} -1 \\ 1 \end{pmatrix} \quad \text{and} \quad h^{-}(y^{-}(t)) = p = \begin{pmatrix} 1 \\ 2 \end{pmatrix}. \quad (10.11)$$

We have that $y^{+}(t)$ is a local solution of the considered neutral problem (10.9) if $y_1^{+}(t) - 1 > 0$ for $t \in (1, 1+\delta]$ and that $y^{-}(t)$ is a local solution of (10.9) if $y_1^{-}(t) - 1 < 0$ for $t \in (1, 1+\delta]$. Since $y_1^{+}(t) = 2 - t$ and $y_1^{-}(t) = t$, none of the previous conditions is fulfilled (see Figure 10.2).

Since $y'(y_1(s) - 1) = H(y_1(s) - 1)$ until $y_1(s) \leq 2$, H being the Heaviside function, the problem (10.9) can be reformulated, at least locally, as the discontinuous differential equation

$$y'(t) = \begin{cases} p & \text{if } g(y) < 0, \\ q & \text{if } g(y) > 0, \end{cases} \quad (10.12)$$

where $g(y) = y_1 - 1$ (see Figure 10.3).

This is the case of a system of ODEs with a vector field which is discontinuous on a linear manifold $\mathcal{M} = \{y \mid g(y) = 0\} = \{y \mid y_1 = 1\}$ of codimension 1. Furthermore, the vector field is constant on the two half-spaces separated by the manifold.

10.2. Regularization by a time average of the discontinuous vector field

Let ξ be a termination point. Following Fusco and Guglielmi (2009), we consider for $t > \xi$ the regularized problem

$$y'^\varepsilon(t) = \frac{1}{\varepsilon} \int_{t-\varepsilon}^{t} [H(g(y^\varepsilon(s)))h^+(y^\varepsilon(s)) + (1 - H(g(y^\varepsilon(s))))h^-(y^\varepsilon(s))] \, ds, \tag{10.13}$$

where H denotes the Heaviside function and $y^\varepsilon(t) = \phi(t)$, $t \leq \xi$, ϕ being a C^1-function. Our goal is that of studying the existence of a solution for sufficiently small $\varepsilon > 0$ and that of investigating the limit behaviour of such solutions as $\varepsilon \longrightarrow 0^+$.

Theorem 10.1. (Existence) Let g, h^-, h^+ be smooth functions. Then there exists $\varepsilon_0 > 0$ such that $\forall \varepsilon \in (0, \varepsilon_0)$, there exist $T > 0$ and $C > 0$, independent of ε, such that the problem (10.13) has a C^1-solution $y^\varepsilon : [\xi, \xi + T] \to \mathbb{R}^d$ with

- $|g(y^\varepsilon(t))| \leq C\varepsilon$, $t \in [\xi, \xi + T]$.

Moreover, there exists a C^1-function y^0 such that

- $g(y^0(t)) \equiv 0$;
- $\lim_{\varepsilon \to 0} \|y^\varepsilon - y^0\|_{C^0[\xi,\xi+T]} = 0$;
- $y'^0(t) = \mu(t)h^+(y^0(t)) + (1 - \mu(t))h^-(y^0(t)) \in T_{y^0}\mathcal{M}$ with $\mu(t) \in [0, 1]$, $t \in [0, T]$;

where $\mathcal{M} = \{y \mid g(y) = 0\}$ is the manifold delimiting the two smooth regions of the vector fields and $T_{y^0}\mathcal{M}$ is the linear manifold tangent to $\mathcal{M}$ at $y^0(t)$.

According to Theorem 10.1, the dynamics of the limit solution y^0 takes place in the manifold $\mathcal{M}$.

Then it is natural to define a *weak solution* of the problem (10.1) after a termination point as the solution of the *limit problem*

$$\begin{pmatrix} I & 0 \\ 0 & 0 \end{pmatrix} \begin{pmatrix} y'(t) \\ \mu'(t) \end{pmatrix} = \begin{pmatrix} h(t, y(t), \mu(t)) \\ \zeta - \alpha(y(t)) \end{pmatrix}, \tag{10.14}$$

I being the identity matrix and

$$h(t, y(t), \mu(t)) = \mu(t) f(t, y(t), y(\zeta), x'^+(\zeta)) \\ + (1 - \mu(t)) f(t, y(t), y(\zeta), x'^-(\zeta)),$$

with consistent initial data $(y(\xi), \mu(\xi))$ at $t = \xi$.

This system includes the constraint $\alpha(y(t)) = \zeta$ and, hence, is a differential-algebraic equation of index 2. It replaces the original problem (10.1) for $t \geq \xi$ until a classical solution is recovered.

10.3. Neutral problems as implicit delay equations

In view of the numerical integration of (10.1), by introducing a new variable $z(t) = y'(t)$, we rewrite it as the equivalent implicit system

$$\begin{pmatrix} I & 0 \\ 0 & 0 \end{pmatrix} \begin{pmatrix} y'(t) \\ z'(t) \end{pmatrix} = \begin{pmatrix} z(t) \\ -z(t) + f(y(t), y(\alpha(y(t))), z(\alpha(y(t)))) \end{pmatrix}, \qquad (10.15)$$

with the initial conditions $y(t) = \phi(t)$, $z(t) = \phi'(t)$, $t \le t_0$, where I denotes the identity matrix.

Note that (10.15) falls in the general class of implicit problems

$$M u'(t) = f(u(t), u(\alpha(t, u(t)))), \quad t_0 \le t \le t_f,$$
$$u(t) = \psi(t), \quad t \le t_0,$$

where the $d \times d$ matrix M is constant and possibly singular, which will be studied in Section 11.

Regularization by a singular perturbation

Following Bellen and Guglielmi (2009), as a regularization of (10.15) we consider the singularly perturbed problem

$$\begin{pmatrix} I & 0 \\ 0 & \varepsilon I \end{pmatrix} \begin{pmatrix} y'_\varepsilon(t) \\ z'_\varepsilon(t) \end{pmatrix} = \begin{pmatrix} z_\varepsilon(t) \\ -z_\varepsilon(t) + f(y_\varepsilon(t), y_\varepsilon(\alpha(y_\varepsilon(t))), z_\varepsilon(\alpha(y_\varepsilon(t)))) \end{pmatrix}, \qquad (10.16)$$

which coincides with (10.15) for $\varepsilon = 0$.

Under standard assumptions on f, problem (10.16) admits a solution on a bounded interval for any fixed $\varepsilon > 0$. If the initial datum $(y_\varepsilon, z_\varepsilon) = (\phi, \psi)$ is continuous, then the corresponding solution is also continuous.

Although a theoretical analysis of the limit of the solution of (10.16) as $\varepsilon \to 0$ is still missing, the numerical experiments provided by Bellen and Guglielmi (2009) suggest that a limit solution exists.

Example 10.2. We consider problem (10.9) again, *i.e.*,

$$y'_1(t) = 1 - 2y'_1(y_1(t) - 1),$$
$$y'_2(t) = 2 - \frac{1}{2}y'_2(y_1(t) - 1),$$

with initial data $y_1(t) = y_2(t) \equiv 0$ for $t \le 0$, $y_1(t) = t$, $y_2(t) = 2t$ for $0 \le t \le 1$. We have seen that $\xi = 1$ is a termination point.

The regularized problem of the form (10.13) is

$$y'^\varepsilon(t) = \frac{1}{\varepsilon} \int_{t-\varepsilon}^{t} [H(g(y^\varepsilon(s))) q + (1 - H(g(y^\varepsilon(s)))) p] \, ds, \qquad (10.17)$$

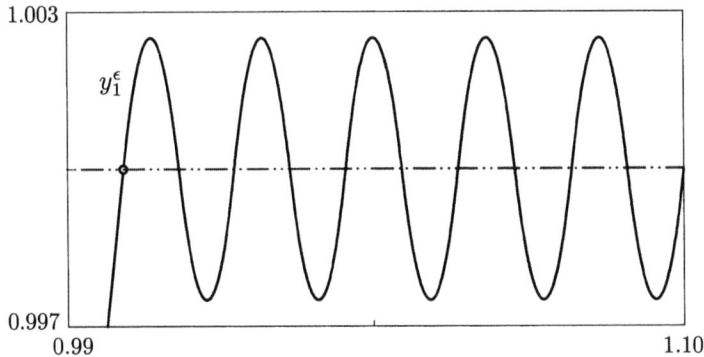

Figure 10.4. The solution component y_1^ε for $\varepsilon = 10^{-2}$.

where the vector fields p and q are constant and the corresponding manifold $\mathcal{M}$ is linear.

The solution of (10.17) can be found explicitly. The first component y_1^ε is periodic of period 2ε and continuously differentiable for $t \geq 1$ (see Figure 10.4). It is given by

$$y_1^\varepsilon(t) = \begin{cases} 1 + (t-1) - \frac{(t-1)^2}{\varepsilon}, & 1 \leq t \leq 1+\varepsilon, \\ 1 - (t-1-\varepsilon) + \frac{(t-1-\varepsilon)^2}{\varepsilon}, & 1+\varepsilon \leq t \leq 1+2\varepsilon, \end{cases}$$

in the interval $[1, 1+2\varepsilon]$ and is repeated periodically for $t \geq 1+2\varepsilon$. The second component y_2^ε is given by the sum of a periodic function and of a linear function

$$y_2^\varepsilon(t) = \frac{3}{2}t + \frac{1}{2}y_1^\varepsilon(t), \quad t \geq 1.$$

Note that, for all $t \geq 1$, the solution remains ε-close to the manifold

$$\mathcal{M} = \{y \mid y_1 = 1\}.$$

We also observe that y_1 and y_2 converge in the C^0-topology as $\varepsilon \to 0$, that is, there exist

$$y_1^0(t) = \lim_{\varepsilon \to 0} y_1^\varepsilon(t) = 1, \qquad (10.18)$$

$$y_2^0(t) = \lim_{\varepsilon \to 0} y_2^\varepsilon(t) = \frac{3}{2}t + \frac{1}{2}. \qquad (10.19)$$

Hence, y_1^0 and y_2^0 naturally represent the weak solution to the original problem (10.9).

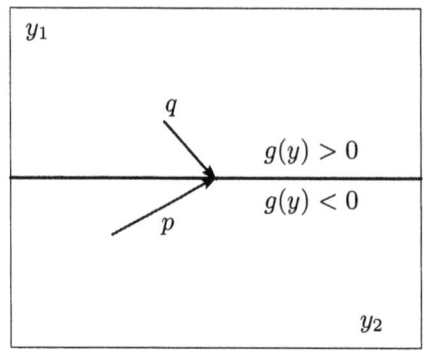 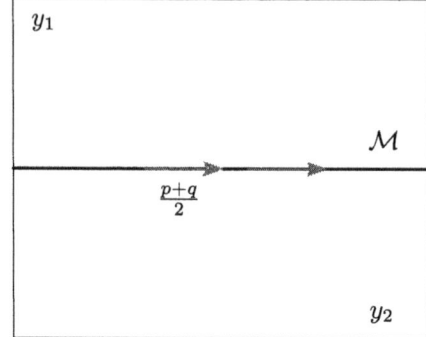

Figure 10.5. The limit problem associated to (10.17).

By (10.14), for $t \geq 1$ we get

$$\begin{pmatrix} I & 0 \\ 0 & 0 \end{pmatrix} \begin{pmatrix} y'(t) \\ \mu'(t) \end{pmatrix} = \begin{pmatrix} \mu(t)\, q + (1-\mu(t))\, p \\ y_1(t) - 1 \end{pmatrix}. \tag{10.20}$$

The consistent initial value for μ is $\mu(1) = \frac{1}{2}$.

By differentiating twice the constraint $y_1(t) \equiv 1$, we get

$$\mu'(t)(p_1 - q_1) = 0 \implies \mu'(t) = 0,$$

which yields

$$\mu(t) = \frac{1}{2}.$$

As a consequence, for $t \geq 1$ the weak solution satisfies the equation

$$y'(t) = \frac{p}{2} + \frac{q}{2} = \begin{pmatrix} 0 \\ 3/2 \end{pmatrix}. \tag{10.21}$$

According to Theorem 10.1, the right-hand side of (10.20) gives the unique convex combination of p and q which lies on the manifold $\mathcal{M}$ (see Figure 10.5). This agrees with the definition of generalized solution of the discontinuous problem given in Filippov (1964, 1988).

Now, we consider the second proposed regularization (10.16), i.e.,

$$\begin{aligned} y_1'^{\varepsilon}(t) &= z_1^{\varepsilon}(t), \\ y_2'^{\varepsilon}(t) &= z_2^{\varepsilon}(t), \\ \varepsilon z_1'^{\varepsilon}(t) &= 1 - 2\, z_1^{\varepsilon}\!\left(y_1^{\varepsilon}(t) - 1\right) - z_1^{\varepsilon}(t), \\ \varepsilon z_2'^{\varepsilon}(t) &= 2 - \frac{1}{2} z_2^{\varepsilon}\!\left(y_1^{\varepsilon}(t) - 1\right) - z_2^{\varepsilon}(t), \end{aligned} \tag{10.22}$$

with initial data $y_1^{\varepsilon}(t) = y_2^{\varepsilon}(t) = z_1^{\varepsilon}(t) = z_2^{\varepsilon}(t) \equiv 0$ for $t \leq 0$.

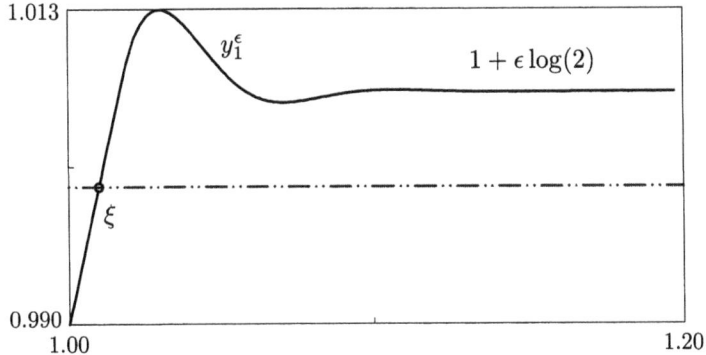

Figure 10.6. The first component of the solution of (10.22) in a neighbourhood of $\xi = 1$ (computed for $\varepsilon = 1/100$).

The first component y_1^ε of the solution has the behaviour shown in Figure 10.6.[1] In particular, it is still ε-close to the manifold $\mathcal{M}$ and its limit as $\varepsilon \to 0$ still coincides with (10.18). In any case, y_1^ε does not exhibit the oscillations of the previous regularizaton (compare to Figure 10.4). From a numerical point of view, the oscillations whose wavelength is of order ε constitute a challenging difficulty.

10.4. The numerics of weak solutions

Concerning the numerical approximation of weak solutions, we have two options: one is that of integrating, when required, the limit problem (10.14) and the other is that of approximating it by integrating the regularized problems (10.13) or (10.16).

Alternating the integration of (10.14) and (10.1) implies a repeated event detection and a possible change of problem whenever (10.7) occurs, as well as when the weak solution is such that

$$\langle n(y(t^*)), h^+(y(t^*))\rangle \cdot \langle n(y(t^*)), h^-(y(t^*))\rangle = 0$$

at some instant t^*. On the contrary, the numerical integration of the regularized problems does not require this. Nevertheless, due to the high oscillations, the numerical integration of (10.13) appears quite expensive, whereas integrating the singularly perturbed problem (10.16) seems to be cheaper. In any case, step-size restrictions are to be expected whenever some component of the z-variables (the solution derivatives) has a jump. Indeed, such jumps correspond to steep transitions of the same component of the z_ε variables.

---

[1] The asymptotic value $1 + \varepsilon \log(2)$ of y_1^ε was obtained using an asymptotic expansion of the solution, by E. Hairer, to whom we are grateful.

10.5. Numerical integration of neutral problems in implicit form

In order to integrate (10.15), it is natural to consider the method

$$\begin{pmatrix} Y_{n+1}^i - y_n \\ 0 \end{pmatrix} = h_{n+1} \begin{pmatrix} \sum_{j=1}^{\nu} a_{ij} Z_{n+1}^j \\ \sum_{j=1}^{\nu} a_{ij} (-Z_{n+1}^j + f(Y_{n+1}^j, \tilde{Y}_{n+1}^j, \tilde{Z}_{n+1}^j)) \end{pmatrix},$$

for $i = 1, \ldots, \nu$, (10.23)

where

$$\tilde{Y}_{n+1}^j = \begin{cases} \phi(\alpha_{n+1}^j) & \text{if } \alpha_{n+1}^j < t_0, \\ \eta(\alpha_{n+1}^j) & \text{if } t_m \le \alpha_{n+1}^j < t_{m+1}, \end{cases}$$

for some $m \le n$ and

$$\tilde{Z}_{n+1}^j = \begin{cases} \phi'(\alpha_{n+1}^j) & \text{if } \alpha_{n+1}^j < t_0, \\ \lambda(\alpha_{n+1}^j) & \text{if } t_m \le \alpha_{n+1}^j < t_{m+1}, \end{cases}$$

with $\alpha_{n+1}^j = \alpha(Y_{n+1}^j)$.

The continuous approximation $\eta(t)$ is still given by (8.32) and, for the continuous approximation $\lambda(t)$, one can consider two options. The first is chosen if t_m is not a computed breaking point and is given by

$$\lambda(t_m + \theta h_{m+1}) = \sum_{i=0}^{\nu} \ell_i(\theta) Z_{m+1}^i, \quad \theta \in [0, 1], \tag{10.24}$$

where $Z_{m+1}^0 = z_m$. The second option is chosen when t_m is a computed breaking point and is given by

$$\lambda(t_m + \theta h_{m+1}) = \sum_{i=1}^{\nu} \hat{\ell}_i(\theta) Z_{m+1}^i, \quad \theta \in [0, 1], \tag{10.25}$$

where $\hat{\ell}_i(\theta)$, $i = 1, \ldots, \nu$, are the Lagrange polynomials of degree $\nu - 1$ involving the collocation abscissae $c_1, \ldots, c_\nu$ only.

Observe that the choice (10.24)–(10.25) also provides a generally discontinuous approximation of the solution derivative $z(t)$ at the computed breaking point t_m, according to the fact that, in general, the solution $y(t)$ is only C^0-continuous at breaking points.

In the case where $\alpha_{n+1}^j \in (t_n, t_{n+1}]$, i.e., when the corresponding delay is smaller than the current step-size, $\eta(\alpha_{n+1}^j)$ and $\lambda(\alpha_{n+1}^j)$ are not known a priori, but only implicitly through the current stage values which are still to be computed.

For a non-singular coefficient matrix A (this is the case, for example, for

Gauss and Radau IIA methods), (10.23) gives
$$Z^j_{n+1} = f(Y^j_{n+1}, \tilde{Y}^j_{n+1}, \tilde{Z}^j_{n+1}), \quad j = 1, \ldots, \nu,$$
whereas the first row remains
$$Y^i_{n+1} = y_n + h_{n+1} \sum_{j=1}^{\nu} a_{ij} Z^j_{n+1}, \quad i = 1, \ldots, \nu.$$

In the case of overlapping, for some j we may have
$$\tilde{Y}^j_{n+1} = \eta(t_n + \theta^j_{n+1} h_{n+1}) = \sum_{i=0}^{\nu} \ell_i(\theta^j_{n+1}) Y^i_{n+1}$$
and
$$\tilde{Z}^j_{n+1} = \lambda(t_n + \theta^j_{n+1} h_{n+1}) = \sum_{i=0}^{\nu} \ell_i(\theta^j_{n+1}) Z^i_{n+1},$$
where $\theta^j_{n+1} = \alpha^j_{n+1}/h_{n+1}$. Therefore, all the approximated delayed terms $\tilde{Y}^j_{n+1}$ and $\tilde{Z}^j_{n+1}$ may be written in terms of Y^j_{n+1} and Z^j_{n+1}. It turns out that these are the same values provided by the approach described in Section 8 with the option (8.5) for the neutral equations in the form (10.1). Consequently, the two approaches are equivalent and the method converges according to Theorem 8.9, which holds under the crucial assumption that exact breaking points are included in the mesh Δ.

10.6. Checking existence and uniqueness numerically

Whenever the solution ceases to exist, a code which has not been designed to check termination typically stops the integration after the step-size has been reduced to a minimal value. This is certainly inconvenient, since the cause of this arrest would remain unclear. Hence it is important to check the possible termination numerically.

For $\bar{y} = y(\xi) \in \mathbb{R}^d$ such that $g(\bar{y}) = \alpha(\bar{y}) - \zeta = 0$, we must compute the sign of scalar products (10.6), or equivalently of

$$\langle \nabla g(\bar{y}), h^+(\bar{y}) \rangle = \sum_{i=1}^{d} \frac{\partial \alpha}{\partial y_i}(y(\xi)) f_i(y(\xi), y(\zeta), x'^+(\zeta)), \qquad (10.26)$$

$$\langle \nabla g(\bar{y}), h^-(\bar{y}) \rangle = \sum_{i=1}^{d} \frac{\partial \alpha}{\partial y_i}(y(\xi)) f_i(y(\xi), y(\zeta), x'^-(\zeta)), \qquad (10.27)$$

where f_i denotes the ith component of f.

The idea for a numerical investigation is based on the observation that, for a point $\bar{y} \in \mathcal{M}$, we can approximate (10.6) in the following way. According to Hairer and Wanner (1996) and Guglielmi and Hairer (2008), by

considering a first-order approximation of $g(\bar{y} + \delta h^{\pm}(\bar{y}))$, we get
$$g(\bar{y} + \delta h^{\pm}(\bar{y})) = g(\bar{y}) + \delta \langle \nabla g(\bar{y}), h^{\pm}(\bar{y}) \rangle + \mathcal{O}(\delta^2).$$
For a small $\delta > 0$, exploiting the property $g(\bar{y}) = 0$ yields
$$\langle \nabla g(\bar{y}), h^{+}(\bar{y}) \rangle \approx \frac{1}{\delta} g(\bar{y} + \delta h^{+}(\bar{y})), \qquad (10.28)$$
$$\langle \nabla g(\bar{y}), h^{-}(\bar{y}) \rangle \approx \frac{1}{\delta} g(\bar{y} + \delta h^{-}(\bar{y})). \qquad (10.29)$$
Note that this corresponds to applying a step of the Euler method to the pair of problems
$$y'(t) = h^{+}(y(t)) \quad \text{and} \quad y'(t) = h^{-}(y(t))$$
with step-size δ.

Let $t_n = \xi^*$ (approximating ξ) and $t_m = \zeta^*$ (approximating ζ) be a numerical breaking point and its ancestor, respectively. Then let
$$\lambda^{-}(t) = \lambda(t), \quad t \in [t_{m-1}, t_m),$$
$$\lambda^{+}(t) = \lambda(t), \quad t \in [t_m, t_{m+1}),$$
be the polynomial extensions of the derivative of the solution on the right-hand and left-hand side of the breaking point t_m, respectively. Such polynomials are clearly well defined in a whole neighbourhood of t_m. Observe that, in general, we expect that $\lambda^{+}(t_m) \neq \lambda^{-}(t_m)$.

Now, in order to proceed, it is sufficient to replace $x'^{+}(s)$ and $x'^{-}(s)$ by $\lambda^{+}(s)$ and $\lambda^{-}(s)$ in (10.1) (see Guglielmi and Hairer (2008)). Then, with
$$y_n^{+} = y_n + \delta f(y_n, y_m, \lambda^{+}(t_m)) \quad \text{and} \quad y_n^{-} = y_n + \delta f(y_n, y_m, \lambda^{-}(t_m)),$$
by using (10.4)–(10.5) and (10.28)–(10.29), at y_n we obtain
$$\langle \nabla g(y_n), h^{+}(y_n) \rangle \approx a_{\delta}^{+} = \frac{\alpha(t_n + \delta, y_n^{+}) - t_m}{\delta},$$
$$\langle \nabla g(y_n), h^{-}(y_n) \rangle \approx a_{\delta}^{-} = \frac{\alpha(t_n + \delta, y_n^{-}) - t_m}{\delta}. \qquad (10.30)$$

If $a_{\delta}^{+} \cdot a_{\delta}^{-} > 0$, so that the solution continues to exist, the integration proceeds with the right-hand limit of $z(t)$ at t_n. On the contrary, if $a_{\delta}^{+} < 0$ and $a_{\delta}^{-} > 0$, the solution ceases to exist at t_n. Finally, if $a_{\delta}^{+} \approx 0$ or $a_{\delta}^{-} \approx 0$, a further analysis could determine whether the solution bifurcates at t_n.

10.7. Accuracy and breaking points

As in the non-neutral case, we want to overcome the need to include the exact breaking points in the mesh Δ. Moreover, with the use of the differential-algebraic formulation (10.15), we also intend to control the error in the z variable, that is, in the derivative of the solution y.

In line with (9.4)–(9.5), we couple the RK equations (10.23) to the equation for the breaking point ξ (with ancestor ζ) whenever the presence of a breaking point has been detected, so as to obtain the system

$$0 = \alpha\big(\eta(t_n + h_{n+1})\big) - \zeta, \qquad (10.31)$$

$$\begin{pmatrix} Y^i_{n+1} - y_n \\ 0 \end{pmatrix} = h_{n+1} \begin{pmatrix} \sum_{j=1}^{\nu} a_{ij} Z^j_{n+1} \\ \sum_{j=1}^{\nu} a_{ij}\big(-Z^j_{n+1} + f(Y^j_{n+1}, \tilde{Y}^j_{n+1}, \tilde{Z}^j_{n+1})\big) \end{pmatrix},$$

$$\text{for } i = 1, \ldots, \nu, \quad (10.32)$$

for the unknowns $Y^1_{n+1}, \ldots, Y^\nu_{n+1}, Z^1_{n+1}, \ldots, Z^\nu_{n+1}$ and h_{n+1}.

We still assume that the breaking points are simple, i.e.,

$$\frac{d}{dt}(\alpha(y(t)))\big|_{t=\xi} \neq 0, \qquad (10.33)$$

and we get the following analogous result to Theorem 9.2 (see Guglielmi and Hairer (2008)).

Theorem 10.2. Consider a smooth problem (10.1) with simple breaking points (i.e., (10.33) holds) and with non-vanishing delay satisfying the hypothesis (H_1) (see Section 9) and assume that, instead of the exact breaking points, those obtained by solving (10.31)–(10.32) are inserted into the mesh.

If the underlying collocation method has discrete order p (and uniform order $q = \nu$), then the resulting method for the NDDE (10.1) still has discrete global order and uniform global order q', where $q' = \min\{p, \nu + 1\}$.

If one considers the ν-stage Radau IIA methods, whose classical order is $p = 2\nu - 1$, with interpolants $\eta(t)$ and $\lambda(t)$ of uniform order $q = \nu$, the NDDE method converges with global uniform order $q' = \nu + 1$ for any $\nu \geq 2$. Such order results hold for any step-size, including the case of overlapping. The 3-stage method is used, for example, in the code Radar5.

Concerning the accuracy of the z variable, we cannot obtain any uniform estimate if we do not use a simple trick. As a matter of fact, for problems with state-dependent delays it is not possible to obtain uniform bounds to the global error of z. In fact, if a mesh point t_n is a numerically computed breaking point, the corresponding exact breaking point is slightly different in general. If the solution derivative has a jump discontinuity at this point, here the global error might be large independently of h.

It is possible to bypass this difficulty by comparing the solution derivative and its numerical approximation at slightly different times. In particular, Guglielmi and Hairer (2008) proved that

$$\lambda(t) - z(s) = \mathcal{O}(h^{q'}),$$

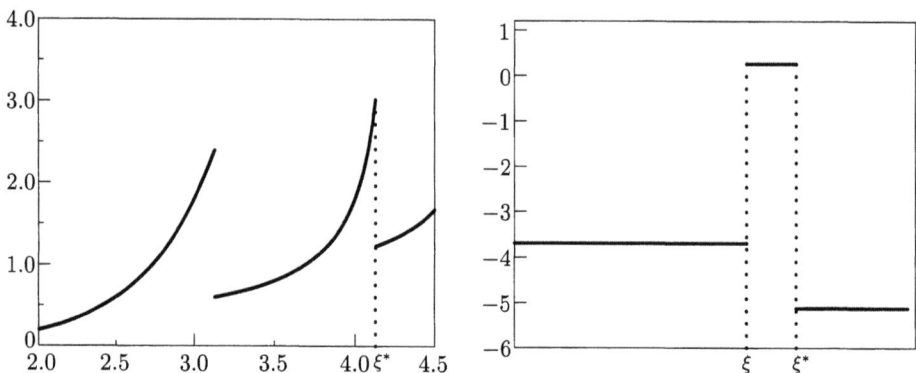

Figure 10.7. The solution component z of equation (10.34) and its numerical approximation (*left*); logarithm of the error of the numerical approximation of z in a neighbourhood of the breaking point ξ (*right*).

where $s = s(t)$ is a suitable smooth function which satisfies $s = t + \mathcal{O}(h^{q'})$ and $h = \max_{n \geq 1} h_n$.

Example 10.3. Let us consider the equation

$$y'(t) = y'(y(t)) + \frac{1}{5}y(t), \quad 2 \leq t \leq 5,$$
$$\phi(t) = (t-1)^2, \quad t \leq 2. \tag{10.34}$$

Figure 10.7 shows that the derivative of the solution is completely inaccurate in a neighbourhood of the breaking point $\xi^* = 4.130469677\cdots$ of amplitude proportional to the error tolerance (which, in the specific case, is $R_{\text{tol}} = A_{\text{tol}} = 10^{-4}$). In fact, the computed breaking point is $\xi = 4.130454\cdots$.

11. Implicit problems with state-dependent delays

As anticipated in the previous section, here we face the study of implicit systems of DDEs of the general form

$$M u'(t) = f\big(u(t), u(\alpha(t, u(t)))\big), \quad t_0 \leq t \leq t_f,$$
$$u(t) = \psi(t), \quad t \leq t_0, \tag{11.1}$$

where the $d \times d$ matrix M is constant and possibly singular.

Note that, for the sake of simplicity, we consider the autonomous case with a single deviated argument.

Besides NDDEs, this class of problems also includes singularly perturbed problems and, since we allow M to be singular, a variety of delay differential-algebraic equations (see, e.g., the models in Shampine and Gahinet (2006)).

11.1. The numerical scheme

Consider an *implicit collocation method*

$$M(U_{n+1}^i - u_n) = h_{n+1} \sum_{j=1}^{\nu} a_{ij} f(U_{n+1}^j, \tilde{U}_{n+1}^j), \quad i = 1, \ldots, \nu, \quad (11.2)$$

$$u_{n+1} = u_n + h_{n+1} \sum_{i=1}^{\nu} b_i f(U_{n+1}^i, \tilde{U}_{n+1}^i),$$

with ν distinct abscissae $c_1, \ldots, c_\nu$ such that $c_i \neq 0$, $i = 1, \ldots, \nu$, where $\tilde{U}_{n+1}^i$ is an approximation to $u(\alpha(u(t_n + c_i h_{n+1})))$ defined by

$$\tilde{U}_{n+1}^i = \begin{cases} \phi(\alpha_{n+1}^i) & \text{if } \alpha_{n+1}^i < t_0, \\ \eta(\alpha_{n+1}^i) & \text{if } \alpha_{n+1}^i \geq t_0, \end{cases} \quad (11.3)$$

where, in turn,

$$\alpha_{n+1}^i = \alpha(U_{n+1}^i).$$

As in (8.32), the continuous approximate solution is given by

$$\eta(t_n + \theta h_{n+1}) = \sum_{i=0}^{\nu} \ell_i(\theta) U_{n+1}^i, \quad \theta \in [0, 1), \quad (11.4)$$

where $U_{n+1}^0 = u_n$ and $c_0 = 0$.

If the mesh point t_n is a computed breaking point, in the step $[t_n, t_{n+1}]$ the polynomial (11.4) can optionally be replaced by

$$\eta(t_n + \theta h_{n+1}) = \sum_{i=1}^{\nu} \ell_i(\theta) U_{n+1}^i, \quad \theta \in [0, 1],$$

which interpolates the internal stage values only, but not u_n (see Guglielmi and Hairer (2001)). The use of this option is important in the presence of a jump discontinuity in some component of the solution, since it permits to have also a discontinuity in the continuous approximation of the solution. Hence, in general, we have $\eta(t_n) \neq u_n$.

As usual, for overlapping, that is, when $\alpha_{n+1}^j \in (t_n, t_{n+1}]$ for some j, the term $\eta(\alpha_{n+1}^j)$ is not known *a priori*, but only implicitly through the unknown stage values $U_{n+1}^1, \ldots, U_{n+1}^\nu$.

11.2. Computing the breaking points

As for the neutral case examined in the previous section, the computation of the breaking points is based on the coupling of the system of the Runge–Kutta equations (11.2) and of the equation for the breaking point

$$0 = \alpha(t_n + h_{n+1}, \eta(t_n + h_{n+1})) - \zeta. \quad (11.5)$$

Generalizing what we have explained in Section 10.7, in this case it is possible to prove the following result (which is stated in Guglielmi and Hairer (2008)).

Theorem 11.1. Consider a smooth problem (11.1) with simple breaking points, *i.e.*, such that

$$\frac{\mathrm{d}}{\mathrm{d}t}(\alpha(u(t)))|_{t=\xi^*} \neq 0,$$

and with non-vanishing delay satisfying the hypothesis (H_1) (see Section 9). Moreover, assume that the uniform global error of the RK method (11.2) is of size $\mathcal{O}(h^r)$ ($h = \max_{n \geq 1} h_n$) if the exact breaking points of order $\leq r$ are inserted into the mesh.

Then, if, instead of the exact breaking points, those computed by solving (11.2)–(11.5) are used, the global error of the resulting method satisfies

$$\max_{t_0 \leq t \leq t_f} |u(s) - \eta(t)| = \mathcal{O}(h^r), \qquad (11.6)$$

where $s = s(t)$ is a suitable smooth function such that $s(t) = t + \mathcal{O}(h^r)$.

The proof is based on an error estimate for the computed breaking points which is analogous to that proved for explicit problems (see Theorem 9.1) and on the classical convergence proof. The only additional difficulty lies in the fact that, for the discontinuous components, it is necessary to align the computed and the exact breaking points in order to obtain a significant error bound. Such an alignment is based on the error estimate $|\xi - \xi^*| = \mathcal{O}(h^r)$ between the computed and the exact breaking point. As a consequence, the global error $|u(t) - \eta(t)|$ is still of size $\mathcal{O}(h^r)$ in $[t_0, t_f]$, except for all the small intervals of the type $[\xi, \xi^*]$, whose size is $\mathcal{O}(h^r)$. This is consistent with what we have shown in Figure 10.7 for Example 10.3.

11.3. Solving the RK equations

In this section we focus on the solution of the RK equations in the particular case when overlapping occurs. The RK system (11.2) has the form

$$F_{n+1}^i(U_{n+1}^1, \ldots, U_{n+1}^\nu, \tilde{U}_{n+1}^1, \ldots, \tilde{U}_{n+1}^\nu) = 0, \quad i = 1, \ldots, \nu, \qquad (11.7)$$

for the unknowns $U_{n+1}^1, \ldots, U_{n+1}^\nu$, where

$$F_{n+1}^i = M(U_{n+1}^i - u_n) - h_{n+1} \sum_{j=1}^\nu a_{ij} f(U_{n+1}^j, \tilde{U}_{n+1}^j).$$

We recall that

$$\tilde{U}_{n+1}^j = \eta(\alpha_{n+1}^j) \quad \text{if } \alpha_{n+1}^j > t_0.$$

We are interested in solving (11.7) by means of a suitable iterative Newton process. For sake of conciseness, we omit the dependence of F_{n+1}^i, U_{n+1}^j, $\tilde{U}_{n+1}^j$ and α_{n+1}^j on n. Moreover, we denote by $f(u, \tilde{u})$ the function on the right-hand side of (11.1).

In order to obtain an accurate computation of the derivatives of F^i, we consider the approximation

$$\frac{\partial F^i}{\partial U^k} \approx M\,\delta_{ik} - h_{n+1}\,(a_{ik}\,D^k + \hat{D}^k), \tag{11.8}$$

where δ_{ik} is the Kronecker delta and

$$D^k = \frac{\partial f}{\partial u}(U^k, \tilde{U}^k) + \frac{\partial f}{\partial \tilde{u}}(U^k, \tilde{U}^k)\,\eta'(\alpha^k)\frac{\partial \alpha}{\partial u}(U_k),$$

$$\hat{D}^k = \sum_{j=1}^{\nu} a_{ij}\,\frac{\partial f}{\partial \tilde{u}}(U^k, \tilde{U}^k)\,\frac{\partial \tilde{U}^j}{\partial U^k}.$$

Note that the term $\frac{\partial \tilde{U}^j}{\partial U^k} = 0$ if the deviated argument $\alpha^j \le t_n$. More precisely, since

$$\eta(t_n + \theta h_{n+1}) = \sum_{k=0}^{\nu} \ell_k(\theta) U^k, \quad \theta \in [0,1),$$

in the current interval, we get

$$\frac{\partial \tilde{U}^j}{\partial U^k} = \mathcal{U}_{jk}\,I_d,$$

where I_d denotes the $d \times d$ identity matrix and

$$\mathcal{U}^{jk} = \begin{cases} \ell_k(\theta_j) & \text{if } \theta_j > 0, \\ 0 & \text{otherwise,} \end{cases} \tag{11.9}$$

with

$$\theta_j = (\alpha^j - t_n)/h_{n+1}.$$

In order to reduce computation, we simplify (11.8) by approximating all derivatives of the functions f and α by

$$\frac{\partial \alpha}{\partial u}(U^k) \approx \frac{\partial \alpha}{\partial u}(u_n), \tag{11.10}$$

$$\frac{\partial f}{\partial u\,\partial \tilde{u}}(U^k, \tilde{U}^k) \approx \frac{\partial f}{\partial u\,\partial \tilde{u}}(u_n, \tilde{u}_n), \tag{11.11}$$

where $\tilde{u}_n = \eta(\alpha^0)$ and $\alpha^0 = \alpha(u_n)$. As a consequence, we expect the Newton process to be linearly convergent.

11.4. General form of the Jacobian

In order to solve (11.7) by means of a Newton process, consider the Jacobian

$$J = I \otimes M - h_{n+1} A \otimes \left(\frac{\partial f}{\partial u} + \frac{\partial f}{\partial \tilde{u}} \eta'(\alpha^0) \frac{\partial \alpha}{\partial u} \right) - h_{n+1} A \cdot \mathcal{U} \otimes \frac{\partial f}{\partial \tilde{u}}. \quad (11.12)$$

In general, J has a full structure and does not admit any transformation allowing for a reduction in the cost of its LU-factorization.

Structure of the Jacobian in the case without overlapping

However, when no overlapping occurs, that is, when $\alpha(U^j) < t_n$ for all $j = 1, \ldots, \nu$, we have $\mathcal{U} = O$ and, therefore,

$$J = J_0 = I \otimes M - h_{n+1} A \otimes B_0, \quad (11.13)$$

where

$$B_0 = \frac{\partial f}{\partial u} + \frac{\partial f}{\partial \tilde{u}} \eta'(\alpha^0) \frac{\partial \alpha}{\partial u}.$$

Then, as in the ODE case (see, e.g., Butcher (1976)), if the RK matrix A is invertible, the matrix J_0 can be pre-multiplied by $(h_{n+1} A)^{-1} \otimes I_d$. In such a case it is useful to transform A^{-1}, so as to obtain a block-diagonal matrix D

$$T^{-1} A^{-1} T = D$$

(see, e.g., Hairer and Wanner (1996)). By introducing the transformed variables $W = (T^{-1} \otimes I_d) U$, we obtain an equivalent Newton iteration with Jacobian

$$\hat{J}_0 = h_{n+1}^{-1} D \otimes M - I \otimes B_0. \quad (11.14)$$

Such a matrix has block-diagonal structure and, thus, the computational cost for its factorization is much cheaper than that of J.

11.5. Preserving the tensor structure of the Jacobian

Unfortunately, in the case of overlapping the previous transformation is not possible. However, an analogous transformation of the Jacobian J to block-diagonal structure is possible if we approximate the matrix $\mathcal{U}$ by

$$\mathcal{U} \approx \gamma I_\nu \quad \text{for an optimal } \gamma \in \mathbb{R},$$

where I_ν stands for the $\nu \times \nu$ identity matrix.

The iteration with no overlapping corresponds to having $\gamma = 0$. In general, simply approximating J by J_0 may prevent the Newton iteration from convergence or make it very slow (see the examples by Castleton and Grimm (1973) and by Waltman (1978) studied in Guglielmi (2005)).

A better choice consists in choosing an optimal $\gamma \in \mathbb{R}$, according to some approximation criteria. The idea is then to make use of an inexact Jacobian which can be block-diagonalized and, if the corresponding inexact Newton process does not converge, either to reduce the step-size or to switch to the exact iteration, that is, to make use of (11.12).

We adopt the optimization criterion

$$\gamma^* \longrightarrow \min_{\gamma \in \mathbb{R}} \|\mathcal{U} - \gamma I_\nu\|_F^2, \qquad (11.15)$$

where $\|\cdot\|_F$ is the Frobenius norm. This choice leads to an explicit formula for the optimal parameter γ^* and has been supported by several numerical experiments.

In the special case $\alpha(y(t)) \equiv t$ (no delay case), we have $\alpha(U^j) = t_n + c_j h_{n+1}$, $j = 1, \ldots, \nu$. Hence, $\mathcal{U} = I_\nu$ and, consequently, $\gamma^* = 1$, which can also be considered a good approximation of the optimal parameter for those cases when the step-size is much larger than the delay.

Example 11.1. Let us consider the Radau IIA 2-stage method, whose tableau is given by

$$\begin{array}{c|cc} \frac{1}{3} & \frac{5}{12} & -\frac{1}{12} \\ 1 & \frac{3}{4} & \frac{1}{4} \\ \hline & \frac{3}{4} & \frac{1}{4} \end{array}.$$

The jth row of the 2×2 matrix $\mathcal{U}$ is zero if $\alpha(U^j) < t_n$, $j = 1, 2$. We have that

$$\mathcal{U} = \begin{pmatrix} H(\theta_1) & 0 \\ 0 & H(\theta_2) \end{pmatrix} \cdot \begin{pmatrix} \ell_{11} & \ell_{12} \\ \ell_{21} & \ell_{22} \end{pmatrix},$$

where $\ell_{jk} = \ell_k(\theta_j)$ and $H(\cdot)$ is the unit Heaviside function.

Since the function to minimize in (11.15) is quadratic with respect to γ, the minimizer

$$\gamma^* = \frac{-9\,H(\theta_1)\,(-1+\theta_1)\,\theta_1 + H(\theta_2)\,\theta_2\,(-1+3\,\theta_2)}{4}$$

is a global one.

Synthesis of the inexact Newton process

With the previous procedure we obtain an approximation of the Jacobian (11.12) given by

$$J \approx J_{\gamma^*} = I \otimes M - h_{n+1} A \otimes B_{\gamma^*}, \qquad (11.16)$$

where

$$B_\gamma = \frac{\partial f}{\partial u} + \frac{\partial f}{\partial \tilde{u}} \left(\eta'(\alpha^0) \frac{\partial \alpha}{\partial u} + \gamma I_\nu \right).$$

By making use of the same transformation used to obtain $\hat{J}_0$ (see (11.14)), we get

$$\hat{J}_{\gamma^*} = (h_{n+1})^{-1} D \otimes M - I \otimes B_{\gamma^*}, \qquad (11.17)$$

which has the same block-diagonal structure as $\hat{J}_0$.

The experimental results obtained on the examples from the test set by Paul (1994) and on the test problems included in the code Radar5 have shown that the use of the inexact Jacobian J_{γ^*} allows us to obtain a more efficient integration of problems with vanishing or small delays since, in most cases, the use of the exact Jacobian J can be avoided.

REFERENCES

C. T. H. Baker (1996), Numerical analysis of Volterra functional and integral equations, in *The State of the Art in Numerical Analysis* (I. S. Duff and G. A. Watson, eds), Clarendon Press, Oxford.

C. T. H. Baker (2000), 'Retarded differential equations', *J. Comput. Appl. Math.* **125**, 309–335.

C. T. H. Baker and C. A. H. Paul (2006), 'Discontinuous solutions of neutral delay differential equations', *Appl. Numer. Math.* **56**, 284–304.

C. T. H. Baker, C. A. H. Paul and D. R. Willé (1995a), 'Issues in the numerical solution of evolutionary delay differential equations', *Adv. Comput. Math.* **3**, 171–196.

C. T. H. Baker, C. A. H. Paul and D. R. Willé (1995b), A bibliography on the numerical solution of delay differential equations. NA Report 269, Department of Mathematics, University of Manchester.

A. Bellen (1985), Constrained mesh methods for functional differential equations, in *Delay Equations, Approximation and Application*, Vol. 74 of *Internat. Ser. Numer. Math.*, pp. 52–70.

A. Bellen and N. Guglielmi (2009), 'Solving neutral delay differential equations with state dependent delays', *J. Comput. Appl. Math.*, in press.

A. Bellen and M. Zennaro (2003), *Numerical Methods for Delay Differential Equations*, Numerical Mathematics and Scientific Computation, Oxford University Press, Oxford.

R. Bellman and K. L. Cooke (1963), *Differential-Difference Equations*, Academic Press.

H. Brunner (2004), *Collocation Methods for Volterra Integral and Related Functional Differential Equations*, Cambridge University Press, Cambridge.

J. C. Butcher (1976), 'On the implementation of implicit Runge–Kutta methods', *BIT* **6**, 237–240.

R. N. Castleton and L. J. Grimm (1973), 'A first order method for differential equations of neutral type', *Math. Comput.* **27**, 571–577.

K. L. Cooke and J. Wiener (1984), 'Retarded differential equations with piecewise constant delays', *J. Math. Anal. Appl.* **99**, 265–297.

C. W. Cryer (1972), Numerical methods for functional differential equations, in *Delay and Functional Differential Equations and their Applications* (K. Schmitt, ed.), Academic Press, New York, pp. 17–101.

C. W. Cryer and L. Tavernini (1972), 'The numerical solution of Volterra functional differential equations by Euler's method', *SIAM J. Numer. Anal.* **9**, 105–129.

O. Diekmann, S. A. van Gils, S. M. Verduyn Lunel and H. O. Walther (1995), *Delay Equations: Functional-, Complex-, and Nonlinear Analysis*, AMS series, Springer, Berlin.

R. D. Driver (1977), *Ordinary and Delay Differential Equations*, Springer, Berlin.

L. E. El'sgol'ts and S. B. Norkin (1973), *Introduction to the Theory and Application of Differential Equations with Deviating Arguments*, Academic Press, New York.

W. H. Enright and H. Hayashi (1997), 'A delay differential equation solver based on a continuous Runge–Kutta method with defect control', *Numer. Algorithms* **16**, 349–364.

W. H. Enright, K. R. Jackson, S. P. Nørsett and P. G. Thomsen (1988), 'Effective solution of discontinuous IVPs using a Runge–Kutta formula pair with interpolants', *Appl. Math. Comput.* **27**, 313–335.

A. Feldstein (1964), Discretization methods for retarded ordinary differential equation. PhD Thesis, Department of Mathematics, UCLA, Los Angeles.

A. Feldstein and K. W. Neves (1984), 'High order methods for state-dependent delay differential equations with nonsmooth solutions', *SIAM J. Numer. Anal.* **21**, 844–863.

A. Feldstein, K. W. Neves and S. Thompson (2006), 'Sharpness results for state dependent delay differential equations: An overview', *Appl. Numer. Math.* **56**, 472–487.

A. F. Filippov (1964), 'Differential equations with discontinuous right-hand sides', *Trans. Amer. Math. Soc.* **42**, 199–231.

A. F. Filippov (1988), *Differential Equations with Discontinuous Righthand Sides*, Vol. 18 of *Mathematics and its Applications* (Soviet Series), Kluwer Academic, Dordrecht (translated from the Russian).

G. Fusco and N. Guglielmi (2009), A regularization for discontinuous differential equations with application to state-dependent delay differential equations of neutral type. In preparation.

N. Guglielmi (2005), 'On the Newton iteration in the application of collocation methods to implicit delay equations', *Appl. Numer. Math.* **53**, 281–297.

N. Guglielmi and E. Hairer (2001), 'Implementing Radau IIA methods for stiff delay differential equations', *Computing* **67**, 1–12.

N. Guglielmi and E. Hairer (2008), 'Computing breaking points in implicit delay differential equations', *Adv. Comput. Math.* **29**, 229–247.

E. Hairer and G. Wanner (1996), *Solving Ordinary Differential Equations II: Stiff and Differential Algebraic Problems*, Springer Series in Computational Mathematics, Springer, Berlin.

J. K. Hale (1977), *Theory of Functional Differential Equations*, Springer, New York.

J. K. Hale and S. M. Verduyn Lunel (1993), *Introduction to Functional Differential Equations*, Applied Mathematical Sciences, Springer, New York.

R. Hauber (1997), 'Numerical treatment of retarded differential-algebraic equations by collocation methods', *Adv. Comput. Math.* **7**, 573–592.

V. Kolmanovskii and A. Myshkis (1992), *Applied Theory of Functional Differential Equations*, Kluwer, Dordrecht.

V. Kolmanovskii and V. Nosov (1986), *Stability of Functional Differential Equations*, Academic Press, London.

J. Kuang and Y. Cong (2005), *Stability of Numerical Methods for Delay Differential Equations*, Science Press, Beijing.

Y. Kuang (1993), *Delay Differential Equations with Applications in Population Dynamics*, Academic Press, Boston.

S. Maset (2009), Theoretical and numerical analysis of retarded functional differential equations. In preparation.

S. Maset, L. Torelli and R. Vermiglio (2005), 'Runge–Kutta methods for retarded functional differential equations', *Math. Models Methods Appl. Sci.* **15**, 1203–1251.

G. Meinardus and G. Nürnberger (1985), Approximation theory and numerical methods for delay differential equations, in *Delay Equations, Approximation and Application*, Vol. 74 of *Internat. Ser. Numer. Math.*, pp. 13–40.

K. W. Neves (1975), 'Automatic integration of functional differential equations: An approach', *ACM Trans. Math. Software* **1**, 357–368.

C. A. H. Paul (1994), A test set of functional differential equation. NA Report 243, Department of Mathematics, University of Manchester.

L. F. Shampine and P. Gahinet (2006), 'Delay-differential-algebraic equations in control theory', *Appl. Numer. Math.* **56**, 574–588.

L. F. Shampine and S. Thompson (2000), 'Event location for ordinary differential equations', *Comput. Math. Appl.* **39**, 43–54.

L. Tavernini (1971), 'One-step methods for the numerical solution of Volterra functional differential equations', *SIAM J. Numer. Anal.* **4**, 786–795.

P. Waltman (1978), A threshold model of antigen-stimulated antibody production, in *Theoretical Immunology*, Vol. 8 of *Immunology* series, Dekker, New York, pp. 437–453.

W. Wang and S. Li (2004), 'Stability analysis of nonlinear delay differential equations of neutral type', *Math. Numer. Sin.* **26**, 303–314.

D. R. Willé and C. T. H. Baker (1992), 'The tracking of derivative discontinuities in systems of delay differential equations', *Appl. Numer. Math.* **9**, 299–222.

M. Zennaro (1995), Delay differential equations: Theory and numerics, in *Theory and Numerics of Ordinary and Partial Differential Equations* (M. Ainsworth, J. Levesley, W. A. Light and M. Marletta, eds), Clarendon Press, Oxford, pp. 291–333.

Adaptivity with moving grids

Chris J. Budd
Centre for Nonlinear Mechanics,
University of Bath, Bath BA2 7AY, UK
E-mail: mascjb@bath.ac.uk

Weizhang Huang
Department of Mathematics,
University of Kansas,
Lawrence, Kansas 66045, USA
E-mail: huang@math.ku.edu

Robert D. Russell
Department of Mathematics,
Simon Fraser University,
Burnaby V5A 1S6, Canada
E-mail: rdr@cs.sfu.ca

In this article we survey r-adaptive (or moving grid) methods for solving time-dependent partial differential equations (PDEs). Although these methods have received much less attention than their h- and p-adaptive counterparts, particularly within the finite element community, we review the substantial progress that has been made in developing more robust and reliable algorithms and in understanding the basic principles behind these methods, and we give some numerical examples illustrative of the wide classes of problems for which these methods are suitable alternatives to the traditional ones.

More specifically, we first examine the basic geometric properties of moving meshes in both one and higher spatial dimensions, and discuss the discretization process for PDEs on such moving meshes (both structured and unstructured). In particular, we consider the issues of mesh regularity, equidistribution, alignment, and associated variational methods. An overview is given of the general interpolation error analysis for a function or a truncation error on such an adaptive mesh. Guided by these principles, we show how to design effective moving mesh strategies. We then examine in more detail how these strategies can be implemented in practice. The first class of methods which we consider are based upon controlling mesh density and hence are called position-based methods. These make use of a so-called moving mesh PDE (MMPDE) approach and variational methods, as well as optimal transport methods. This is followed by an analysis of methods which have a more Lagrange-like interpretation, and due to this focus are called velocity-based methods. These include the moving finite element method (MFE), the geometric conservation law (GCL) methods, and the deformation map method. Finally, we present a number of specific types of examples for which the use of a moving mesh method is particularly effective in applications. These include scale-invariant problems, blow-up problems, problems with moving fronts and problems in meteorology. We conclude that, whilst r-adaptive methods are still in their relatively early stages of development, with many outstanding questions remaining, they have enormous potential and indeed can produce an optimal form of adaptivity for many problems.

CONTENTS

1	Introduction	112
2	Moving mesh basics	121
3	Location-based moving mesh methods	154
4	Velocity-based moving mesh methods	193
5	Applications of moving mesh methods	201
	References	231

1. Introduction

1.1. Motivation

Time-dependent systems of partial differential equations (PDEs) often have structures that evolve significantly as the integration of the PDEs proceeds. These can be interfaces, shocks, singularities, changes of phase, high vorticity or regions of complexity. Associated with such structures are the evolution of small length (and time) scales, rapid movement of the solution features and the possibility of finite time blow-up of a component of the solution. Frequently associated are also conservation laws, usually linked to underlying symmetries. Examples of these phenomena occur in many applications, such as gas and fluid dynamics, conservation laws, nonlinear optics, free boundary problems, combustion, detonation, meteorology, mathematical biology and nonlinear optics. To solve such PDEs numerically it is typical to impose some form of spatial mesh and then to discretize the solution on this mesh by using a finite element, finite volume, finite difference, or collocation method. However, this strategy may not be effective in the case of structures that involve small length scales, leading to large localized errors. In such cases it is often beneficial to use some form of non-uniform mesh, adapted to the solution, on which to perform all of the computations. The advantages of doing this can be a reduced overall error, better conditioning of the system, and better computational efficiency. Unfortunately, introducing the extra level of complexity to the system through adaptivity can also lead to additional computational cost and possible numerical instability. Mesh adaptation should thus be used with care and appropriate analysis where possible.

1.2. Adaptivity on a moving mesh

Adaptive methods for solving partial differential equations broadly fit into three categories. The most extensively developed are *static regridding methods*, in which a mesh is updated at each time level. The most widely used of these are *h-refinement methods*, which form the basis of many commercial codes. Usually such codes start with an initially uniform mesh, and then locally coarsen or refine this by the inclusion or deletion of mesh points. The strategy for doing this is normally guided by some *a posteriori* estimate of the solution error, and may consider problems in which the error is due to the solution geometry (such as re-entrant corners) or high derivatives. In *p-refinement methods* some finite element discretization of the PDE is used with local polynomials of some particular order. This order is then increased or decreased in accordance with the solution error. These methods may be combined with h-refinement methods and with careful *a posteriori* estimates to give hp methods (Ainsworth and Oden 2000). The principal objective of the hp methods is to obtain solutions within prescribed error bounds by such refinement procedures. There is not usually an upper bound on the number of points used in the calculation. Such methods have now been developed to a high degree of sophistication. However, they are necessarily rather complex, need not take advantage of any dynamic properties of the underlying solution, and the *a posteriori* error estimates rely heavily on certain assumptions on the solution which may be hard to verify for strongly nonlinear problems.

The *r-refinement* (relocation refinement) moving mesh methods which will form the substance of this article are a more recent development than hp methods. Whilst not as widely used as h- or p-adaptive methods, r-adaptivity has been used with success in many applications including computational fluid mechanics (Tang 2005), phase field models and crystal growth (Mackenzie and Mekwi 2007a), and convective heat transfer (Ceniceros and Hou 2001). It also has a natural application to problems with a close coupling between spatial and temporal length scales, such as in problems with symmetry, scaling invariance and self-similarity (Barenblatt 1996, Budd and Williams 2006), where the mesh points become the *natural coordinates* for an appropriately rescaled problem. Less is known about the behaviour of r-adaptive methods than of the much more extensively developed hp methods, and (at least in higher dimensions) they have yet to become part of established large numerical codes. In particular, as we shall see in this article, many outstanding open questions remain on

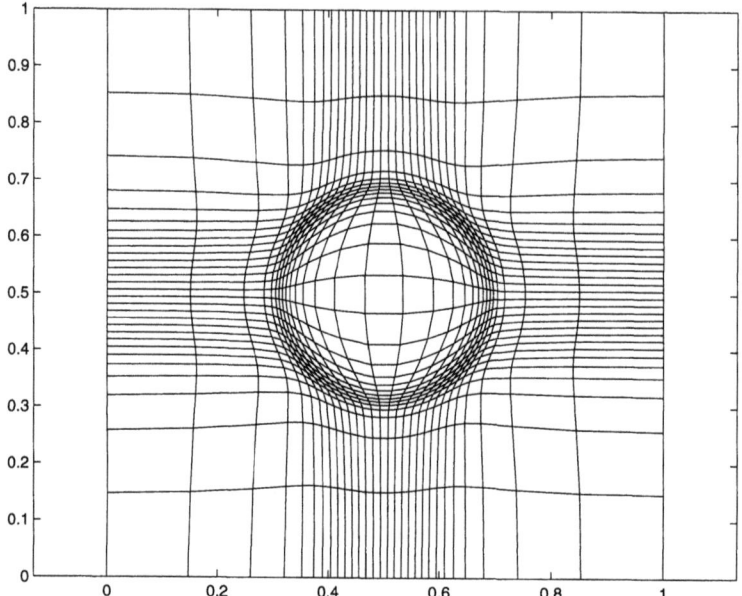

Figure 1.1. A logically rectangular mesh, moved to concentrate points around a ring in an evolving singular solution of the nonlinear Schrödinger equation. Note the good radial symmetry of the adapted mesh around the ring.

their convergence, the nature of the meshes that they generate and the error estimates that can be obtained when using them to solve PDEs with rapidly evolving structures. As a consequence, much of the analysis of such methods has been for one-dimensional problems, and the one-dimensional PDE solver MOVCOL (Huang and Russell 1996, Russell, Williams and Xu 2007) and the celebrated continuation code AUTO (for solving two-point boundary value problems amongst others) both make use of r-adaptive methods. However, r-refinement methods show great potential for solving a much greater range of problems, as we hope to demonstrate in this article.

The r-refinement methods start with a uniform mesh and then *move* the mesh points, keeping the mesh topology and number of mesh points fixed as the solution evolves. Hence the use of the alternative name of *moving mesh methods* for such procedures. The mesh points are then concentrated into regions where the solution has 'interesting behaviour', usually typified by a rapid variation of either the solution or one of its (higher) derivatives. The objective of this approach is to get the smallest error possible for the number N of mesh points used, and to try to obtain error estimates which depend upon the value of N but *not* on the solution itself. For example, if the solution evolves a boundary layer of width ϵ (with ϵ decreasing as time advances), then ideally the mesh points should concentrate into this boundary layer so that the solution error is *independent of* ϵ. The moving mesh methods typically work by generating a mapping from a regular (logical or computational) domain into a physical domain in which the underlying equation is posed. The location, or the velocity, of the mesh points is then determined by solving a system of *auxiliary partial differential equations*, often called the *moving mesh equations*. In all cases a vector or a scalar *monitor function* (or functions) is used to guide the position of the mesh. The monitor function is usually designed to give an estimate of some measure of the solution error which is then *equidistributed* over each mesh cell. The monitor function is usually constructed in one of three ways. It may depend upon *a priori solution estimates* (such as arclength or curvature), on *a posteriori error estimates* (such as the solution residual, as used in moving finite element methods (Baines 1994), or estimates of the derivative jump across element boundaries (Tang 2005)), or on some underlying *physics* related to the solution, such as the potential temperature or the vorticity in a meteorological problem (Budd and Piggott 2005). In the case of *scale-invariant* problems, such physical estimates are often optimal. When using such methods, much care has to be taken in preventing mesh tangling and ensuring mesh regularity and isotropy (where relevant). A discussion of this will form a significant part of Section 2. We also require that discretizations of the underlying PDE on such meshes (in either the computational or the physical domain) should retain important properties of the underlying physical solution, such as conservation laws and scaling structures (Tang 2005). Provided that these conditions are carefully considered, r-adaptive methods can be used with considerable success for many time-evolving systems. Examples of these include computational fluid dynamics (Yanenko, Kroshko, Liseikin, Fomin, Shapeev and Shitov 1976), groundwater flow (Huang and Zhan 2004, Huang, Zheng and Zhan 2002), blow-up problems (Budd, Huang and Russell 1996, Budd and Williams 2006, Ceniceros and Hou 2001, Ren and Wang 2000), chemotaxis systems (Budd, Carretero-Gonzalez and Russell 2005), reaction–diffusion systems (Zegeling and Kok 2004), the nonlinear Schrödinger equation (Sulem and Sulem 1999, Ceniceros 2002, Budd, Chen and Russell 1999a), phase change problems (Mackenzie and Robertson 2002, Mackenzie and Mekwi 2007a, Tan, Lim and Khoo 2007), shear layer calculations (Tang 2005), gas dynamics (Li and Petzold 1997, Li, Petzold and Ren 1998), hyperbolic conservation laws with high Mach number (Li and Petzold 1997, Tang 2005, Stockie, Mackenzie and Russell 2000, Tang and Tang 2003), problems with high vorticity (Ceniceros and Hou 2001), magneto-hydrodynamics (Tan 2007) and meteorological problems (Budd and Piggott 2005). More details of such applications are given in Section 5. In Figure 1.1 we give an example of an r-adaptive mesh which has evolved to capture the structure of a singular solution of the nonlinear Schrödinger equation which has its support concentrated around a ring.

All r-adaptive methods have intimate connections with the geometry of mapping one domain to another. They thus have intimate links with problems in differential geometry such as optimal transport (Brenier 1991, Gangbo and McCann 1996), mean curvature flows (Huang 2007) and harmonic mappings (Dvinsky 1991). A natural application of such ideas arises in image processing, and r-adaptivity has close connections with such image processing procedures as image segmentation and image de-noising (Sapiro 2003).

There are advantages and disadvantages to each of the strategies outlined above. As discussed earlier, the hp methods have been in use for a long time and are now well established in many commercial codes. There is a significant body of analysis supporting their use. In contrast, r-adaptive methods are more recent and are less well understood. A significant criticism which has often been made of them is that their implementation usually requires the solution of *auxiliary* partial differential equations for the mesh, which must be solved in parallel to the underlying partial differential equation. This requires significant additional computational cost. Furthermore, the equations to be solved to determine a suitable mesh can often be very stiff, and thus expensive to solve. Furthermore, the methods get the best estimates for a given N rather than errors necessarily lower than a specified tolerance. However, r-adaptive methods do have significant advantages in certain applications. Firstly, from a computational point of view, it is convenient to work with the same number of mesh points and the same mesh topology. This makes the linear algebra rather easier, as the matrices considered have a constant sparsity structure, and there is no need for any form of nested data structure to keep track of the node points (an issue which always complicates the use of h-refinement methods). The discretization strategy on the mesh is also easier, especially with a finite element method, as the constancy in the mesh topology and connectivity implies that there is no possibility of hanging nodes. There are further, structural advantages to r-refinement methods. One of these is that the movement of the mesh points may well correspond to natural structures of the PDE itself. An obvious example is Lagrangian-based methods for fluid flow problems, in which mesh points move with the fluid flow. A further such example is given by the use of r-refinement methods for PDEs with natural scaling symmetries, in which the mesh points automatically follow the motion of natural similarity variables (and indeed the use of the r-refinement method becomes equivalent to the use of an appropriate coordinate transformation). A third advantage of r-refinement is that, under certain circumstances, the adaptive strategy when coupled with the PDE can be regarded as one (large) dynamical system, which may then be amenable to a combined analysis. One limitation of having a *fixed* number of points means that it may never be possible to resolve all of the fine structures of a PDE as it evolves (although it is surprising what can be done with often a relatively small number of mesh points). Also, all r-adaptive methods are, in principle, prone to *mesh tangling*, in which lines connecting the mesh points can cross over during the evolution. This generally leads to severe instabilities in the system and a failure of the solution routine. Mesh tangling is often associated with *mesh racing*, where some mesh points move very fast during the evolution, frequently leading to a stiff set of equations to solve. The disadvantages of having to solve an auxiliary set of partial differential equations are less severe than they might originally appear to be. Firstly, the combined system of mesh and underlying equations may be much smaller than the original system defined on a uniform mesh for the *same level of accuracy*. Indeed, in Section 5 we will give an example of the solution of focusing behaviour in the one-dimensional nonlinear Schrödinger (Sulem and Sulem 1999) equation (which can be written as four real first-order PDEs), where a discretization of the PDE on a set of $N = 81$ moving mesh points is able to resolve singular structures with a length scale of 10^{-5}, and outperforms discretizations on uniform meshes with 10^5 mesh points. Hence the additional 81 auxiliary equations for the mesh gives a method which outperforms one with 10^5 equations, giving a very significant cost reduction. Secondly, although the equations describing the mesh are indeed stiff in general, we do not need (in general) to solve them exactly. After all, it is the underlying solution of the PDE that we are interested in, and not the mesh on which it is solved. Thus a quite rough approximation to the solution of the moving mesh equations will often deliver a mesh more than adequate for the task of resolving the structures of the underlying PDE. Indeed, we will argue in Section 3 (and demonstrate by example in Section 5) that a relaxed version of the moving mesh equations can be solved using a simple explicit method, will deliver similar performance for much stiffer equations than the meshes obtained by more computationally expensive methods. Indeed, they may well be stable, more regular, and deliver a better mesh quality than solving the exact equations for the mesh. Finally, one of the main applications of hp methods is to solve otherwise regular PDEs on irregular domains, typically with re-entrant corners, that introduce significant errors due to a lack of solution regularity at the corner. The r-adaptive methods as described in this paper are not really the right tool for this job (though see the results in Touringy (1998)). However, a combination of h and r methods may well prove optimal in this case, where the h method is used to mesh around the corner and the r method to follow any evolving solution structure. Future attempts to combine these two types of adaptive refinement in a general context should prove to be most interesting.

1.3. Computation on moving meshes

The problem of computing solutions of PDEs using a moving mesh method separates into three related problems.

(1) As described, we need some monitor function to guide the mesh evolution, which is typically constrained either to equidistribute this function, or to relax towards an equidistributed state. In practice, whatever the choice of monitor function, some spatial (and temporal) smoothing is usually employed.

(2) Having determined the monitor function, we must determine a mesh that equidistributes it in some way. The equidistribution problem itself is a nonlinear algebraic problem, and several techniques have been developed to solve this problem, *e.g.*, a variational method, the geometric conservation law, moving mesh PDEs and optimal transport methods.

(3) The underlying PDE is then discretized, either on the mesh in the computational domain or in the original physical domain (in the latter case a finite element or finite volume method is usually employed (Tang 2005)). The underlying partial differential equation and the mesh equations can then be solved either simultaneously, typically by using the method of lines (Huang, Ren and Russell 1994), or alternatively (often by using a predictor–corrector method). The first method avoids the need for any interpolation from one mesh to another, but is usually associated with having to solve stiff differential equations. Alternating solutions can be implemented using either the quasi-Lagrange approach (Huang and Russell 1997b) or the rezoning approach (Tang 2005). The former transforms time derivatives to those along mesh trajectories and avoids interpolation of the physical solution from the old mesh to the new one. However, it has the disadvantages that it has to deal with extra convection terms caused by mesh movement and may cause a time lag in mesh movement. On the other hand, the rezoning approach solves the physical PDE on a fixed mesh over a time step but requires interpolation from one mesh to another (which often has to be done very carefully to preserve conservation laws). We will consider both methods in detail in this article.

We are currently in a situation where the mesh formulation problem, mesh generation and the solution of PDEs on a moving mesh are generally well understood in one spatial dimension. Reliable and efficient moving mesh methods exist (and are implemented in a number of packages) which are based on such formulations and can be used to solve time-evolving PDEs in one spatial dimension, with associated error estimates in certain cases. Indeed, for such problems the use of moving mesh PDEs to evolve the mesh coupled with a method of lines approach has proved to be very effective, and also amenable to analysis. In this article we will be able to give a detailed description of the theory, implementation and application of such methods. However, the problem of mesh movement, and the discretization of PDEs on such meshes, is much less understood in higher dimensions, and this will form the bulk of the discussion in this paper.

1.4. A historical survey

Moving meshes and the use of adaptive strategies to minimize estimates of the solution error have a rich and diverse literature. Moving mesh methods can be classified according to the mesh movement strategy into two groups (Cao, Huang and Russell 2003): *velocity-based* methods and *location-based* methods. The first group is referred to as velocity-based since the methods directly target the mesh velocity and obtain mesh point locations by integrating the velocity field. Methods in this group are more or less motivated by the Lagrange method in fluid dynamics, where the mesh coordinates, defined to follow fluid particles, are obtained by integrating flow velocity. A major effort in the development of these methods has been to avoid mesh tangling, an undesired property of the Lagrange method. This type of method includes those developed in Anderson and Rai (1983), Cao, Huang and Russell (2002), Liao and Anderson (1992), Miller and Miller (1981), Miller (1981), Petzold (1987) and Yanenko *et al.* (1976). The method of Yanenko *et al.* is of Lagrange type. In the work of Anderson and Rai, mesh movement is based on attraction and repulsion pseudo-forces between nodes motivated by a spring model in mechanics. The moving finite element method (MFE) of Miller and Miller has aroused considerable interest. It computes the solution and the mesh simultaneously by minimizing the residual of the PDEs written in a finite element form. Penalty terms are added to avoid possible singularities in the mesh movement equations; see Carlson and Miller (1998*a*, 1998*b*). A way of treating the singularities but without using penalty functions has been proposed by Wathen and Baines (1985). Liao and Anderson (1992) and Cai, Fleitas, Jiang and Liao (2004) use a deformation map approach. Cao *et al.* (2002) develop the GCL method based on the geometric conservation law (see Section 4). Similar ideas have been used by Baines, Hubbard and Jimack (2005) and Baines, Hubbard, Jimack and Jones (2006) for fluid flow problems.

The second group is referred to as *location-based* because the methods directly control the location of mesh points. Methods in this group typically employ an adaptation functional and determine the mesh or the coordinate transformation as a minimizer of the functional. For example, the method of Dorfi and Drury (1987) can be linked to a functional associated with equidistribution principle (Huang *et al.* 1994). The moving mesh PDE (MMPDE) method developed in Cao, Huang and Russell (1999*b*), Huang *et al.* (1994)

and Huang and Russell (1997a, 1999) moves the mesh through the gradient flow equation of an adaptation functional, which includes the energy of a harmonic mapping (Dvinsky 1991) as a special example. A combination of the MMPDE method with local refinement is studied in Lang, Cao, Huang and Russell (2003). Li, Tang and Zhang (2002) and Tang and Tang (2003) also use the energy of a harmonic mapping as their adaptation functional, but discretize the physical PDE in the rezoning approach.

So far a number of moving mesh methods and a variety of variants have been developed and successfully applied to practical problems; see the review articles of Cao et al. (2003), Eisman (1985, 1987), Hawken, Gottlieb and Hansen (1991), Thompson (1985), Thompson, Warsi and Mastin (1982) and Thompson and Weatherill (1992), and the books of Baines (1994), Carey (1997), Knupp and Steinberg (1994), Liseikin (1999), Thompson, Warsi and Mastin (1985) and Zegeling (1993). In particular, Hawken et al. (1991) give an extensive overview and references on moving mesh methods before 1990. In addition to the references cited above, we would also like to bring the reader's attention to the recent interesting work of Bank and Smith (1997), Beckett, Mackenzie and Robertson (2001a), Budd et al. (1996), Calhoun, Helzel and LeVeque (2008), Ceniceros and Hou (2001), Chacón and Lapenta (2006), Lapenta and Chacón (2006), Di, Li, Tang and Zhang (2005), Huang and Zhan (2004), Mackenzie and Robertson (2002), Ren and Wang (2000), Stockie et al. (2000), Tang and Tang (2003) and Zegeling and Kok (2004) on moving mesh methods and their applications.

1.5. Outline of this article

The purpose of this Introduction has been to give an underlying motivation for the theory and application of (adaptive) moving meshes. In Section 2 we will consider in detail the geometry of possible meshes (with special regard to equidistribution and isotropy), and the nature of discretizations of differential equations on them. In Section 3 we then look in detail, and with reference to many examples, at 'location-based' meshes in which the *local density* of the mesh points is controlled by a monitor function. These include *moving mesh PDE* (MMPDE) methods, variational methods and optimal-transport-based methods. This discussion will look at moving meshes in both one and higher dimensions and compare the strategies used for these two cases. In Section 4 we will then look at *velocity-based* methods, such as the geometric conservation law (GCL) methods and the moving mesh finite element methods, in which the *velocity* of the mesh points, rather than their position, is controlled. The concluding section, Section 5, will then look at some examples in much more detail, considering scale invariance, blow-up problems, problems with convection and moving fronts, phase change and combustion problems, and problems arising in meteorology.

2. Moving mesh basics

In this section we will give an overview of the main aspects of adaptive moving mesh generation, and will concentrate on the nature of the *geometry* of an adapted mesh, the equidistribution and variational approaches to defining a mesh, and the relation of the mesh to solution (truncation and interpolation) errors. The *movement* of the mesh and the way that it can be *coupled to a partial differential equation* will be discussed briefly, but will mainly be the subject of Sections 3 and 4.

As described in the Introduction, in an r-adaptive procedure a fixed number of mesh points are *moved* in response to some user-designed condition. Any r-adaptive method has two main features, a description of the *optimal geometry* of the mesh (which is related both to intrinsic properties of the mesh regularity and to the structure of the underlying solution of the PDE) and a *strategy for evolving the mesh towards this optimal geometry*. Optimal mesh geometries are typically expressed in terms of equidistribution measures (related to the solution of the underlying PDE by monitor functions) or using variational principles, and we review these here. *Movement strategies* are generally methods for determining either the *location* of the mesh points or the *velocity* of the mesh points. We discuss both briefly in this section, and then in more detail in Sections 3 and 4.

Essential to mesh adaptation is the ability to control the shape, size and orientation of mesh elements, and hence to control the error of the solution of the underlying PDE. This is done in three steps. Firstly an estimate of the solution error and/or mesh quality is made. Secondly the mesh is aligned and moved according to this estimate. Thirdly the solution of the underlying PDE is advanced on the new mesh. Typically this can be in response to some structure of the solution of a PDE which is evolving in the space supporting the mesh; however, there are more general circumstances (such as in image processing) where we might wish to evolve a mesh in a manner independent of any PDE.

In this section we will study the geometry of the meshes that arise from various adaptive strategies, considering such aspects as local element size, skewness and orientation as well as considering both isotropic and anisotropic meshes, and looking at solution error estimation and control.

2.1. Mesh-mapping functions

To describe an r-adaptive mesh we consider a *fixed computational domain* $\Omega_C \subset \mathbb{R}^n$, in which most of the computations associated with the PDE will be made. The domain Ω_C will have the usual Lebesgue measure and may have a non-trivial topology. We now consider there to be a *fixed mesh* τ_C on the computational domain. This can either be uniform or non-uniform, depending on the nature of the underlying problem, and in the simplest case will be a uniform set of logical rectangles. It can also be triangular, and usually takes this form when a finite element or finite volume method is used to discretize the PDE in the *physical* domain. Alternatively, if a finite difference or a spectral method is used to discretize the PDE in the *computational domain* then a regular rectangular mesh may be more appropriate. Note that we have a lot of *a priori* freedom in the choice of the computational domain Ω_C, and hence when Ω_C is *simply connected* it is often convenient to consider it to be a logically rectangular domain, so that

$$\Omega_C = [0,1]^n.$$

To describe a computational mesh in the case of such simply connected domains we typically divide $\Omega_C \subset \mathbb{R}^n$ into N^n uniform, *regular tetrahedra or cuboids* of side proportional to $1/N$ and volume proportional to $1/N^n$, and we will initially assume that this is the case. In the r-adaptive procedure considered in this section we consider the mesh points to be joined in a simple (logically rectangular or triangular) network, the topology of which (and consequently the ordering of the nodes in the network) is fixed for most (if not all) of the time during the computation. Indeed, it is this constancy of ordering which makes the r-adaptive procedure very attractive for finite element and related computations.

To derive a moving mesh, the computational domain with its associated mesh is then mapped to a *physical domain* $\Omega_P \in \mathbb{R}^n$, in which the underlying PDE is posed. We assume that there is an invertible, *adaptive mesh generating* function

$$\mathbf{F} : \Omega_C \to \Omega_P$$

describing this map, so that $\mathbf{F}$ is *smooth* on the interior of Ω_C and continuous on Ω_C. Throughout this article we will denote variables in Ω_C by Greek letters, e.g., $\boldsymbol{\xi}$, and in Ω_P by Roman letters, $\mathbf{x}$, and consider the function $\mathbf{F}(\boldsymbol{\xi}, t)$ to be *time-dependent*. The action of the function $\mathbf{F}$ on the fixed mesh τ_C generates a *moving mesh* τ_P in the physical domain. An example of such a mesh is given in Figure 2.1, in which a uniform rectangular mesh in Ω_C is mapped to a mesh τ_P. (This map was constructed by using the optimal transport method described in Section 3.)

In the case where τ_C is a uniform rectangular mesh, the resulting mesh τ_P in the *physical space* is then (in the representative example of a two-dimensional system) given by the points $(X_{i,j}, Y_{i,j})$, where $\mathbf{F} = (x, y)$ and

$$X_{i,j} = x\left(\frac{i}{N}, \frac{j}{N}\right), \quad Y_{i,j} = y\left(\frac{i}{N}, \frac{j}{N}\right). \tag{2.1}$$

We assume further that the boundary of $\partial\Omega_C$ of Ω_C is mapped by $\mathbf{F}$ to the boundary of $\partial\Omega_P$ of Ω_P. In some r-adaptive strategies (such as the multi-equidistribution and/or variational strategies described in Huang and

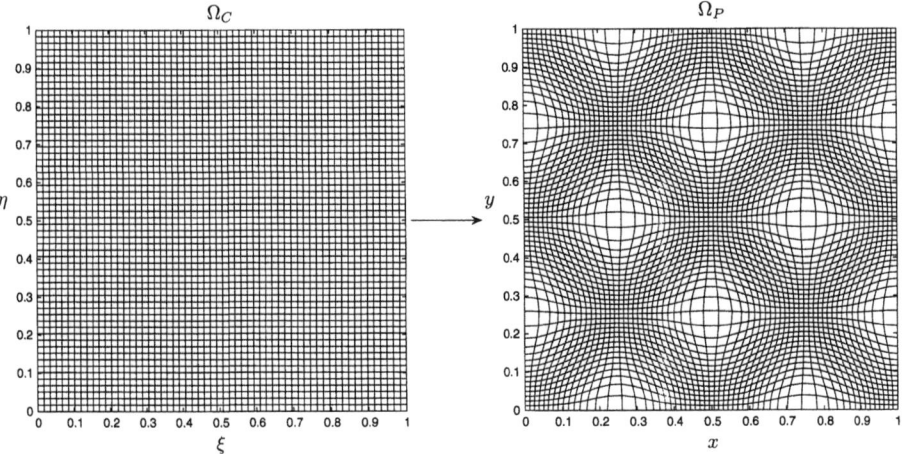

Figure 2.1. A typical map $(x(\xi,\eta), y(\xi,\eta))$ from a computational domain Ω_C to a physical domain Ω_P.

Russell (1999), the map $\mathbf{F}$ is augmented with a second map,

$$\partial F : \partial \Omega_C \to \Omega_P,$$

which explicitly describes the map from one boundary to another. This has the advantage of close control of the meshing strategy right up to the boundary, but has the disadvantage of introducing extra complexity into the system. In other algorithms, such as the optimal mapping strategy (Delzanno, Chacón, Finn, Chung and Lapenta 2008, Budd and Williams 2006), the boundary map is obtained automatically as part of the algorithm. This is an attractive feature from the perspective of algorithmic complexity, although it does lead to a reduction of control of the boundary points.

The great merit of this approach is that it transforms the problem of finding (and describing) a mesh in Ω_P (which in the case of h-adaptive methods can require subtle data structures, including hierarchical trees) into the much simpler problem of describing the function $\mathbf{F}$. Much of this article is devoted to deriving suitable equations for $\mathbf{F}$ and seeking effective solution strategies for them. It goes without saying that it should not be more difficult to find $\mathbf{F}$ than to solve the underlying PDE, and indeed that many such functions $\mathbf{F}$ may give appropriate meshes on which to solve the PDE. The properties of the mesh τ_P then follow immediately from the structure of the map $\mathbf{F}$. This simple observation is a key to the success of r-adaptive methods, as it allows the use of powerful mathematical tools to describe, construct and control the mesh behaviour. These include the application of methods from differential geometry (especially the theory of optimal transport) to describe the static structure of τ_P, and methods from the theory of dynamical systems to describe its evolution. The latter is especially appropriate when coupled to the partial differential equations which are often solved to find **F**. The unity that r-adaptive meshes give for both solving the underlying PDE, and finding the mesh, is a significant advantage of r-adaptivity over other adaptive approaches.

2.2. Static mesh properties: skewness, regularity and smoothness

We consider first some immediate properties of the mesh τ_P which are related to the function **F**. Broadly speaking these divide into *local* and *global* properties. The *global* properties relate to the isotropy of the mesh, orthogonality issues and the behaviour close to boundaries. We discuss these presently.

The *local* properties relate to the size and shape of the elements of τ_P. If τ_C is divided into logical regular rectangular or triangular elements τ_C^e, then these are mapped to elements τ_P^e in τ_P. These elements will then be distorted rectangles/triangles, possibly with small angles at the vertices. Locally we can characterize each such element by the size h_e of the largest side and (in two dimensions) the radius ρ_e of the largest inscribed circle. If a partial differential equation is discretized over τ_P using (say) a finite element method in two dimensions, then the error in the solution has contributions from the size and the shape of the elements, as well as from the derivatives of the solution itself. For example, if the solution u is interpolated over τ_p^e by a piecewise linear interpolant $\Pi(u)$, then the following *a priori* error estimates are standard (Johnson 1987):

$$\max_{\tau_p^e} |\Pi(u) - u| \leq 2h_e^2 \max \left| \frac{\partial^2 u}{\partial x_i \partial x_j} \right|,$$

$$\max_{\tau_p^e} \left| \frac{\partial}{\partial x_i}(\Pi(u) - u) \right| \leq 6 \frac{h_e^2}{\rho_e} \max \left| \frac{\partial^2 u}{\partial x_i \partial x_j} \right|. \tag{2.2}$$

An adapted mesh will usually aim to control the size and the shape of each element so that, for any particular solution u, the overall error is controlled. Thus, for example, if the second derivatives of u are large over τ_p^e, then the error (2.2) can be controlled locally by taking h_e to be small, and ensuring that h_e/ρ_e remains bounded. We discuss these issues in detail later in this section. (See Cao (2005, 2007a) and Chen, Sun and Xu (2007) for a very complete analysis of this problem.)

Both the size and the shape of the mesh elements can be described in terms of the local properties of F, and in particular its Jacobian J, given by

$$J = \frac{\partial F}{\partial \xi}. \tag{2.3}$$

The following is immediate.

Condition 2.1. For the map to be locally well-posed we require that J should be both bounded and invertible at all points in Ω_C.

2.2.1. Mesh scaling

The local scaling factor $1/\rho$ of the transformation (also called the *adaptation factor*) is given by

$$\Lambda \equiv 1/\rho = \det(J) \equiv |J|.$$

Assuming that J has a full set of singular values $\lambda_1, \ldots, \lambda_n$, the local stretching is given by the determinant

$$\Lambda = |\lambda_1||\lambda_2|\cdots|\lambda_n|. \tag{2.4}$$

The adaptation factor controls the (possibly higher-dimensional) *area* $|\tau_P^e|$ of the element τ_P^e so that

$$\rho|\tau_C^e| = |\tau_P^e|. \tag{2.5}$$

The area $|\tau_p^e|$ implicitly enters into the expression (2.2). Indeed, in so-called *shape-regular* two-dimensional meshes there exist constants α and β such that, for all elements $\tau_p^e \in \tau_p$, we have

$$\alpha|\tau_p^e| \leq h_e^2 \leq \beta|\tau_p^e|.$$

Accordingly, many moving mesh methods (such as those based on equidistribution or variational methods) aim to control the adaptation factor. Scale-invariant methods relate the adaptation factor to local length scales of the underlying PDE. It is easily possible for the adaptation factor to vary over *many orders of magnitude*, particularly when the adaptive method is being used to compute singular structures in the underlying PDE in which the solution u and/or its derivatives vary over similar orders of magnitude.

2.2.2. Mesh skewness

In the case of one-dimensional meshes, control of the adaptation factor for each element completely describes the mesh. In higher dimensions many more mesh properties are important, such as the local rotation or the skewness of the mesh. A special class of *irrotational* meshes control the local element rotation by requiring that J is symmetric, so that

$$J^T = J,$$

or equivalently that

$$\nabla_\xi \times \mathbf{F} = \mathbf{0}.$$

This is by no means true of all such mappings, but it can be shown (Delzanno *et al.* 2008, Brenier 1991) that meshes in an averaged sense closest to uniform meshes have this property.

The shape of the element τ_p^e, in particular the existence of any small angles, is also important in the error estimate (2.2). A measure of this is the local *mesh skewness*. A measure for the *local skewness* s of the mesh is then given by

$$s = \frac{\max |\lambda_i|}{\min |\lambda_j|}. \tag{2.6}$$

Other measures of mesh skewness are also referred to as shape or quality measures. Liu and Joe (1994) investigate several shape measures for tetrahedra and show that they are equivalent to each other. Denote the four vertices of a tetrahedron τ_P^e by $a_0, \ldots, a_3$, and define the so-called edge matrix as $E = [a_1 - a_0, a_2 - a_0, a_3 - a_0]$. Let $\hat{e}$ be a regular tetrahedron having the same volume as τ_P^e. Denote the corresponding vertices and the edge matrix of $\hat{e}$ by $\hat{a}_0, \ldots, \hat{a}_3$ and $\hat{E}$, respectively. Then one of the shape measures for τ_P^e is defined by

$$\eta(\tau_P^e) = \frac{3\left[\det((E\hat{E}^{-1})^T(E\hat{E}^{-1}))\right]^{\frac{1}{3}}}{\operatorname{trace}((E\hat{E}^{-1})^T(E\hat{E}^{-1}))}.$$

Notice that the $\eta(\tau_P^e)$ ranges from 0 to 1, with $\eta(\tau_P^e) = 1$ for a regular tetrahedron and $\eta(\tau_P^e) = 0$ for a flat tetrahedron. A geometric quality measure is introduced by Huang (2005a) for measuring the shape of a simplicial element in any dimension. Let τ_P^e be a simplicial element in n dimensions and let $\hat{K}$ be an n-simplex with unit edge length. There exists a unique invertible affine mapping

$$F_e : \hat{K} \to \tau_P^e, \quad \tau_P^e = F_e(\hat{K}).$$

Denote the Jacobian matrix of F_e by F'_e. Then the geometric measure is defined by

$$Q_{\text{geo}}(\tau_P^e) = \frac{\operatorname{trace}((F'_e)^T F'_e)}{d[\det((F'_e)^T F'_e)]^{\frac{1}{d}}}.$$

Notice that $Q_{\text{geo}}(\tau_P^e)$ ranges from 1 to ∞, with $Q_{\text{geo}}(\tau_P^e) = 1$ for a regular n-simplex and $Q_{\text{geo}}(\tau_P^e) = \infty$ for a flat d-simplex. Interestingly, for tetrahedra these two shape measures have the relation $Q_{\text{geo}}(\tau_P^e) = 1/\eta(\tau_P^e)$. To see this, we first notice that $\hat{K}$ and $\hat{e}$ are similar. Thus, the mapping $G_e : \hat{e} \to \tau_P^e$ is related to F_e by

$$G_e = cF_e, \quad G'_e = cF'_e$$

for some positive constant c. Then the edge matrices E and $\hat{E}$ are related by

$$E = G'_e \hat{E} = cF'_e \hat{E}.$$

Using this relation we can rewrite $\eta(\tau_P^e)$ as

$$\eta(\tau_P^e) = \frac{3\left[\det((F'_e)^T F'_e)\right]^{\frac{1}{3}}}{\operatorname{trace}((F'_e)^T F'_e)} = \frac{1}{Q_{\text{geo}}(\tau_P^e)}.$$

It is shown by Huang (2005a) that measures s defined in (2.6) and Q_{geo} are mathematically equivalent.

Some other quality measures can be found in Liseikin (1999, Chapter 3), Knupp (2001) and Shewchuk (2002). Again, a good adaptive method aims to control some or all of these measures of skewness, either explicitly or implicitly throughout the calculation, and we discuss this later in this section. In the case of scale-invariant meshes we shall show that, whilst the adaptation factor changes a great deal, *the skewness hardly varies*. More generally, it should be noted that whereas the adaptation factor often changes a great deal in a mesh, the skewness generally does not. To control terms in the error expression (2.2) arising from large solution gradients, it is generally more important to vary the adaptation factor. If this results in a locally larger value of the skewness then this can usually be tolerated.

2.2.3. Mesh smoothness and regularity

The smoothness or regularity of a mesh is a measure of how much the mesh elements vary over the mesh. This can be important since the accuracy and error in the numerical solution of partial differential equations generally depend upon the type of discretization, the quality of the mesh, the treatment of boundary conditions, and so on. A uniform mesh has the highest degree of regularity, which can lead to particularly low error estimates on such meshes. It is sometimes claimed that only uniform meshes have such low estimates, but in fact, as we shall see, they share this with sufficiently regular meshes. Although there is generally no simple relationship between the smoothness of the mesh and the error (see Veldman and Rinzema (1992)), for most problems and most discretization methods, abrupt variations in the mesh will cause a deterioration in the convergence rate and an increase in the error (Thompson *et al.* 1985), or indeed in the accuracy of the approximation of a function over the mesh. Moreover, most discrete approximations of spatial differential operators have much larger condition numbers on an abruptly varying mesh than they do on a gradually varying one, and these ill-conditioned approximations may result in stiffness in the time integration for time-dependent problems.

The smoothness of a mesh can be expressed in terms of the regularity of the underlying mesh function $\mathbf{F}$.

Definition. A mesh τ_P has degree of regularity r if $\mathbf{F} \in C^r(\Omega_C)$.

The regularity of $\mathbf{F}$ can often be achieved by determining $\mathbf{F}$ as a solution of a PDE system or a minimizer of a functional as in variational mesh generation methods. In many cases it is possible to have strong control over the derivatives of $\mathbf{F}$ allowing guaranteed regularity of the mesh τ_P. It should be noted that an obvious way to determine a mesh is to prescribe the Jacobian function J exactly (Brackbill and Saltzman 1982). However, it is in general very hard to do this, and instead some property of J (such as its determinant) is prescribed, and this is then used to determine the mesh.

A mesh can also be made smoother by some direct methods. For example, (weighted) Laplace smoothing is often used in hp adaptation (see Carey (1997)). In this strategy, the coordinates of an interior mesh point are adjusted so that they become the (weighted) average of the coordinates of its neighbouring points. Typically this is carried out in a Jacobian or Gauss–Seidel fashion. When this is the case, Laplace smoothing can be viewed as the application of the Jacobian or Gauss–Seidel iteration to the solution of a discretization of the partial differential equation

$$-\Delta_\xi(\hat{x}, \hat{y}) = (x, y), \qquad (2.7)$$

where Δ_ξ is the Laplacian operator applied in the computational domain. In r-adaptive methods based on equidistribution, on the other hand, a smoother mesh is often obtained indirectly by smoothing the monitor function M used for controlling mesh adaptation and movement. We will describe this strategy presently.

When calculating solutions to a (partial) differential equation on a non-uniform mesh it is essential that there is a strong control on the mesh variation. For one-dimensional meshes for which we have a mesh function $x(\xi)$, mesh points $X_i = x(i\Delta\xi)$ and local mesh spacing given by $\Delta_i = X_{i+1} - X_i$ then the *grid size ratio* or *local stretching factor* r is given by

$$r = \frac{\Delta_i}{\Delta_{i-1}}. \qquad (2.8)$$

In a uniform mesh we have that $r = 1$. For many calculations on a non-uniform mesh, we require instead that

$$r = 1 + \mathcal{O}(\Delta_i). \qquad (2.9)$$

Such grids are termed *quasi-uniform* (Li et al. 1998, Zegeling 2007, Kautsky and Nichols 1980, Kautsky and Nichols 1982), and normally lead to truncation (and approximation) errors of the same order as uniform meshes (Veldman and Rinzema 1992). We note that

$$r = 1 + \frac{\Delta_i - \Delta_{i-1}}{\Delta_i} = 1 + \frac{X_{i+1} - 2X_i + X_{i-1}}{X_i - X_{i-1}}.$$

Consequently, as $\Delta_i \approx \Delta\xi x_\xi$, etc., we have

$$r = 1 + \Delta_i \frac{x_{\xi\xi}}{x_\xi^2} + \mathcal{O}(\Delta_i^2).$$

Thus the mesh is quasi-uniform provided that

$$\Lambda \equiv \frac{x_{\xi\xi}}{x_\xi^2} = \mathcal{O}(1). \qquad (2.10)$$

The condition (2.10) plays an important role in our subsequent analysis of the errors of computations on both static and moving non-uniform meshes. The ratio between lengths of adjacent elements is also used in Dorfi and Drury (1987) and studied by Verwer, Blom, Furzeland and Zegeling (1989).

The concept of quasi-uniformity has natural extensions to higher dimensions augmented with small angle conditions. For example, in two dimensions, if we have a triangulation τ_P then this is *shape-regular*, ensuring control over small angles, if, for each element $\tau_e \in \tau_p$ with area $|\tau_e|$, longest side of length h_e and interior circle of diameter ρ_e, we have a constant σ_1 such that

$$\max_{\tau_e} \frac{h_e}{\rho_e} \leq \sigma_1. \tag{2.11}$$

Such a shape-regular mesh is then *quasi-uniform* if there is a second constant σ_2 for which

$$\frac{\max_{\tau_e \in \tau_p} |\tau_e|}{\min_{\tau_e \in \tau_P} |\tau_e|} \leq \sigma_2. \tag{2.12}$$

As in the one-dimensional case, quasi-uniform meshes have similar error estimates to uniform ones (Johnson 1987). However, it is often much harder to achieve this for time-dependent problems.

2.3. Mesh calculation, mesh tangling and mesh racing

The function **F** must be determined as part of the process of calculating τ_P. This map can be calculated either explicitly or implicitly. In the explicit method, an equation is derived for **F** which is expressed in terms of the *position* of the mesh points. This (usually large and nonlinear) system is then solved to find **F** and hence to determine the location of the mesh. This procedure lies at the heart of a number of equidistribution *position-based* methods for calculating the mesh, such as the moving mesh partial differential equation, optimal transport and variational methods. Typically such methods *cluster* the mesh points where high precision is required, and the location of points of *density* of the mesh points moves as the solution evolves (in a similar manner to a longitudinal wave passing down the length of a spring, whilst the coils of the spring do not move very far from their equilibrium positions).

In an alternative procedure, the *velocity* **v** of the mesh points in τ_P is determined. This is given by

$$\mathbf{v} = \mathbf{F}_t. \tag{2.13}$$

The mesh point positions are updated using this velocity. This approach is very closely linked with particle and Lagrangian methods and includes methods such as GCL, MFE and the deformation map method. (Using the analogy above, such methods are like moving the whole spring.)

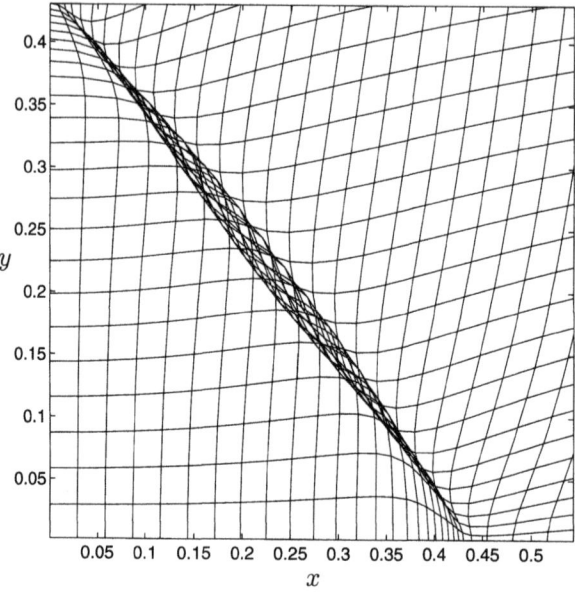

Figure 2.2. An example of a mesh which has tangled whilst attempting to resolve a front.

In general, position-based methods tend to produce smoother meshes, and are much less prone to the problem of *mesh tangling* than velocity-based methods. Mesh tangling occurs when the lines connecting adjacent mesh points intersect. An example of this is given in Figure 2.2, in which we see an attempted calculation of a solution front for Burgers' equation which has led to a tangled mesh through the use of an inappropriately large time step in evolving the mesh.

Mesh tangling can occur either locally or globally and can often arise in Lagrangian-type methods computing solutions with high vorticity. It is associated with a local loss of inevitability of the map $\mathbf{F}$ or, equivalently, at a point for which $\Lambda = \det(J) = 0$. If Λ is controlled throughout the evolution of the mesh, then mesh tangling can be avoided. Position-based methods usually try to do this (see the calculations using the optimal transport and MMPDE methods), hence their robustness to mesh tangling. Of course, control of Λ for all time is impossible, as any equations describing the time evolution of $\mathbf{F}$ will inevitably be discretized in time. If this discretization is too coarse then mesh tangling may result.

Mesh racing is related to mesh tangling and occurs if $\mathbf{v}$ is too large relative to the evolutionary behaviour of the underlying system being solved (so that the mesh evolves more rapidly than the solution of the underlying PDE). Mesh racing can occur for a variety of reasons, such as an inappropriate choice of adaptivity strategy, gross mesh distortion or problems when the moving mesh interacts with a fixed boundary. In a purely *Lagrangian* setting a moving mesh used to calculate (for example) a fluid flow might seek to have **v** equal to the local velocity of the fluid particles. In practice, as we will demonstrate, such a procedure can lead to mesh tangling in the presence of flows with high vorticity, and mesh racing when the fluid particles leave the boundary and the mesh labelling has to be reassigned. In practice (and in a manner to be made precise presently) it is often optimal for the mesh points to move in a similar manner to the particles, but not to follow their motions too precisely.

The connectivity of a mesh reflects how adjacent nodes are connected together. In an h-adaptive mesh connectivity can be a significant issue, and changes every time a local mesh refinement step is implemented. This causes additional overheads in setting up the equations of any discretization on this mesh, as the connectivity matrix needs to be constantly updated. In contrast, in an r-adaptive method, the connectivity of the moving mesh is usually determined by the connectivity of the underlying computational mesh, which usually does not change during the calculation, and we can presume to be relatively simple. A significant benefit of this approach is that various mesh-smoothing methods can use this constant connectivity and can (for example) exploit fast spectral methods which take advantage of the constant mesh connectivity in the computational domain. As an example, if the functions $(x(\xi, \eta), y(\xi, \eta))$ determine a particular mesh, then it is possible to construct a smoother mesh from this. One example is given by

$$(\hat{x}, \hat{y}) = (I - \gamma \Delta_\xi)^{-1}(x, y), \qquad (2.14)$$

where Δ_ξ is the Laplacian operator applied in the *computational domain* and γ is chosen appropriately. An important reason for constructing a smoother mesh is to avoid significant variation in mesh size between adjacent elements. We presently consider the effect of this on the solution error. This procedure was introduced by Huang and Russell (1997a). The Laplacian operator can be inverted very rapidly on a simply connected uniform rectangular mesh by using a fast spectral method, for example the fast cosine transform. This smoothing procedure also damps out the creation of certain chess board modes that can lead to a deterioration of the mesh quality.

2.4. Mesh topology

The discussion so far has been restricted to the use of moving meshes mapping one convex region, indeed logically rectangular regions, to logically rectangular regions. There is no real problem mapping a logically rectangular region Ω_C to another convex region. For example, the article by Calhoun *et al.* (2008) describes in detail how a logically rectangular mesh can be mapped to both circular and spherical domains. This is especially useful for calculations in meteorology involving whole Earth models. However, moving mesh methods are not ideal for mappings to and from non-convex regions, due to the inherent singularities associated with re-entrant corners. See Dvinsky (1991) for a brief discussion of this point. The issue of refining a mesh close to such a corner where the geometry of the solution and the associated singularity is (of course) known *a priori* has been extensively covered in the literature of h-adaptive methods: see, for example, Ainsworth and Oden (2000), Johnson (1987) and many other texts. This approach can be very naturally coupled with a moving mesh approach by using an h-adaptive method to construct a mesh in the *computational domain*, which is refined close to the re-entrant corner. This mesh can then be mapped, in a similar manner to that described earlier, to a moving mesh in the *physical domain*. We will not pursue this further here as this article is largely concerned with the construction of meshes adapted to the evolving structures of time-dependent PDEs.

2.5. Equidistribution and monitor functions

Having considered the general aspects of the mesh geometry and the mesh function **F**, we now consider the issues associated with calculating appropriate functions **F** to give meshes with certain properties. There are several general approaches to this, and we consider two closely related methods: equidistribution-based and variational-based. Both methods generate meshes determined by suitable *monitor functions*, which are typically determined both by properties of the solution of the underlying partial differential equation and by other considerations of the mesh regularity.

2.5.1. Equidistribution

At the heart of many r-adaptive methods is the concept of *equidistribution*, introduced as a computational device by de Boor (1973). Equidistribution is a widely used means of prescribing the *optimum geometry* of the mesh, but many different strategies have been devised to move the mesh towards this optimum state, leading to a variety of moving mesh methods. In a certain sense, all meshes equidistribute some function, and to motivate equidistribution we consider the fundamental Radon–Nikodym theorem from measure theory. To do this we consider an invertible mesh mapping function F which maps an arbitrary set A in Ω_C to an image set $B = F(A)$ in Ω_P. We can induce a measure $\nu(B)$ on Ω_P by $\nu(B) = |A|$, where $|A|$ is the usual Lebesgue measure on Ω_C. We then have the following.

Theorem 2.2. (Radon–Nikodym) If ν is a well-defined Borel measure on Ω_P, then there is a non-negative measurable function $M : \Omega_P \to \mathbb{R}$ such

that
$$\nu(B) = \int_B M(\mathbf{x})\, d\mathbf{x},$$

for any Borel subset B of Ω_P, where $d\mathbf{x}$ is the usual Lebesgue measure on Ω_P. Furthermore M is unique up to sets of Lebesgue measure zero.

Proof. See Capiński and Kopp (2004). □

The Radon–Nikodym theorem shows that for any invertible map $\mathbf{F}$ we can find a unique function $M(\mathbf{x})$ such that, for any set $A \subset \Omega_C$, we have

$$\int_A d\mathbf{x} = \int_{\mathbf{F}(A)} M(\mathbf{x})\, d\mathbf{x}. \tag{2.15}$$

Note, however, that (other than the special case of one dimension) the same function M may be associated with many different maps.

The function $M(\mathbf{x}) > 0$ is a function of $\mathbf{x}$ and t, but is more usually defined in terms of the solution $u(\mathbf{x}, t)$ of the underlying PDE, so that we might have

$$M(\mathbf{x}, t) \equiv M(\mathbf{x}, u(\mathbf{x}, t), \nabla u(\mathbf{x}, t), \ldots, t).$$

In this context M is usually called a *scalar monitor function*, and is chosen to be large when the mesh points need to be clustered, for example if the solution of the underlying problem has a high gradient. In this case the Lebesgue measure of the set B may be small even if the measure $\nu(B)$ is not. This may occur, for example, in the neighbourhood of a solution singularity or a sharp front. An obvious example of a monitor function is some estimate of the truncation error in the calculation of the solution of the underlying PDE, and this was the original motivation of the equidistribution approach of de Boor (1973). Loosely speaking, equidistributing the error in calculating the solution of a PDE over all mesh elements is a necessary condition for finding a global minimum of that error (Johnson 1987)

Assume now that we know the scalar function M and consider how we might determine an appropriate map $\mathbf{F}$. To do this we introduce an arbitrary non-empty set $A \subset \Omega_C$ in the computational domain, with a corresponding image set $\mathbf{F}(A, t) \subset \Omega_P$. The map $\mathbf{F}$ *equidistributes the respective scalar monitor function* M if the Stieltjes measures of A and $\mathbf{F}(A, t)$, normalized over the measure of their respective domains, are the same. This implies that

$$\frac{\int_A d\boldsymbol{\xi}}{\int_{\Omega_C} d\boldsymbol{\xi}} = \frac{\int_{\mathbf{F}(A,t)} M(\mathbf{x}, t)\, d\mathbf{x}}{\int_{\Omega_P} M(\mathbf{x}, t)\, d\mathbf{x}}. \tag{2.16}$$

It follows from a change of variable that

$$\frac{\int_A d\boldsymbol{\xi}}{\int_{\Omega_C} d\boldsymbol{\xi}} = \frac{\int_A M(\mathbf{x}(\boldsymbol{\xi},t),t)|J(\boldsymbol{\xi},t)|\,d\boldsymbol{\xi}}{\int_{\Omega_P} M(\mathbf{x}(\boldsymbol{\xi},t),t)\,d\mathbf{x}}. \quad (2.17)$$

As the set A is arbitrary, the map $\mathbf{F}(\boldsymbol{\xi},t)$ must (for all $(\boldsymbol{\xi},t)$) obey the identity

$$M(\mathbf{x}(\boldsymbol{\xi},t),t)|J(\boldsymbol{\xi},t)| = \theta(t), \quad \text{where } \theta(t) = \frac{\int_{\Omega_P} M(\mathbf{X}(\boldsymbol{\xi},t),t)\,d\mathbf{x}}{\int_{\Omega_C} d\boldsymbol{\xi}}. \quad (2.18)$$

We shall refer to (2.18) as the *equidistribution equation*. This equation must always be satisfied by the map $\mathbf{F}(\boldsymbol{\xi},t)$. It is the central equation of much of mesh generation, and we shall show presently that it is strongly connected to a variational representation of the mesh transformation.

2.5.2. Choice of a scalar monitor function

The choice of a scalar monitor function M appropriate to the accurate solution of a PDE is difficult, problem-dependent, and the subject of much research. We do not consider this in detail here but give a brief review of various choices used for certain problem classes. The function M can be determined by *a priori* considerations of the geometry or of the physics of the solution. An example is the generalized solution arclength given by

$$M = \sqrt{1 + c^2|\nabla_x u(x)|^2}, \quad \text{or alternatively} \quad M = \sqrt{1 + c^2|\nabla_{\boldsymbol{\xi}} u(x(\boldsymbol{\xi}))|^2}. \quad (2.19)$$

The first of these is often used to construct meshes which can follow moving fronts with locally high gradients (Winslow 1967, Huang 2007). A careful analysis of the application of arclength-based monitor functions to the resolution of the solution of singularly perturbed PDEs is given in Kopteva and Stynes (2001). Ceniceros and Hou (2001) successfully used the second monitor function (with u being the temperature) to resolve small scale singular structures in Boussinesq convection. It is also common to use monitor functions based on the (potential) vorticity, or curvature, of the solution (Beckett and Mackenzie 2000), and these have been used in computations of weather front formation (Budd and Piggott 2005, Budd, Piggott and Williams 2009, Walsh, Budd and Williams 2009). In certain problems, moving fronts are associated with changes in the physics of the solution. An example is problems with phase changes, where the phase front occurs at those points $(x_m)_i$ at a temperature $T = T_m$. In such cases it is possible to construct meshes which resolve behaviour close to the phase boundary by using the monitor function $M = a/\sqrt{b|x - x_m| + c}$, where $|x - x_m| = \min|x - (x_m)_i|$ (Mackenzie and Mekwi 2007a). Alternatively, M can be linked to estimates of the solution error. A significant calculation in which M was determined in terms of *a priori* error estimates (typically proportional to the higher derivatives of the solution or estimates of these) was given in Dorfi and Drury (1987), and is discussed in more detail presently. More recently, monitor functions determined by *a posteriori* error estimates have been constructed. An example of these, in the context of a piecewise linear finite element approximation u_h to a function u, is $M = \sqrt{1 + \alpha \zeta^2}$, where

$$|u - u_h|^2_{1,\Omega_P} \sim \zeta^2(u_h) \equiv \sum_{l:\text{ interior edge}} \int_l [\nabla u_h . n_l]^2_l \, dl \qquad (2.20)$$

and $[.]_l$ is the jump in the computed solution along the element edges. This monitor function is used by Tang (2005) to compute solutions adaptively to the Navier–Stokes equations with thin shear layers and/or high Mach numbers. Similarly, in a series of papers studying both isotropic and anisotropic meshes (Huang 2001a, 2001b, 2005a, 2005b, 2007), Huang explicitly considers monitor functions which are designed to control the regularity, alignment and quality of the mesh. These include monitor functions which are based as estimates of the interpolation error of the computed solution, and we consider them presently. Other measures of mesh quality can be incorporated into the monitor function including maximum and minimum angle conditions (Zlamal 1968, Babuška and Rheinboldt 1979), conditions on aspect ratio and quantities that combine both shape and solution behaviour (Berzins 1998). Finally, it is sometimes possible in the case of PDEs with *strong scaling structures* (such as problems related to combustion and gas dynamics) to find suitable monitor functions which give meshes reflecting the natural scales of the problem (Budd and Williams 2006). We give an example of these in Section 5, looking at a PDE which has solutions which blow up in a finite time. In this case we need a fine mesh when the solution is large, and take $M(u) = \sqrt{a^2 + b^2 u^{2p}}$, $p > 0$.

We note at this stage that most choices of monitor function need a degree of smoothing and regularization to perform effectively, and we will consider this presently.

2.5.3. Matrix-valued monitor functions

The monitor function defined above is a scalar measure and is effective in the specification and generation of certain isotropic meshes. However, much greater freedom in mesh calculation may be required when calculating anisotropic meshes, and in this case a *matrix-valued monitor function* M can be used. In this case the meshes are defined via the metric determined by an $n \times n$ matrix-valued monitor function that specifies the shape, size and orientation of the elements throughout the physical domain Ω_P. Huang (2007) defines a matrix-valued monitor function $\mathbf{M}(\mathbf{x})$ using the identity

$$J^{-T}J^{-1} = \left(\frac{\int_{\Omega_P} \sqrt{\det(\mathbf{M})}\, d\mathbf{y}}{|\Omega_C|}\right)^{-\frac{2}{n}} \mathbf{M}(\mathbf{x}), \qquad (2.21)$$

which is closely linked to the equidistribution principle for the scalar function. Indeed, the mesh satisfies an *equidistribution equation*

$$|J|\sqrt{\det(\mathbf{M})} = \frac{\int_{\Omega_p} \sqrt{\det(\mathbf{M})}\, d\mathbf{y}}{|\Omega_C|}. \qquad (2.22)$$

It also follows an *alignment condition*

$$\frac{1}{n}\mathrm{trace}(J^{-1}\mathbf{M}^{-1}J^{-T}) = \det(J^{-1}\mathbf{M}^{-1}J^{-T})^{\frac{1}{n}}. \qquad (2.23)$$

A matrix monitor function **M**, together with a proper boundary correspondence, then specifies a mesh via the conditions (2.22) and (2.23). Presently we relate these conditions of alignment and equidistribution to mesh quality and interpolation error, and show how to construct monitor functions that give explicit control over mesh quality.

2.6. Moving mesh PDEs, variational principles and harmonic maps

2.6.1. Moving the mesh to an equidistributed state

The equidistribution equation must be solved to find a mesh which equidistributes the monitor function M, and this equation must be augmented with additional conditions to obtain a unique map **F**. Indeed, the equidistribution principle has different consequences in one dimension from higher dimensions. In one dimension it (together with boundary conditions) uniquely defines the map $\mathbf{F}(\boldsymbol{\xi}, t)$. In this case the strategies for moving the mesh are all similar (or indeed trivially equivalent), and rely on either exactly solving the equidistribution equation (2.18), or relaxing towards a solution of some differentiated form of (2.18).

An important set of examples of the relaxation methods, which we will consider in detail (both in one and higher dimensions) in Section 3, are the variety of *moving mesh PDE* (MMPDE) methods. An example is MMPDE6, given by

$$-\epsilon x_{\xi\xi} = (Mx_\xi)_\xi. \qquad (2.24)$$

Here $0 < \epsilon \ll 1$ is a *relaxation time* over which the mesh evolves to the equidistributed state.

However, uniqueness of the solution of the equidistribution equation is lost in higher dimensions. Informally, this is because there is a unique interval (up to translation) of prescribed length in one dimension, but there is an uncountable number of sets of prescribed area in higher dimensions. Thus the equidistribution principle needs to be augmented with some additional conditions if it is to be applied in dimension $n > 1$. The determination of these additional conditions is neither straightforward nor unique, and leads to a variety of different methods for mesh generation, for which control of mesh skewness and other geometric properties must also be considered. Two examples of the additional conditions might be to impose an irrationality condition in the computational domain so that $\nabla_\xi \times \mathbf{F} = 0$ (Budd and Williams 2006) – which is the basis of the optimal transport methods – or to require that the mesh velocity is irrotational in the physical domain so that $\nabla_\mathbf{x} \times \mathbf{v} = 0$ (Cao et al. 2002) – which is the basis of the GCL methods. The augmented equations can then be solved in a number of ways to find the mesh: by directly solving the nonlinear system, which can be expensive; by differentiating the condition and solving the resulting differential equations which leads to the GCL methods we will consider in Section 4; by relaxing towards a solution of the system, which leads to the MMPDE methods in one and higher dimensions; or to have a global variational principle associated with the error and to find the gradient flow equations associated with it. We consider the latter now, with more details in Section 3.

2.6.2. Variational methods in one dimension and links to equidistribution

An alternative strategy for determining a mesh, also based on an appropriate monitor function, is the variational method. In such a method the stationary points determine the optimal mesh, and the associated gradient flow equations towards the stationary points determine a suitable mesh motion strategy.

·Suppose that $I(\boldsymbol{\xi})$ is a certain functional and that the mesh generation strategy is equivalent to minimizing I over a certain function space. Finding the Euler–Lagrange equations then leads to a gradient flow equation to evolve the mesh towards the equilibrium state (a stationary point of I), which is given by

$$\frac{\partial \xi}{\partial t} = -\frac{\delta I}{\delta \xi}. \tag{2.25}$$

This can then lead directly to an MMPDE to move the mesh by introducing some additional local control on the mesh movement in the form

$$\frac{\partial \xi}{\partial t} = -\frac{P}{\tau}\frac{\delta I}{\delta \xi}, \tag{2.26}$$

where P is a positive differential operator and $\tau > 0$ is a parameter for adjusting the time scale of the mesh movement. In one dimension, equidistributing the scalar monitor function M is exactly equivalent to minimizing the functional

$$I(\xi) = \frac{1}{2}\int_0^1 \frac{1}{M}\left(\frac{\partial \xi}{\partial x}\right)^2 \mathrm{d}x. \tag{2.27}$$

If the function P in (2.26) is taken to be

$$P = \left(\frac{M}{\xi_x}\right)^2,$$

then we obtain MMPDE5 (see Section 3 for an alternative derivation),

$$\frac{\partial x}{\partial t} = \frac{1}{\tau}\frac{\partial}{\partial \xi}\left(M\frac{\partial x}{\partial \xi}\right). \tag{2.28}$$

This equation can be used to evolve the one-dimensional mesh towards an equidistributed state. It also has a natural generalization to the PMA equation derived from the *optimally transported meshes* we will consider in Section 3.

2.6.3. Variational methods in higher dimensions

Motivated by (2.27) we can consider a generalization to two dimensions, which is essentially a form of *equidistribution in each coordinate direction* (Huang and Russell 1997b, 1999). This is given by

$$I(\xi, \eta) = \frac{1}{2}\int_{\Omega_P}\left[\nabla \xi^T \mathbf{M}^{-1}\nabla \xi + \nabla \eta^T \mathbf{M}^{-1}\nabla \eta\right] dx\, dy, \tag{2.29}$$

where $\mathbf{M}$ is now a symmetric positive definite matrix-valued monitor function, which is a generalization of the original scalar monitor function. The Euler–Lagrange equations which define the coordinate transformation at the steady state are then given by

$$\nabla \cdot (\mathbf{M}^{-1}\nabla \xi) = 0, \quad \nabla \cdot (\mathbf{M}^{-1}\nabla \eta) = 0, \tag{2.30}$$

where *all derivatives are expressed in terms of the physical variables* so that $\nabla = (\partial_x, \partial_y)^T$. A moving mesh PDE can then be obtained via the gradient flow equations, given by

$$\frac{\partial \xi}{\partial t} = -\frac{P}{\tau}\frac{\delta I}{\delta \xi}, \quad \frac{\partial \eta}{\partial t} = -\frac{P}{\tau}\frac{\delta I}{\delta \eta}, \tag{2.31}$$

and we will give more details of this procedure in Section 3. A special case of this system is given by

$$\mathbf{M} = wI, \tag{2.32}$$

where w is known as the (scalar) weight function. This corresponds to one-dimensional equidistribution and in the steady state gives the equations

$$\nabla \cdot \left(\frac{1}{w}\nabla \xi\right) = 0, \quad \nabla \cdot \left(\frac{1}{w}\nabla \eta\right) = 0. \tag{2.33}$$

Finding a mesh which satisfies this is called *Winslow's variable diffusion method* (Winslow 1967, 1981).

2.6.4. Harmonic maps

Another method closely related to (2.29) is the method based on *harmonic maps* (Dvinsky 1991). It defines the coordinate transformation used for mesh adaptation as a harmonic map minimizing the functional

$$I(\xi,\eta) = \frac{1}{2}\int_{\Omega_P} \sqrt{\det(\mathbf{M})}\left[\nabla\xi^T \mathbf{M}^{-1}\nabla\xi + \nabla\eta^T \mathbf{M}^{-1}\nabla\eta\right] dx\, dy, \quad (2.34)$$

where, once again, $\mathbf{M}$ is a matrix-valued monitor function. We note that in this case the matrix-valued function $\mathbf{M}$ cannot be chosen to be a scalar monitor function (see Winslow (1967)) as this would lead to no mesh adaptivity in two dimensions. Brackbill and Saltzman (1982) generalize Winslow's idea and define the needed coordinate transformation by minimizing a combination of three functionals characterizing adaptivity, smoothness, and orthogonality, respectively. Its final functional takes the form

$$I(\xi,\eta) = \theta_a \int_{\Omega_P} w|J|\, dx\, dy + \theta_s \int_{\Omega_P} (\nabla\xi^T\nabla\xi + \nabla\eta^T\nabla\eta)\, dx\, dy$$
$$+ \theta_o \int_{\Omega_P} (\nabla\xi^T\nabla\eta)^2\, dx\, dy, \quad (2.35)$$

where w is the (scalar) weight function and θ_a, θ_s, and θ_o are positive parameters. Notice that the three integrals on the right-hand side have different dimensions. As a consequence, the choice of the parameters may depend on specific applications. Directional control is further considered by Brackbill (1993). Variational methods have also been developed based on mechanical models; see Jacquotte (1988), Jacquotte and Coussement (1992) and de Almeida (1999). Dvinsky (1991) also discusses the advantages and disadvantages of formulating the harmonic map method in the physical domain and in the computational domain. However, numerical results show that the method formulated in the computational domain produces crossover meshes for a non-convex physical domain, whereas the method formulated in the physical domain leads to non-singular meshes.

2.7. Mesh quality, isotropy and alignment

The methods discussed in Section 2.6 are primarily based on physical and/or geometric considerations. Although they have been applied with a degree of success to numerical solution of a variety of PDEs, it is unclear how mesh concentration is controlled precisely through the monitor function for these methods. This is important because a clear understanding of the effect of the monitor function on mesh concentration will lead to a better choice of the monitor function as well as a better design of the mesh adaptation method itself. Moreover, neither of the methods, or their choice of the monitor function, is directly connected to any sort of error analysis. (A qualitative analysis of the effect of the monitor function on mesh concentration is given by Cao, Huang and Russell (1999b) for the functional (2.29).)

A variational method based on appropriate functionals which addresses these issues has been developed based on the equidistribution and alignment conditions (2.22) and (2.23) in Huang (2001b), Huang and Sun (2003) and Huang (2007). Recall that, for a given matrix-valued monitor function $\mathbf{M}(\mathbf{x})$, the condition (2.22) specifies the size of elements while (2.23) determines the shape and orientation of elements. The main idea of the variational method in this context is to then generate a coordinate transformation that closely satisfies these two conditions.

2.7.1. A functional for mesh alignment

First consider the alignment condition (2.23). Let the eigenvalues of the matrix $J^{-1}\mathbf{M}^{-1}J^{-T}$ be $\lambda_1, \ldots, \lambda_n$. By the arithmetic-mean/geometric-mean inequality, the desired coordinate transformation can be obtained by minimizing the difference between the two sides of the inequality

$$\left(\prod_i \lambda_i\right)^{\frac{1}{n}} \le \frac{1}{n}\sum_i \lambda_i.$$

Notice that

$$\sum_i \lambda_i = \operatorname{trace}(J^{-1}\mathbf{M}^{-1}J^{-T}) = \sum_i (\boldsymbol{\nabla}\xi_i)^T \mathbf{M}^{-1} \boldsymbol{\nabla}\xi_i,$$

$$\prod_i \lambda_i = \det(J^{-1}\mathbf{M}^{-1}J^{-T}) = \frac{1}{(|J|\sqrt{\det(\mathbf{M})})^2}.$$

Then we have

$$\left(\frac{1}{(|J|\sqrt{\det(\mathbf{M})})^2}\right)^{\frac{1}{n}} \le \frac{1}{n}\sum_i (\boldsymbol{\nabla}\xi_i)^T \mathbf{M}^{-1} \boldsymbol{\nabla}\xi_i,$$

or equivalently

$$\frac{n^{\frac{n}{2}}}{|J|} \le \sqrt{\det(\mathbf{M})}\left(\sum_i (\boldsymbol{\nabla}\xi_i)^T \mathbf{M}^{-1} \boldsymbol{\nabla}\xi_i\right)^{\frac{n}{2}}. \qquad (2.36)$$

Integrating the above inequality over the physical domain yields

$$n^{\frac{n}{2}}\int_{\Omega_C} d\boldsymbol{\xi} \le \int_{\Omega_P} \sqrt{\det(\mathbf{M})}\left(\sum_i (\boldsymbol{\nabla}\xi_i)^T \mathbf{M}^{-1} \boldsymbol{\nabla}\xi_i\right)^{\frac{n}{2}} d\mathbf{x}.$$

Hence, the adaptation functional associated with mesh alignment for the inverse coordinate transformation $\boldsymbol{\xi} = \boldsymbol{\xi}(\mathbf{x})$ can be defined as

$$I_{\text{ali}}(\boldsymbol{\xi}) = \frac{1}{2}\int_\Omega \sqrt{\det(\mathbf{M})}\left(\sum_i (\boldsymbol{\nabla}\xi_i)^T \mathbf{M}^{-1} \boldsymbol{\nabla}\xi_i\right)^{\frac{n}{2}} d\mathbf{x}. \qquad (2.37)$$

We remark that the functional (2.37) can also be derived from the concept of conformal norm in the context of differential geometry (Huang 2001b). Moreover, in two dimensions ($n = 2$), (2.37) gives the energy of a harmonic mapping (Dvinsky 1991). In this sense, the harmonic map method can be understood as a functional associated with alignment. Similarly, we can take squares on both sides of (2.36) and integrate the resulting inequality over Ω_P. We get

$$n^n \int_{\Omega_P} \frac{\sqrt{\det(\mathbf{M})}}{(|J|\sqrt{\det(\mathbf{M})})^2} \, d\mathbf{x} \leq \int_{\Omega_P} \sqrt{\det(\mathbf{M})} \left(\sum_i (\nabla \xi_i)^T \mathbf{M}^{-1} \nabla \xi_i \right)^2 d\mathbf{x}.$$

The resulting functional for alignment then takes the form

$$\tilde{I}_{\text{ali}}(\boldsymbol{\xi}) = \int_{\Omega_P} \sqrt{\det(\mathbf{M})} \left(\sum_i (\nabla \xi_i)^T \mathbf{M}^{-1} \nabla \xi_i \right)^2 d\mathbf{x}$$

$$- n^n \int_{\Omega_P} \frac{\sqrt{\det(\mathbf{M})}}{(|J|\sqrt{\det(\mathbf{M})})^2} \, d\mathbf{x}. \qquad (2.38)$$

2.7.2. A functional for equidistribution

We now consider the equidistribution condition (2.22). From Hölder's inequality we have

$$\left(\int_{\Omega_P} \frac{\sqrt{\det(\mathbf{M})}}{|J|\sqrt{\det(\mathbf{M})}} \, d\mathbf{x} \right)^2 = \left(\int_{\Omega_C} d\boldsymbol{\xi} \right)^2 \leq \int_{\Omega_P} \frac{\sqrt{\det(\mathbf{M})}}{(|J|\sqrt{\det(\mathbf{M})})^2} \, d\mathbf{x},$$

which leads to the functional for equidistribution given by

$$I_{\text{eq}}(\boldsymbol{\xi}) = \int_{\Omega_P} \frac{\sqrt{\det(\mathbf{M})}}{(|J|\sqrt{\det(\mathbf{M})})^2} \, d\mathbf{x}. \qquad (2.39)$$

2.7.3. An adaptation functional based on equidistribution and alignment

We note that neither of the adaptation functionals defined in the previous subsections can alone lead to a robust mesh adaptation method because each of them represents only one of the mesh control conditions (2.23) and (2.22). It is necessary and natural to combine them. A way to achieve this goal is to take an average of the functionals (2.38) and (2.39), i.e.,

$$I(\boldsymbol{\xi}) = \theta \int_{\Omega_P} \sqrt{\det(\mathbf{M})} \left(\sum_i (\nabla \xi_i)^T \mathbf{M}^{-1} \nabla \xi_i \right)^n d\mathbf{x}$$

$$+ (1 - 2\theta) n^n \int_{\Omega_P} \frac{\sqrt{\det(\mathbf{M})}}{(|J|\sqrt{\det(\mathbf{M})})^2} \, d\mathbf{x}, \qquad (2.40)$$

where $\theta \in [0, 1]$ is a parameter. Notice that the two terms in the functional have the same dimension. The balance between them is controlled by a dimensionless parameter θ. When $\theta = 1/2$, only the first term remains. Regarding well-posedness, it is noted that the first term of the functional is convex, and the existence, uniqueness, and the maximal principle for its minimizer are guaranteed; *e.g.*, see Reshetnyak (1989). It is unclear if this result can apply to the whole functional.

2.7.4. Mesh quality measures: alignment and equidistribution

Mesh quality measures can also be developed based on the alignment and equidistribution conditions (2.23) and (2.22). Indeed, for a given matrix-valued monitor function $\mathbf{M} = \mathbf{M}(\mathbf{x})$ and a coordinate transformation $\mathbf{x} = \mathbf{x}(\boldsymbol{\xi})$ (or its inverse), we can use

$$Q_{\text{ali}} = \left[\frac{\text{trace}(J^T \mathbf{M} J)}{n \det(J^T \mathbf{M} J)^{\frac{1}{n}}}\right]^{\frac{n}{2(n-1)}}, \tag{2.41}$$

$$Q_{\text{eq}} = \frac{|J| \sqrt{\det(\mathbf{M})} |\Omega_c|}{\int_{\Omega_P} \sqrt{\det(\mathbf{M})}\, dy} \tag{2.42}$$

to measure how closely the coordinate transformation (*i.e.*, mesh) satisfies the alignment and equidistribution conditions (2.23) and (2.22), respectively. We note that Q_{ali} is equivalent to

$$\hat{Q}_{\text{ali}} = \left[\frac{\text{trace}(J^{-1} \mathbf{M}^{-1} J^{-T})}{n \det(J^{-1} \mathbf{M}^{-1} J^{-T})^{\frac{1}{n}}}\right]^{\frac{n}{2(n-1)}}. \tag{2.43}$$

The quantity Q_{ali} ranges from 1 to ∞, with $Q_{\text{ali}} \equiv 1$ for the identity mapping, while Q_{eq} takes values in $(0, \infty)$, with $\max_x Q_{\text{eq}} = 1$ implying an equidistributing mesh. Interestingly, Q_{ali} reduces to an equivalence of Q_{geo} when $\mathbf{M} = \text{Id}$. In this sense, Q_{ali} can be viewed as a geometric quality measure in the metric specified by $\mathbf{M}$.

2.8. Error control and associated monitor functions

The measures for mesh quality and geometry described in Section 2.7 have largely been constructed in the absence of a clear application. For the majority of this article we are considering the effectiveness of a mesh for computing the solution of a partial differential equation. In this case we are expecting to impose some form of discretization of the system on the mesh. From this discretization we hope to solve a (typically rapidly evolving) partial differential equation. The mesh so constructed should attempt to minimize error (such as the truncation or the interpolation error) in some way. In this subsection we consider three forms of error, namely static truncation error, static interpolation error and dynamic errors, and in the first two cases look at monitor functions which lead to reduced errors. The difficulty in implementing such a procedures is, of course, that not only is it difficult to measure (or indeed to precisely define) the error during the calculation and adapting the mesh accordingly, but also some of the 'best' adaptive meshes for solving time-dependent problems lead to very stiff differential equations, thus significantly increasing the computational cost of the process and in some cases making the whole calculation significantly unstable. However, it is intuitively reasonable that for solutions with small length scales over part of the domain, and larger length scales elsewhere, we might expect to gain significant efficiency by using a smaller mesh in the region of high variation. Exactly this observation motivated the important early work of Dorfi and Drury (1987).

2.8.1. Static truncation error and 'optimal' meshes
The most natural reason for using an adaptive mesh in the context of solving a PDE is to control the overall error in any discretization. It is a surprisingly difficult problem to obtain such an estimate in the context of a (non-uniform) moving mesh. Typically there are contributions to the local truncation error from the local mesh scale, the variation of the mesh from one element to the next, and also the effects of the mesh motion, which all need to be taken into consideration. It is also difficult to then extrapolate from a local to a global error estimate in such cases. Accordingly, we will confine ourselves to giving a flavour of this analysis by looking at some simple one-dimensional problems for which we can perform the technical calculations needed to analyse the error. Following this we will return to more general ideas shortly. Accordingly, as examples of two *steady-state* problems for which adaptivity may be required, we may wish to solve the Poisson equation

$$-\Delta u = f(x, y, z, \ldots) \qquad (2.44)$$

for a potentially singular right-hand side f. Alternatively we may seek to solve the singular diffusion equation

$$-\epsilon u'' - c(x)u' = f(x), \quad \epsilon \ll 1. \qquad (2.45)$$

An ideal mesh with N points, used to compute the function u, is one which leads to low errors – ideally, in the case of (2.45), to errors which are ϵ-independent and depend only upon N. Such a mesh should also keep computational costs low. Essentially we can consider two types of non-uniform mesh for the computation. One type is an *optimal, fitted or a priori mesh*, which is prescribed in advance of the calculation and gives best possible errors for that computation in some appropriate norm. Important examples of this class are the *Shishkin and Bakhvalov meshes* for the singularly perturbed problems (2.45) and optimal meshes for Poisson-type problems. In Babuška and Rheinboldt (1979), a general analysis of such meshes is made in the context of finite element calculations in one dimension. An adaptive mesh, on the other hand, attempts to approximate an optimal mesh through equidistributing a suitable monitor function determined during the computation.

It is important to note at this stage that the error in discretizing a differential equation (or indeed in interpolating the solution to that equation or a function in general) is a combination of the error that would occur on a uniform mesh together with further errors that arise from the non-uniformity of the mesh (variation in the size of the elements) and (in more than one dimension) the mesh skewness. The latter errors have to be treated with great care as they can easily dominate the truncation error on the uniform mesh and make an adapted mesh worse than useless in solving the underlying problem. However, in contrast, a common error in many numerical analysis texts is to assume either that these errors add together to give a larger error, or that, for example, the error due to the non-uniformity of the mesh is always at a lower order than the error on the uniform mesh and thus dominates the overall calculation. In fact, provided that the mesh function is *suitably smooth*, for example if (in one dimension) the mesh is quasi-uniform and obeys the condition (2.9), then the three errors can be *at the same order* when expressed in terms of $1/N^p$. In an 'optimal' mesh it may be possible for the errors to cancel each other out to leading order. However, such meshes are usually very hard to construct and require a lot of *a priori* information about the solution. Adaptive meshes generally work by bounding the leading order error (regardless of the behaviour of the underlying solution).

We start by looking at both optimal and adaptive non-uniform meshes on which we can pose finite difference discretization of the Poisson equation:

$$-\frac{d^2 u}{dx^2} = f(x). \qquad (2.46)$$

If the mesh is a function $x(\xi)$ of the computational variable, then in the computational domain we have

$$-\frac{1}{J}\left(\frac{u_\xi}{J}\right)_\xi = f(x(\xi)), \quad J = x_\xi. \qquad (2.47)$$

The equation (2.47) can then be discretized in the computational domain for which we use the approximations

$$U_j \approx u(X_j), \quad X_j = x(j\Delta_\xi), \quad f_j = f(X_j).$$

A natural centred difference approximation to (2.47) then takes the form

$$-\frac{2}{(\Delta\xi)^2}\frac{\left(\frac{U_{j+1}-U_j}{X_{j+1}-X_j} - \frac{U_j-U_{j-1}}{X_j-X_{j-1}}\right)}{X_{j+1}-X_{j-1}} = f_j, \qquad (2.48)$$

with

$$\Delta_i = X_{i+1} - X_i.$$

We now assume that the mesh function $x(\xi)$ has regularity C^2 and exactly equidistributes a scalar monitor function M.

Lemma 2.3.

(i) The local truncation error T of the above discretization is given in the original variables by

$$T = \frac{\Delta_i^2}{3}\left[\frac{x_{\xi\xi}}{x_\xi^2}u_{xxx} + \frac{u_{xxxx}}{4}\right] + \mathcal{O}(\Delta_i^3), \qquad (2.49)$$

and in the computational variables by

$$T = \frac{\Delta\xi^2 x_\xi^2}{3}\left[-\frac{M_\xi u_{xxx}}{Mx_\xi} + \frac{u_{xxxx}}{4}\right] + \mathcal{O}(\Delta\xi^3). \qquad (2.50)$$

(ii) The truncation error is of second order if the mesh is quasi-uniform so that condition (2.10) is satisfied.

(iii) The truncation error is *zero to leading order* if the mesh equidistributes the monitor function

$$M_{\text{opt}} = (u_{xxx})^{1/4} = (-f_x)^{1/4}. \qquad (2.51)$$

Proof. Let $\Delta_j = X_{j+1} - X_j$. A simple Taylor expansion gives

$$U_{j+1} = U_j + \Delta_j u' + \frac{\Delta_j^2}{2}u'' + \frac{\Delta_j^3}{6}u''' + \frac{\Delta_j^4}{24}u'''' + \mathcal{O}(5),$$

$$U_{j-1} = U_j - \Delta_{j-1}u' + \frac{\Delta_{j-1}^2}{2}u'' - \frac{\Delta_{j-1}^3}{6}u''' + \frac{\Delta_{j-1}^4}{24}u'''' + \mathcal{O}(5),$$

where all derivatives of u are expressed in terms of x. Hence, expanding the left-hand side of (2.48), we obtain (after some manipulation) that the truncation error is given by

$$T = \frac{1}{3}(\Delta_j - \Delta_{j-1})u''' + \frac{1}{12}\frac{\Delta_j^3 + \Delta_{j-1}^3}{\Delta_j + \Delta_{j-1}}u'''' + \mathcal{O}(3).$$

This error has two components. The second is the usual component (of order Δ_j^2) which is seen on a uniform mesh. The first is an additional error due to the variation in the size of the mesh. In many texts this is considered to be large (as it is apparently of higher order); however, if Δ_j varies smoothly over the domain then it is actually of the same order as the second error. Since, to leading order,

$$\Delta_j = \Delta\xi x_\xi,$$

we have, to leading order,

$$T = (\Delta\xi)^2\left(\frac{1}{3}x_{\xi\xi}u''' + \frac{1}{12}x_\xi^2 u''''\right) + \mathcal{O}(\Delta\xi^3).$$

Setting (to leading order) $\Delta_j = \Delta \xi x_\xi$ gives (2.49). Now, from the equidistribution equation we have

$$x_\xi = \frac{\theta}{M}.$$

Hence $M_\xi x_\xi + M x_{\xi\xi} = 0$. Substituting for M in the above gives (2.50).
Result (ii) follows immediately from the expression (2.49)

The optimal form of M in (iii) arises from setting the leading-order term to zero and integrating. Note that this latter calculation can break down if u_{xxx} vanishes at some point. □

An almost identical calculation to the above leads to the following result.

Lemma 2.4.

(i) If we consider using the standard central difference approximation to u_x given by

$$u_x = \frac{U_{i+1} - U_{i-1}}{X_{i+1} - X_{i-1}},$$

then the truncation error is given in the physical coordinates by

$$T = \frac{\Delta_i^2}{2}\left[\frac{x_{\xi\xi}}{x_\xi^2}u_{xx} + \frac{1}{3}u_{xxx}\right] + \mathcal{O}(\Delta_i^3), \tag{2.52}$$

or in the computational coordinates by

$$T = \frac{\Delta \xi^2 x_\xi^2}{2}\left[-\frac{M_\xi u_{xx}}{M x_\xi} + \frac{1}{3}u_{xxx}\right] + \mathcal{O}(\Delta \xi^3). \tag{2.53}$$

(ii) This error is of second order on a quasi-uniform mesh, and is zero to leading order on an 'optimal mesh' given when

$$M = (u_{xx})^{1/3}. \tag{2.54}$$

These calculations, both of the errors in approximating u_x and u_{xx} and of the possible optimal meshes, are revealing in a number of ways. Firstly, they show that in all such calculations there is a subtle interplay between the mesh variability and the mesh size. This is even more marked in the case of singular perturbation problems. By choosing M very carefully we can exploit this to give very high accuracy and an *optimal mesh*. In general this is not usually possible. Indeed this calculation requires an accurate knowledge of the third derivative of the function u.

Secondly, we can also see from (2.50) the effect of choosing other types of monitor function as part of an adaptive calculation. The optimal mesh eliminates the truncation error to leading order. However, the truncation error is actually an estimate for the second derivative (with respect to x) of the solution error between the calculated solution U_j and $u(X_j)$. To obtain a true estimate, this expression needs to be integrated. A useful expression for the error for both (2.46) and (2.45) (see Andreev and Kopteva (1998)) is then given by the next lemma.

Lemma 2.5.
$$\|U_j - u(X_j)\|_\infty \leq C \max\left[\Delta_i^2 \max_{[X_i,X_{i+1}]} |u''| + \Delta_i^2\right]. \tag{2.55}$$

Given suitable *a priori* estimates, this error can be bounded by using a monitor function M which controls this via the expression

$$\Delta_i^2\left(\max_{[X_i,X_{i+1}]} |u''| + 1\right) = \Delta\xi^2 x_\xi^2 \left(\max_{[X_i,X_{i+1}]} |u''| + 1\right)$$
$$= \Delta\xi^2 \left(\max_{[X_i,X_{i+1}]} |u''| + 1\right) \theta^2/M^2.$$

This motivates the choice of the curvature-dependent monitor function given by
$$M = \sqrt{1 + |u''|^2}.$$

Such a function has been used by Blom and Verwer (1989); see also Mackenzie and Robertson (2002), Chen (1994), Huang and Sun (2003) and Kopteva (2007). In this case the error *becomes a function of $\Delta\xi$ only* and does not depend upon the solution. Hence it has the great advantage of yielding a mesh for which large variations in u'' (for example at boundary layers) do not affect accuracy. This is good enough for most calculations. Observe, however, the difference between using the curvature-based monitor function to bound the error, and the optimal monitor function for the Poisson equation which eliminates this error to leading order.

Similar issues arise in the case of the singularly perturbed problems (2.45). For example, it is possible to get very sharp estimates on the solution in certain cases (such as when $c(x) = 1$). In the latter case (Andreev and Kopteva 1998) we have the following result.

Lemma 2.6.
$$\|U_j - u(X_j)\|_\infty \leq C\left[\|\min\{\Delta_i^2/\epsilon^2, 1\} e^{-x_{i-1}/\epsilon}\|_\infty + \max \Delta_i^2\right]. \tag{2.56}$$

This error can be completely controlled to be proportional only to $\Delta\xi^2$ using a Bakhvalov mesh. For such problems it can also be shown (Kopteva 2007) that, if the monitor function is chosen to be a discrete arclength of the form
$$M = \sqrt{1 + u_x^2},$$
then, provided that the solution has converged closely to an equidistributed one, the computed solution is first-order accurate with errors $\mathcal{O}(\Delta\xi)$, independent of the value of ϵ.

It is interesting to point out that uniform convergence has been obtained by a number of researchers for equidistributing meshes determined *a priori* by the exact solution or the singularity information of the exact solution for singularly perturbed differential equations: *e.g.*, see Sloan *et al.* (Qiu and Sloan 1999, Qiu, Sloan and Tang 2000), Mackenzie *et al.* (Mackenzie 1999, Beckett and Mackenzie 2000, Beckett, Mackenzie, Ramage and Sloan 2001*b*, Beckett and Mackenzie 2001*a*, 2001*b*, Mackenzie and Mekwi 2007*b*), and Huang (2005*c*). Convergence results are also obtained for *a posteriori* equidistributing meshes determined by computational solutions for differential equations in Babuška and Rheinboldt (1979), Kopteva and Stynes (2001), He and Huang (2009) and Huang, Kamenski and Lang (2009).

2.8.2. Static interpolation error

In higher dimensions it is very hard to obtain reliable estimates for the truncation error when solving a general PDE. A somewhat easier, but still very important, question to address is whether a mesh is suitable to approximate the solution of the PDE, in particular to interpolate the solution. We now consider this question.

Suppose that the solution of the differential equation (or indeed any appropriate function defined in the physical domain) is given by $u(x, y, \ldots)$. For the case of a problem in two dimensions we can define the point values of u on the non-uniform mesh by

$$U_{i,j} = u(X_{i,j}, Y_{i,j}).$$

A natural measure of error is the *interpolation error* obtained by approximating u on the mesh with suitable functions using the above point values. Significant progress in finding meshes with good properties in reducing the interpolation error of a solution has been made in this direction in the past decade. Formulae giving the optimal monitor function to minimize this error over a suitable mesh have been developed based on interpolation error estimates by Huang and Sun (2003) in the H^m-norm, Chen *et al.* (2007) in the L^q-norm, Huang (2005*a*, 2005*b*) in the $W^{m,q}$-norm, and Cao (2005, 2007*a*, 2007*b*, 2008) for higher-order interpolation in two dimensions. Formulae have also been developed based on an *a posteriori* error estimate for one dimension (He and Huang 2009), a hierarchical basis *a posteriori* error estimate (Huang *et al.* 2009), and semi *a posteriori* error estimates for variational problems (Huang and Li 2009). Formulae for the monitor functions in these cases can be obtained as follows (Huang and Sun 2003, Huang 2005*a*). A so-called anisotropic error bound, taking into consideration the directional effect of the error or solution derivatives, is first developed. This error bound can be regarded as a function of the monitor function M when only meshes satisfying the alignment and equidistribution conditions (2.23) and (2.22) are concerned. Then the optimal monitor function is obtained by minimizing the bound among all possible matrix-valued functions $\mathbf{M}$. Consider a simple situation where a function $u \in H^2(\Omega_P)$ is interpolated by piecewise linear polynomials on a simplicial mesh (of N elements) and the error is measured in L^2-norm. Then an anisotropic asymptotic bound (as $N \to \infty$) can be obtained from the interpolation theory of Sobolev spaces (Huang and Sun 2003), namely

$$\|e_h\|^2_{L^2(\Omega_P)} \leq C\alpha^2 N^{-\frac{4}{n}} \int_{\Omega_P} \left(\text{trace}(J^T[I+\alpha^{-1}|H(u)|]J)\right)^2 \mathrm{d}\mathbf{x} + \text{h.o.t.}, \quad (2.57)$$

where $H(u)$ denotes the Hessian of the function u, $|H(u)| = \sqrt{H(u)^T H(u)}$, and $\alpha > 0$ is an arbitrary number which serves as a regularization parameter, whose value will be determined later. Note that a rigorous bound can be obtained. But in this situation the derivation has to be associated with a discrete form; e.g., see Huang (2007). Since the procedure is the same for both, for simplicity we use the non-rigorous continuous form. Noticing that a mesh satisfying (2.23) and (2.22) (and a proper boundary correspondence) is a function of $\mathbf{M}$, we can regard the integral on the right-hand side of (2.57) as a function of $\mathbf{M}$, namely

$$B(\mathbf{M}) = \int_{\Omega_P} \left(\text{trace}(J^T[I+\alpha^{-1}|H(u)|]J)\right)^2 \mathrm{d}\mathbf{x}. \quad (2.58)$$

In the following analysis, we consider only meshes satisfying (2.22) and (2.23) and derive the optimal monitor function by minimizing $B(\mathbf{M})$ among all possible matrix-valued functions $\mathbf{M}$. First we notice that (2.23) is mathematically equivalent to

$$\frac{1}{n}\text{trace}(J^T\mathbf{M}J) = \det(J^T\mathbf{M}J)^{\frac{1}{n}}. \quad (2.59)$$

A direct comparison of (2.59) suggests that $\mathbf{M}$ can be chosen in the form

$$\mathbf{M} = \theta(\mathbf{x})[I + \alpha^{-1}|H(u)|], \quad (2.60)$$

where $\theta = \theta(\mathbf{x})$ is a scalar function. For matrix-valued monitor functions in this form, (2.59) reduces to

$$\frac{1}{n}\text{trace}(J^T[I+\alpha^{-1}|H(u)|]J) = \det(J^T[I+\alpha^{-1}|H(u)|]J)^{\frac{1}{n}}.$$

Inserting this into (2.58) and using Hölder's inequality, we have

$$B(\mathbf{M}) = n^2 \int_{\Omega_P} |J|^{\frac{4}{n}} \det(I+\alpha^{-1}|H(u)|)^{\frac{2}{n}} \mathrm{d}\mathbf{x}$$

$$= n^2 \int_{\Omega_C} \left[|J|\det(I+\alpha^{-1}|H(u)|)^{\frac{2}{n+4}}\right]^{\frac{n+4}{n}} \mathrm{d}\boldsymbol{\xi}$$

$$\geq n^2 |\Omega_C|^{-\frac{4}{n}} \left[\int_{\Omega_C} |J| \det(I + \alpha^{-1}|H(u)|)^{\frac{2}{n+4}} \, d\boldsymbol{\xi} \right]^{\frac{n+4}{n}} \quad (2.61)$$

$$= n^2 |\Omega_C|^{-\frac{4}{n}} \left[\int_{\Omega_P} \det(I + \alpha^{-1}|H(u)|)^{\frac{2}{n+4}} \, d\mathbf{x} \right]^{\frac{n+4}{n}}. \quad (2.62)$$

We note that equality in (2.61) holds when the mesh satisfies

$$|J| \det(I + \alpha^{-1}|H(u)|)^{\frac{2}{n+4}} = \frac{1}{|\Omega_C|} \int_{\Omega_P} \det(I + \alpha^{-1}|H(u)|)^{\frac{2}{n+4}} \, d\mathbf{y}.$$

Comparing this with the equidistribution condition (2.22), we have

$$\sqrt{\det(\mathbf{M})} = \det(I + \alpha^{-1}|H(u)|)^{\frac{2}{n+4}}.$$

From this and (2.60), the optimal matrix-valued monitor function to minimize the interpolation error is given by

$$\mathbf{M} = \det(I + \alpha^{-1}|H(u)|)^{-\frac{1}{n+4}} \, [I + \alpha^{-1}|H(u)|]. \quad (2.63)$$

Inserting (2.62) into (2.57), the interpolation error bound for a mesh satisfying (2.23) and (2.22) with optimal $\mathbf{M}$ given in (2.63) is then

$$\|e_h\|_{L^2(\Omega_P)}^2 \leq C\alpha^2 N^{-\frac{4}{n}} \left[\int_{\Omega_P} \det(I + \alpha^{-1}|H(u)|)^{\frac{2}{n+4}} \, d\mathbf{x} \right]^{\frac{n+4}{n}} + \text{h.o.t.} \quad (2.64)$$

We now discuss how to choose α. We first notice that conditions (2.57) and (2.62) are invariant under scaling transformations of $\mathbf{M}$ of the form $\mathbf{M} \to c\mathbf{M}$, for any positive constant c. Thus, if $|H(u)|$ is strictly positive definite on Ω_P, we can take $\alpha \to 0$ in (2.63) and (2.64). This gives

$$\mathbf{M} = \det(|H(u)|)^{-\frac{1}{n+4}} \, |H(u)|, \quad (2.65)$$

$$\|e_h\|_{L^2(\Omega_P)}^2 \leq CN^{-\frac{4}{n}} \left[\int_{\Omega_P} \det(|H(u)|)^{\frac{2}{n+4}} \, d\mathbf{x} \right]^{\frac{n+4}{n}} + \text{h.o.t.} \quad (2.66)$$

When $|H(u)|$ vanishes locally, the monitor function cannot be defined by (2.65) since the right-hand side is not positive definite. In this case, a positive α should be used. Huang (2001b) suggests that α be defined implicitly via

$$\int_{\Omega_P} \det(I + \alpha^{-1}|H(u)|)^{\frac{2}{n+4}} \, d\mathbf{x} = 2|\Omega_P|. \quad (2.67)$$

It is easy to show that (2.67) has a unique solution for α. A simple iteration method such as the bisection method can be used for solving this equation. Moreover, when α is defined in this way, $\mathbf{M}$ is invariant for scaling transformation of $|H(u)|$. Furthermore, it is shown in Huang (2001b) that about fifty per cent of the mesh points are then concentrated in regions where

$\det(I + \alpha^{-1}|H(u)|)^{\frac{2}{n+4}}$ is large. Finally, the error bound reads as

$$\|e_h\|_{L^2(\Omega_P)}^2 \leq C\alpha^2 N^{-\frac{4}{n}}. \qquad (2.68)$$

From (2.67) it is not difficult to show that, for $n \leq 4$, α is bounded as

$$\left[\frac{1}{2|\Omega_P|}\int_{\Omega_P} \det(|H(u)|)^{\frac{2}{n+4}}\, d\mathbf{x}\right]^{\frac{n+4}{2n}} \leq \alpha \leq$$

$$\left[\frac{1}{n^{\frac{2n}{n+4}}|\Omega_P|}\int_{\Omega_P}\left(\mathrm{trace}(|H(u)|)\right)^{\frac{2n}{n+4}}\, d\mathbf{x}\right]^{\frac{n+4}{2n}}. \qquad (2.69)$$

2.8.3. Dynamic error

Usually when we apply a moving mesh method we are interested in solving a time-evolving PDE. This leads to additional dynamic errors (Li and Petzold 1997, Li et al. 1998) such as oscillations around rapidly evolving fronts or miscalculations of the front speed. These depend significantly on the way in which the mesh is updated and coupled to the PDE. We consider these in more detail in the next section when we look at how the moving mesh equations are coupled to the underlying PDE.

2.9. Monitor function smoothing and regularization

Having considered the mesh quality, we now return to further considerations of the monitor function and of mesh smoothness. Recall that for one-dimensional problems it is essential that the mesh should be quasi-uniform in order to have a low truncation error. Smoothing a mesh either directly or indirectly through smoothing/averaging aims to achieve this.

2.9.1. The Dorfi and Drury method

A direct approach to smoothing a one-dimensional mesh derived from an equidistribution principle is proposed in Dorfi and Drury (1987), and is often called the Dorfi and Drury method. In this method, if

$$n_i = (\Delta X_i)^{-1} \equiv (X_{i+1} - X_i)^{-1},$$

then a smoother mesh is given by computing $\hat{n}_i$ where

$$\hat{n}_i = n_i - \gamma(n_{i+1} - 2n_i + n_{i-1}) \qquad (2.70)$$

for a suitable constant γ. A variant of this, considered in Li and Petzold (1997), is given by updating the mesh differences by

$$\Delta \hat{X}_i = \sum_{j=i-p}^{i+p} \theta^{|i-j|}\Delta X_i$$

for a suitable constant $0 < \theta < 1$. There are many other strategies for direct mesh smoothing. For example it is possible to use *a posteriori* estimates of the solution on the mesh to do this (Bank and Smith 1997).

2.9.2. Monitor function smoothing

Alternatively we can generate a smoother mesh by averaging the monitor function prior to the mesh calculation. Suppose that point values of (a scalar or matrix-valued) monitor function $M_{i,j}$ are given. A Jacobian-type strategy, also referred to as averaging or low-pass filtering in the literature, is commonly used, *e.g.*, see Dorfi and Drury (1987), Verwer *et al.* (1989) and Huang and Sloan (1994). When a rectangular computational mesh is used, this smoothing can be conveniently expressed as

$$\hat{M}_{i,j} = \frac{\sum_{k=-1}^{1} \sum_{l=-1}^{1} M_{i+k,j+l} \gamma^{|k|+|l|}}{\sum_{k=-1}^{1} \sum_{l=-1}^{1} \gamma^{|k|+|l|}},$$

where $\gamma \in (0,1)$ is a parameter. This type of local smoothing of the monitor function can be viewed as an approximation of Laplace-operator-based smoothing:

$$(I - \lambda^{-2}\Delta_\xi)\hat{M} = M,$$

where λ is a parameter. In one dimension, Huang and Russell (1997*a*) show that, when λ is chosen as a value of order $\mathcal{O}(N)$, where N is the number of sub-intervals, a mesh equidistributing $\hat{M}$ is *locally quasi-uniform*, indeed there exists a reasonably small constant $\nu \geq 1$ such that

$$\frac{1}{\nu} \leq \frac{X_{j+1} - X_j}{X_j - X_{j-1}} \leq \nu \quad \forall j.$$

Another interesting strategy is to use a reference Jacobian matrix (Knupp 1996, Knupp, Margolin and Shashkov 2002) where a new mesh is generated to have a close Jacobian matrix to the reference one that is typically obtained from a reference, often non-smooth mesh.

2.9.3. 50:50 meshes and the Mackenzie regularization

A recurring problem with r-adaptive meshes which equidistribute a poorly chosen monitor function is that they can concentrate points in areas of particular identified interest where high resolution is needed, but leave other regions sparse of points. This can lead not only to low resolution in such areas, but also to a severe lack of mesh regularity and consequent large errors caused by a too rapid mesh variation. An example of such would be the calculation of the solution of a system which is blowing up in finite time with a large peak, in which all of the mesh points are concentrated in the peak alone. Such problems can be significantly reduced if the mesh is designed so that roughly half of the points are concentrated in the areas where high resolution is required, and half where it is not. Such meshes are called 50:50 meshes (see Budd et al. (2005), Huang (2001a), Huang et al. (2002)) and a regularization of M to ensure that such meshes arise in practice has been proposed by Beckett and Mackenzie (2000). To show how such problems arise in a calculation in n dimensions, suppose that we have a scalar monitor function M for which $\int_{\Omega_P} M \, dx = \theta$. Consider now the situation in which there are two subsets A and B of Ω_P, with $\Omega_P = A \cup B$ so that the monitor function is designed to concentrate points in a small region A (so that A may be the support of a singularity or of a front). The preimage $A' = F^{-1}(A) \subset \Omega_C$ represents those points in the computational domain which are mapped to A, with a similar set B'. Suppose now that $|A'|$ and $|B'|$ are the areas of these sets in Ω_C, with respective areas $|A|$ and $|B|$ in Ω_P. These areas measure the proportion of mesh points allocated to the corresponding sets A and B. Note that in most applications of an adaptive method, where mesh points have to be concentrated into a small region we would expect that

$$|A'| = \mathcal{O}(1), \quad |A| = o(1), \quad |B'| = \mathcal{O}(1), \quad |B| \approx |\Omega_P|. \tag{2.71}$$

Problems arise with mesh regularity and solution resolution away from the set A if $|B'| \ll |A'|$. It follows immediately from the equidistribution principle that if $\theta = \int_{\Omega_P \equiv A \cup B} M \, d\mathbf{x}$ then

$$\Lambda = \frac{|A'|}{|B'|} = \frac{\int_A M \, d\mathbf{x}}{\int_B M \, d\mathbf{x}} = \frac{\theta}{\int_B M \, d\mathbf{x}} - 1.$$

Furthermore,

$$|A'| = \frac{\int_A M \, d\mathbf{x}}{\theta}.$$

It follows from the conditions on A' and A in (2.71) that over the set A we have $M \gg \theta$. However, if the monitor function is so constructed such that over the set B we have $M \ll \theta$ and hence $\int_B M \, dx \ll \theta$ (so that the integral of M is concentrated in A), then Λ will be very large and the mesh will lose regularity. Indeed, we will have $|A'| \approx 1$. Exactly such problems arose in some of the blow-up calculations reported in Budd et al. (1996). In such cases we must replace M by the regularized function introduced by Beckett and Mackenzie (2000) and given by

$$\hat{M} = M + \frac{\theta}{|\Omega_P|}. \tag{2.72}$$

Observe that over the set A we have $\hat{M} \approx M$, and over B we have $\hat{M} \approx \theta/|\Omega_P|$, and trivially

$$\int_{\Omega_P} \hat{M} \, d\mathbf{x} = 2\theta.$$

Consequently, when we make use of the regularized monitor function $\hat{M}$ to define the mesh we have

$$|A'| = \frac{\int_A \hat{M} \, d\mathbf{x}}{2\theta} \approx \frac{\int_A M \, d\mathbf{x}}{2\theta} \approx \frac{1}{2},$$

and

$$\Lambda \approx \frac{2|\Omega_P|}{|B|} - 1 \approx 1.$$

This gives the desired 50:50 quality to the mesh.

3. Location-based moving mesh methods

In this section we will look in greater detail at the various methods described in Section 2 under the general heading of *location-based methods*. These are those methods which determine the *location* (or more precisely the density) of the mesh points, typically through solving some form of nonlinear differential equation through some form of gradient flow method. The latter can be hard to solve. However, the advantage of these methods is that they tend to give meshes with good global properties, avoiding excessive skewness. We will consider in detail the various methods outlined in the previous section, such as MMPDE-based methods, variational methods and optimal transport methods.

3.1. MMPDE methods in one dimension

Methods based on moving mesh partial differential equations (MMPDEs) are now universally used as a means of r-adaptivity in one dimension, and have been incorporated into codes such as MOVCOL and AUTO. There are many different MMPDEs, which together encapsulate most of the methods used to derive adaptive meshes in one dimension.

We consider a one-dimensional map $x(\xi,t)$ from $[0,1]$ to $[a,b]$ with associated mesh points $X_i = x(i\Delta\xi, t)$, which equidistributes the monitor function M. This map satisfies the equidistribution equation

$$M x_\xi = \theta, \quad x(0,t) = a, \quad x(1,t) = b, \quad \theta = \int_a^b M \, dx. \tag{3.1}$$

Lemma 3.1. *The equidistribution equation (3.1) has a unique monotone increasing solution $x(\xi,t)$ for all $M > 0$.*

Proof. Integrating (3.1) with respect to ξ and changing variables gives

$$\int_a^x M \, dx' = \theta \xi \quad \text{or} \quad \int_{X_{i-1}}^{X_i} M \, dx = \frac{1}{N} \int_a^b M \, dx. \tag{3.2}$$

Now, as $M > 0$ the left-hand side of this expression is a *monotone increasing* function of x. It immediately follows that x is a unique monotone increasing function of ξ. Observe further that this function is as smooth as the function M. □

This simple observation makes equidistribution relatively easy in one dimension.

The most direct way to enforce equidistribution is to solve (3.1) directly. However, this has the disadvantage that it requires the calculation of the integral θ. This can be avoided by a further differentiation with respect to ξ, and thus solving the moving mesh equation (together with boundary conditions) given by

$$(Mx_\xi)_\xi = 0, \quad x(0,t) = a, \; x(1,t) = b. \tag{3.3}$$

To determine an equidistributed mesh, the equation (3.3) can be discretized over the computational domain and then solved. This discretization does not have be done to high accuracy in order to obtain a regular mesh suitable for solving the underlying PDE. A typical such discretization takes the form

$$E_i \equiv \frac{2}{\Delta\xi^2}\left(M_{i+1/2}(X_{i+1} - X_i) - M_{i-1/2}(X_i - X_{i-1})\right) = 0,$$

$$M_{i+1/2} = \frac{1}{2}(M_i + M_{i+1}). \tag{3.4}$$

However, the solution of the system (3.4) requires solving a system of nonlinear equations, which is usually difficult and requires the use of some form of iterative procedure. See Pryce (1989), Xu, Huang, Russell and Williams (2009), He and Huang (2009), Kopteva and Stynes (2001) and Kopteva (2007) for a discussion of such methods, and conditions for them to converge to a solution.

This problem can be avoided by instead introducing a natural time evolution into the mesh equations. Differentiating the equidistribution equation (3.3) with respect to time gives the (so-called) MMPDE0 (Huang et al. 1994):

$$\frac{d}{dt}((Mx_\xi)_\xi) = 0. \tag{3.5}$$

Instead we may also differentiate (3.1) with respect to time (Adjerid and Flaherty 1986). This leads directly to the GCL method described in the next section. The resulting equation then takes the form

$$\frac{\partial}{\partial\xi}(Mx_t) + M_t x_\xi = \theta_t. \tag{3.6}$$

This equation can then also be differentiated with respect to ξ to eliminate the θ contribution, giving MMPDE1 (Huang et al. 1994), where it is assumed that we can find M_t, although in practice this may not be easy.

Starting from any mesh (uniform or otherwise), we can evolve the mesh by solving MMPDE0 (3.5) or MMPDE1. Unfortunately, a uniform mesh does not necessarily satisfy the equidistribution equation (3.1). Furthermore, even if a mesh does exactly satisfy it at some time, solving a discretized form of (3.5) inevitably leads to meshes that drift away from an equidistributed state. Both of these can lead to problems with mesh crossing (the one-dimensional version of mesh tangling) which occurs when $x_\xi = 0$.

As an example of this, which also demonstrates the general applicability of the method, we consider solving (3.5) starting from an initially *uniform mesh* on $[0,1]$ for which $x_\xi = 1$. In this calculation we will assume that we have a time-evolving monitor function $M(x,t)$ with $M(x,0) \equiv M^0(x)$. It follows from integrating (3.5) with respect to t and applying the initial conditions that, for all time, we have

$$(Mx_\xi)_\xi = M^0_\xi.$$

Hence, integrating again we have

$$Mx_\xi = M^0 + B(t),$$

for some function $B(t)$. This can be determined by integrating this expression with respect to ξ to give

$$Mx_\xi = M^0 + \theta(t) - \theta(0).$$

As $M^0 > 0$ it follows that if θ is increasing in time then $x_\xi > 0$. However, if θ decreases with t then it is quite possible for x_t to vanish (initially at the point where M^0 takes its minimum value) and for mesh crossing (tangling) to result.

Such problems can be avoided (both in one and in higher dimensions) by introducing a relaxation time into the solution of (3.3). The philosophy behind doing this is that the equation (3.3) need not be solved exactly to obtain a mesh which is perfectly reasonable for any computation. What is more important is that the mesh evolves *at least as fast as any significant features of the solution*. Exactly the same philosophy applies to moving meshes in any number of dimensions. Thus it is possible to consider meshes which relax towards an equidistributed mesh, provided *the relaxation time is smaller than the natural time scale of the solution*. Ideally the relaxation time should be of a similar order to that of the solution evolution. This prevents the mesh equations becoming unnecessarily stiff. Various different forms of mesh relaxation are possible.

The most obvious way of relaxing towards an equidistributed state was proposed by Anderson and Rai (1983), who computed the mesh through a relaxation equation, based on considering pseudo-forces between the mesh points, given by

$$\epsilon \dot{x} = (Mx_\xi)_\xi, \tag{3.7}$$

where $\epsilon > 0$ is presumed to be small. Alternatively, we can consider the original equidistribution equation in integral form. If a mesh is not exactly equidistributed then we can determine the residual

$$R = \int_a^x M \, \mathrm{d}x - \xi \int_a^b M \, \mathrm{d}x.$$

If we then set $\epsilon \dot{x} = -R$ and differentiate this expression twice with respect to ξ, we obtain

$$\epsilon(\dot{x})_{\xi\xi} = -(Mx_\xi)_\xi. \tag{3.8}$$

This equation was originally derived in Adjerid and Flaherty (1986). The equations (3.7) and (3.8) are known respectively (Huang et al. 1994) as MMPDE5 and MMPDE6. We can combine them to give the (smoothed) moving mesh equation considered in Huang and Russell (1997a) (see also the discussion in Section 2), which takes the form

$$\epsilon \left(1 - \gamma \frac{\partial^2}{\partial \xi^2}\right) \dot{x} = (Mx_\xi)_\xi. \tag{3.9}$$

Here $\gamma > 0$ can be chosen to give some control over the smoothness of the mesh. The equation (3.9) (and its various discretizations) is very dissipative, and leads to extremely stable meshes under most discretizations. The equation (3.9) also has natural extensions to higher dimensions, both in the context of the methods described in Huang and Russell (1997a) and Ceniceros and Hou (2001) and also in the optimal transport methods we consider later in this section. Other smoothed versions of the MMPDEs have also been considered by Huang and Russell (1997a). One of them is given by

$$\epsilon \left(1 - \gamma \frac{\partial^2}{\partial \xi^2}\right) \frac{\partial}{\partial \xi} \left(-\left(\frac{\partial x}{\partial \xi}\right)^{-2} \frac{\partial \dot{x}}{\partial \xi}\right) = -\left(1 - \gamma \frac{\partial^2}{\partial \xi^2}\right) \frac{\partial}{\partial \xi} \left(\frac{\partial x}{\partial \xi}\right). \tag{3.10}$$

Huang and Russell (1997a) have proved that the solutions (*i.e.*, the coordinate transformation and the mesh) to the continuous equation (3.10) using a central finite difference discretization have the properties both of local quasi-uniformity and no node-crossing.

Note. Many other MMPDEs have been derived, such as MMPDE2, given by

$$(M\dot{x})_{\xi\xi} = -(M_t x_\xi)_\xi - \frac{1}{\epsilon}(Mx_\xi)_\xi.$$

However, we will focus our discussion on the more widely used (3.9).

The moving mesh equation evolves a mesh towards an equidistributed state satisfying (3.3). When implementing the MMPDE method, the equation (3.9) is typically discretized over the computational space, leading to a set of ordinary differential equations for the location of the mesh points X_i.

These can then be solved using standard stiff ODE software, e.g., by using an SDIRK (singly diagonally implicit Runge–Kutta) method. A simple such semi-discretization of (3.9) is given by

$$\epsilon \left(\dot{X}_i - \gamma \frac{\dot{X}_{i+1} - 2\dot{X}_i + \dot{X}_{i-1}}{(\Delta \xi)^2} \right) = E_i(t), \qquad (3.11)$$

where the equidistribution measure $E_i(t)$ is as given in (3.4). This leads (on inversion of the simple tri-diagonal system on the right-hand side of this equation) to a simple set of ODEs for the location of the mesh points. Alternatively, a simple full discretization of (3.9) for a mesh X_i^n evaluated at the time $t_n = n\Delta t$ is proposed in Ceniceros and Hou (2001), and takes the form

$$\epsilon \left(X_i^{n+1} - \gamma \frac{X_{i+1}^{n+1} - 2X_i^{n+1} + X_{i-1}^{n+1}}{(\Delta \xi)^2} \right) =$$
$$\epsilon \left(X_i^n - \gamma \frac{X_{i+1}^n - 2X_i^n + X_{i-1}^n}{(\Delta \xi)^2} \right) + \Delta t E_i^n. \qquad (3.12)$$

These equations for the mesh can then be solved together with a suitable discretization of the underlying PDE, either simultaneously or alternately. We will give more details of this procedure later in this section in the context of moving meshes in higher dimensions, but we note at this stage that the simultaneous solution method is both possible and effective in such one-dimensional problems.

The method for evolving the mesh is typically implemented in two stages.

(1) Starting from an *initially uniform mesh* in the physical space, we evolve this to equidistribute the monitor function *at the initial time* over Ω_P. To do this we set $M_0(x) \equiv M(x,0)$, with $x_\xi = a + (b-a)\xi$, and solve (3.46) with M fixed to equal the function M_0, with $\epsilon = 1$ for $0 < t < T$, where T is a fixed time. In this first calculation t is an *artificial time* during which the uniform mesh evolves toward an equidistributed mesh for which the right-hand side of (3.9) is zero. It follows from the earlier results that, provided $M_0 > 0$, such a mesh exists, and we show presently that it is stable. In this initial calculation the right-hand side of (3.9) is initially relatively large, and taking $\epsilon = 1$ prevents the numerical calculation of the solution of the ODEs for the mesh point locations from being unnecessarily stiff. The value of T is chosen large enough to allow the mesh to relax toward the equidistributed state.

(2) We then solve (3.9) with the true *time-dependent* monitor function $M(x,t)$, with t now *actual time*. For this calculation we typically set $\epsilon = 0.01$. We show presently that the resulting mesh is then ϵ-close to a mesh which exactly equidistributes $M(x,t)$, provided that M does not

change too rapidly with time. As this procedure starts from a mesh which exactly equidistributes $M(\mathbf{X}, 0)$, the right-hand side of (3.9) is always close to zero and the resulting differential equations are not especially stiff.

Observe that this algorithm has the convenience of *starting from a uniform mesh*. This is a significant advantage over methods based on (3.5), or related methods such as the GCL method described in Section 4.

We will discuss later in this section the exact mechanism by which this algorithm for moving the mesh is coupled to the solution method for the underlying PDE.

This procedure has been criticized (for example, see Tang (2005)) for being imprecise about the way that ϵ is defined and the possibility of having to solve a very stiff system of equations. However, it can be given a very precise meaning. In the second stage of this calculation we are trying to find a mesh which is close to an equidistributed mesh. The natural time scale τ over which this mesh evolves is given simply by

$$\tau \approx \frac{\epsilon}{M}. \tag{3.13}$$

The key factor governing the choices of both ϵ and M is then to *ensure that τ is smaller than but of the same order as* the natural evolutionary time scale of the underlying PDE. In Section 5 we will show that in the context of PDEs with a strong scaling structure, this allows a natural choice to be made for both ϵ and M.

We now substantiate some of the claims made above, as well as stating another important property of the solutions of the moving mesh PDE (3.9).

Theorem 3.2.

(i) If $M_t = 0$, then the equidistributed mesh is a solution of (3.9) and is linearly stable.

(ii) If $M_t = \mathcal{O}(1)$, then an initially ϵ-close to equidistributed solution of (3.9) remains ϵ-close for all subsequent times.

(iii) At all times the solution of (3.9) satisfies $x_\xi > 0$, so that mesh crossing (tangling) does not occur.

Note. This applies for exact solutions of (3.9). If an overly coarse discretization is used to approximate these solutions then (iii) above may be violated (Smith 1996).

Proof. (i) Let $\hat{x}$ be an equidistributed mesh satisfying $(M(\hat{x})\hat{x}_\xi)_\xi = 0$, with $\hat{x}_t = M_t = 0$. It follows immediately that $\epsilon(\dot{\hat{x}} - \gamma \dot{\hat{x}}_{\xi\xi}) = (M\hat{x}_\xi)_\xi$ so that $\hat{x}$ satisfies the equidistribution equation. Now set $x = \hat{x} + R(\xi, t)$ with

$R \ll 1$ and $R(a) = R(b) = 0$. To leading order, R satisfies the equation

$$\epsilon(\dot{R} - \gamma R_{\xi\xi}) = (MR_\xi)_\xi + (M_x x_\xi R)_\xi = (MR_\xi + M_\xi R)_\xi = (MR)_{\xi\xi}.$$

Therefore

$$\epsilon \dot{R} = (1 - \gamma \partial_\xi^2)^{-1}(MR)_{\xi\xi} \equiv G(MR)_{\xi\xi}. \tag{3.14}$$

Here G is a *positive compact operator* and, as $M > 0$, $ER \equiv (MR_{\xi\xi})$ is a uniformly elliptic operator with a negative real spectrum. It follows that R must decay to zero. Hence the equidistributed solution is locally stable.

(ii) To prove this result, consider a slowly varying monitor function $M(x,t)$ and an exact solution $\hat{x}$ of the equidistribution equation $(M\hat{x}_\xi)_\xi = 0$. If $x = \hat{x} + \epsilon R$ is a solution of (3.9) then, extending the calculation in (3.14), we see that R satisfies the equation

$$\epsilon(\dot{x} - \gamma \dot{x}_{\xi\xi}) + \mathcal{O}(\epsilon^2) = \epsilon(MR)_{\xi\xi} + \mathcal{O}(\epsilon^2).$$

Hence, we have

$$(MR)_{\xi\xi} = \dot{x} - \gamma \dot{x}_{\xi\xi} + \mathcal{O}(\epsilon).$$

But as $(M\hat{x}_\xi)_\xi = 0$, we have

$$(M\dot{\hat{x}}_\xi)_\xi = -(M_t \hat{x}_\xi)_\xi.$$

As the operator $E\phi \equiv (M\phi)_{\xi\xi}$ is uniformly elliptic, it follows that provided $\dot{M}$ is of order one, then $\dot{\hat{x}}$ and hence R and its derivatives are also of order one. Consequently, the solution x of (3.9) stays ϵ-close to the solution of the equidistribution equation.

(iii) To show this we need to show that x_ξ cannot vanish. This result is a consequence of the maximum principle. In the case when $\gamma = 0$, we have, on differentiating (3.9), that

$$\dot{x}_\xi = M_{\xi\xi} x_\xi + 2M_\xi x_{\xi\xi} + M x_{\xi\xi\xi}.$$

Suppose that x_ξ is initially positive everywhere, and as x evolves it vanishes for a first time at (without loss of generality) the point $\xi = 0$. Then locally close to this point we have $x_\xi = a\xi^2 + \mathcal{O}(\xi^3)$ for some $a > 0$. Hence

$$\dot{x}_\xi = a\xi^2 M_{\xi\xi} + 2a\xi M_\xi + aM + \mathcal{O}(\xi).$$

Thus $\dot{x}_\xi > 0$ at this point and time. Hence x_ξ must remain positive. The more general result follows from the positivity of the compact operator G. □

3.1.1. Coupling a one-dimensional MMPDE method to a PDE

In one dimension, MMPDE methods can be very effectively coupled to an underlying PDE system by using a variety of different methods, including finite difference, finite element, collocation and spectral methods. The mesh equations and the PDE equations can then be solved together or alternately. We will discuss this in more detail presently in the context of moving mesh methods in higher dimensions, but it is appropriate to make some preliminary remarks here.

3.1.2. Finite difference methods

To motivate the discussion of appropriate discretizations, we assume that the underlying PDE system takes the form

$$\mathbf{u}_t = f(t, x, \mathbf{u}, \mathbf{u}_x, \mathbf{u}_{xx}). \tag{3.15}$$

If $x(\xi, t)$ is itself a time-dependent function of a computational variable ξ then (3.15) can be cast into the Lagrangian form in the moving coordinate system given by

$$\frac{d\mathbf{u}}{dt} = \mathbf{f}(t, x, \mathbf{u}, \mathbf{u}_x, \mathbf{u}_{xx}) + \mathbf{u}_x x_t. \tag{3.16}$$

The MMPDE governing the mesh motion gives a direct value for x_t. A method effective for solving (3.16) (in one-dimensional problems) is to use a semi-discretization. In this approach we discretize the differential equation (3.16) in the computational coordinates together with a similar discretization of the MMPDE (3.9). In such a semi-discretization we set

$$X_i(t) \approx x(i\Delta\xi, t) \quad \text{and} \quad U_i(t) \approx u(X_i(t), t).$$

As a simple example of the use of a finite difference method we can then take

$$u_x \approx \frac{U_{i+1} - U_{i-1}}{X_{i+1} - X_{i-1}} \quad \text{and} \quad u_{xx} \approx \frac{\frac{U_{i+1} - U_i}{X_{i+1} - X_i} - \frac{U_i - U_{i-1}}{X_i - X_{i-1}}}{\frac{X_{i+1} - X_{i-1}}{2}}. \tag{3.17}$$

These discretizations can then be substituted into (3.16) and the resulting set of ODEs for X_i and U_i solved along with one of the discretizations of (3.9). We discuss presently, and in more detail, the various alternating and simultaneous approaches for discretizing in time and then solving the resulting combined system.

On a static mesh, the truncation errors in calculating these finite difference approximations were given in the expressions (2.49) and (2.52). Provided that the stretching condition (2.10) is satisfied, then these errors are of second order.

We note, however, that additional errors may arise from the additional convective terms arising from the mesh movement, in particular the term

$$\mathbf{u}_x x_t \tag{3.18}$$

arising in (3.16). This additional term leads to both theoretical and practical difficulties in applying the moving mesh methods. From a theoretical perspective it is very possible that certain desirable properties of the equation (3.15) (such as symmetries, Hamiltonian structure and/or conservation laws) may not be inherited by the Lagrangian form (3.16). A practical difficulty, observed by Li *et al.* (1998), arises from certain discretizations of (3.18). In particular it was shown in this paper that it is possible that these can lead to instabilities and degrade the accuracy of the calculation. For example, if a centred finite difference approximation is used to discretize u_x, then from the expression (2.52) we have an additional truncation error given to leading order by (Li *et al.* 1998)

$$\dot{x}\frac{\Delta_i^2}{2}\left[\frac{x_{\xi\xi}}{x_\xi^2}u_{xx} + \frac{1}{3}u_{xxx}\right]. \tag{3.19}$$

It was observed in Li *et al.* (1998) that as $x_{\xi\xi}$ can be negative and $\dot{X}$ large, then the term $X_{\xi\xi}u_{xx}/x_{\xi^2}$ can be anti-diffusive (even dominating the diffusive terms in the underlying PDE), and hence destabilizing, and also potentially quite large. It was considered in Li *et al.* (1998) that this contributed to some large errors and instabilities arising in their calculation of the front solutions to Fisher's equation. Such problems were also observed in calculations of the nonlinear Schrödinger equation reported in Ceniceros (2002). Various strategies can be used to overcome such problems. These include increasing the mesh density in the unstable (wave front) region (Qiu and Sloan 1998), using a higher-order upwind strategy such as an ENO or Roe scheme (Li and Petzold 1997, Li *et al.* 1998) or a higher-order (fourth-order) centred difference scheme (Ceniceros 2002). (Alternatively, a static rezoning method can be used, as discussed in the next subsections.) An alternative strategy in one dimension is to use collocation, which deals with errors on non-uniform meshes very effectively, and we now describe this.

3.1.3. Collocation methods

Spline collocation gives a powerful method of discretizing the underlying partial differential equation in the physical domain, which has significant advantages over finite difference and finite element methods. In particular, it affords a continuous representation of the solution and its derivatives, provides a higher order of convergence, easily handles boundary conditions, and gives errors independent of local mesh grading, so that by using collocation we are able to avoid the problem of approximating high-order derivatives over a widely non-uniform mesh (Saucez, Vande Vouwer and Zegeling 2005). It also discretizes the PDE in the physical domain Ω_P and avoids the problems with the additional advective terms for the mesh movement described in Section 3.1.2. A very effective spline collocation discretization procedure coupled to various possible MMPDEs is adopted in the moving mesh collocation code MOVCOL, described in Huang and Russell (1996) (with extensions to higher-order systems using higher-degree Hermite polynomials, given in the code MOVCOL4 (Russell et al. 2007)) and this package has been used in many tests of adaptive methods in one dimension: see, for example, Huang and Russell (1996) and Budd et al. (1999a). Spline collocation methods for second-order PDEs (see Ascher, Christiansen and Russell (1981)) typically use a basis of third-degree cubic Hermite polynomials to give a piecewise smooth approximation $\mathbf{U}(x,t)$ over a series of N intervals $x \in [X_i(t), X_{i+1}(t)]$ to the solution $\mathbf{u}(x,t)$ of the underlying partial differential equation and its associated boundary conditions The collocation points are then chosen to be the Gauss points within the intervals. The interval points are precisely the mesh points moved by solving the MMPDE. The physical solution $u(x,t)$ is approximated on the moving mesh by the piecewise cubic Hermite polynomial

$$\mathbf{U}(x,t) = \mathbf{U}_i(t)\phi_1(s^{(i)}) + \mathbf{U}_{x,i}(t)H_i(t)\phi_2(s^{(i)})$$
$$+ \mathbf{U}_{i+1}(t)\phi_3(s^{(i)}) + \mathbf{U}_{x,i+1}(t)H_i(t)\phi_4(s^{(i)}), \qquad (3.20)$$

for $x \in [X_i(t), X_{i+1}(t)]$, $i = 1, 2, \ldots, N-1$, where $\mathbf{U}_i(t)$ and $\mathbf{U}_{x,i}(t)$ denote the approximations to $\mathbf{u}(X_i(t),t)$ and $\mathbf{u}_x(X_i(t),t)$, respectively. The local coordinate $s^{(i)}$ is defined by

$$s^{(i)} := (x - X_i(t))/H_i(t), \quad H_i(t) := X_{i+1}(t) - X_i(t), \qquad (3.21)$$

and the piecewise cubic shape functions are defined by

$$\begin{aligned} \phi_1(s) &:= (1+2s)(1-s)^2, & \phi_2(s) &:= s(1-s)^2, \\ \phi_3(s) &:= (3-2s)s^2, & \phi_4(s) &:= (s-1)s^2. \end{aligned} \qquad (3.22)$$

For $x \in [X_i(t), X_{i+1}(t)]$, $i = 1, \ldots, N-1$, we then have

$$\mathbf{U}_x(x,t) = \frac{1}{H_i}\left(\mathbf{U}_i\frac{d\phi_1}{ds} + \mathbf{U}_{x,i}H_i\frac{d\phi_2}{ds} + \mathbf{U}_{i+1}\frac{d\phi_3}{ds} + \mathbf{U}_{x,i+1}H_i\frac{d\phi_4}{ds}\right), \qquad (3.23)$$

$$\mathbf{U}_{xx}(x,t) = \frac{1}{H_i^2}\left(\mathbf{U}_i\frac{d^2\phi_1}{ds^2} + \mathbf{U}_{x,i}H_i\frac{d^2\phi_2}{ds^2} + \mathbf{U}_{i+1}\frac{d^2\phi_3}{ds^2} + \mathbf{U}_{x,i+1}H_i\frac{d^2\phi_4}{d^2s}\right), \qquad (3.24)$$

$$\begin{aligned} \mathbf{U}_t(x,t) &= \frac{d\mathbf{U}_i}{dt}\phi_1 + \left(\frac{d\mathbf{U}_{x,i}}{dt}H_i + \mathbf{U}_{x,i}\frac{dH_i}{dt}\right)\phi_2 \\ &\quad + \frac{d\mathbf{U}_{i+1}}{dt}\phi_3 + \left(\frac{d\mathbf{U}_{x,i+1}}{dt}H_i + \mathbf{U}_{x,i+1}\frac{dH_i}{dt}\right)\phi_4 \\ &\quad - \mathbf{U}_x(x,t)\left(\frac{dX_i}{dt} + s^{(i)}\frac{dH_i}{dt}\right), \end{aligned} \qquad (3.25)$$

where ϕ_j, $(d\phi_j/ds)$ and $(d^2\phi_j/d^2s)$, $j = 1, \ldots, 4$, are functions of $s^{(i)}$. These expressions for U and its derivatives can then be directly substituted into (3.15), and the expression

$$\mathbf{U}_t = \mathbf{f}(t, x, \mathbf{U}, \mathbf{U}_x, \mathbf{U}_{xx})$$

evaluated at the two Gauss points $X_{ij} = X_i + s_j H_i, j = 1, 2$, where

$$s_1 = \frac{1}{2}\left(1 - \frac{1}{\sqrt{3}}\right), \quad s_2 = \frac{1}{2}\left(1 + \frac{1}{\sqrt{3}}\right).$$

This, when coupled to the boundary conditions of the underlying PDE, leads to a set of ordinary differential equations for $\mathbf{U}_i, \mathbf{U}_{i,x}$ which can then be coupled directly to the ODEs for the moving mesh given, for example, by (3.11). In certain circumstances, such as when the underlying PDE has a conservation form, it is also possible for the collocation scheme to satisfy an analogous discrete conservation law. This procedure is implemented in MOVCOL and is described in detail in Huang and Russell (1996) The resulting ODEs are somewhat stiff, and are typically solved using an appropriate stiff solver such as an SDIRK (singly diagonally implicit Runge–Kutta) or a BDF (backward differentiation formula) method. In MOVCOL they are solved using a BDF method in the code dassl (Petzold 1982).

3.1.4. Spectral methods

Spectral methods provide an attractive alternative to finite difference and finite element methods for numerical solution of PDEs. They involve approximation by global basis functions, such as trigonometric or algebraic polynomials. For problems with smooth solutions the convergence rate of spectral methods is faster than algebraic, as the number of grid points increases, and the significance of this so-called spectral convergence is that a specified accuracy can usually be achieved using fewer grid points than would be required by the algebraically convergent finite difference or finite element approaches. However, if a solution has a steep region such as a boundary layer or an interior layer, spectral methods will achieve high accuracy only if the number of grid points is sufficiently high to permit resolution of the localized phenomena. To overcome this difficulty, a common approach is to apply a coordinate transformation that is designed to smooth out regions of high gradient. Such a transformation can be generated numerically and adaptively through a moving mesh method. Another benefit of using a moving mesh method is that PDEs can be conveniently discretized on the computational domain where a rectangular or cubic mesh is often used. Adaptivity of this type has proved successful, producing highly accurate solutions to problems that have steep, smooth solutions using a reasonably small number of grid points, although some care must be taken that the numerically generated coordinate transformation should be made sufficiently smooth to avoid possible deterioration of accuracy: see, *e.g.*, Mulholland, Huang and Sloan (1998), Wang and Shen (2005), Feng, Yu, Hu, Liu, Du and Chen (2006) and Tee and Trefethen (2006). While preliminary results are promising, much further investigation is needed to determine the full potential of these adaptive spectral methods.

3.2. MMPDEs and variational methods in n dimensions

3.2.1. Description of some variational-based methods

In the variational and MMPDE approaches of mesh adaptation in n dimensions, briefly described in Section 2, adaptive meshes are also generated as images of a computational mesh under a coordinate transformation from the computational domain to the physical domain. Such a coordinate transformation is determined by an adaptation functional, which is commonly designed to measure the difficulty in the numerical approximation of the physical solution. The functional often involves mesh properties and employs a monitor function to control mesh quality and mesh concentration. The key to the development of variational and MMPDE methods is the formulation of the adaptation functional. Direct-use standard-error estimates are often not appropriate since they often lead to non-convex functionals in two and higher dimensions. Instead, most of the existing methods have been developed based on physical, geometric, mesh quality control, and/or other considerations.

The functional can be formulated in terms of either the coordinate transformation $\mathbf{x} = \mathbf{F}(\boldsymbol{\xi}, t)$ or its inverse transformation $\boldsymbol{\xi} = \mathbf{F}^{-1}(\mathbf{x}, t)$. The latter has been used more commonly than the former because it is less likely to produce mesh tangling for non-convex domains (*e.g.*, see Dvinsky (1991)). In the latter case, the adaptation functional takes the general form

$$I(\boldsymbol{\xi}) = \int_{\Omega_P} G(\mathbf{M}, \boldsymbol{\xi}, \boldsymbol{\nabla}\xi_i) \, \mathrm{d}\mathbf{x}, \quad i = 1, \ldots, n, \tag{3.26}$$

where G is a continuous function of its arguments, $\mathbf{M}$ is the (scalar or matrix-valued) monitor function, and $\boldsymbol{\nabla}$ is the gradient operator with respect to the physical coordinate $\mathbf{x}$. Once a functional has been defined, an MMPDE can be obtained as described in (2.26) as the gradient flow equation of the functional (Huang and Russell 1997*b*, 1999), *i.e.*,

$$\frac{\partial \xi_i}{\partial t} = -\frac{P}{\epsilon} \frac{\partial I}{\partial \xi_i}, \quad i = 1, \ldots, n, \tag{3.27}$$

where P is a positive differential operator and $\epsilon > 0$ is a parameter for adjusting the time scale of mesh movement. For the general form (3.26), this becomes

$$\frac{\partial \xi_i}{\partial t} = \frac{P}{\epsilon} \left(\boldsymbol{\nabla} \cdot \frac{\partial G}{\partial (\boldsymbol{\nabla}\xi_i)} - \frac{\partial G}{\partial \xi_i} \right), \quad i = 1, \ldots, n. \tag{3.28}$$

For example, Winslow's variable diffusion method (Winslow 1981), as described in Section 2, takes this form with

$$I(\boldsymbol{\xi}) = \frac{1}{2} \int_{\Omega_P} \frac{1}{w} \sum_i (\nabla \xi_i)^T \nabla \xi_i \, d\mathbf{x}, \qquad (3.29)$$

where w is the weight function prescribed by the user. We can also consider the generalized version of this method described in Huang and Russell (1997b, 1999), given by

$$I(\boldsymbol{\xi}) = \frac{1}{2} \int_{\Omega_P} \sum_i (\nabla \xi_i)^T \mathbf{M}^{-1} \nabla \xi_i \, d\mathbf{x}, \qquad (3.30)$$

where $\mathbf{M}$ is a matrix-valued monitor function in d dimensions. (Obviously, (3.30) reduces to (3.29) when $\mathbf{M} = w\mathbf{I}$.)

Since $\boldsymbol{\xi} = \boldsymbol{\xi}(\mathbf{x}, t)$ does not explicitly define the location of mesh points, a mesh equation for $\mathbf{x}(\boldsymbol{\xi}, t)$ is commonly used in actual computation. Such an equation can be obtained by interchanging the dependent and independent variables in (3.27) or (3.28) and in the case of the variational principle (3.30) in two dimensions, we obtain the following MMPDE:

$$\frac{\partial}{\partial t} \begin{pmatrix} x \\ y \end{pmatrix} = -\frac{1}{\epsilon |J| \sqrt{\det(\mathbf{M})}} \begin{pmatrix} x_\xi \\ y_\xi \end{pmatrix} \left\{ \frac{\partial}{\partial \xi} \left[\frac{1}{|J|\det(\mathbf{M})} \begin{pmatrix} x_\eta \\ y_\eta \end{pmatrix}^T \mathbf{M} \begin{pmatrix} x_\eta \\ y_\eta \end{pmatrix} \right] \right.$$
$$\left. - \frac{\partial}{\partial \eta} \left[\frac{1}{|J|\det(\mathbf{M})} \begin{pmatrix} x_\eta \\ y_\eta \end{pmatrix}^T \mathbf{M} \begin{pmatrix} x_\xi \\ y_\xi \end{pmatrix} \right] \right\}$$
$$- \frac{1}{\epsilon |J| \sqrt{\det(\mathbf{M})}} \begin{pmatrix} x_\eta \\ y_\eta \end{pmatrix} \left\{ -\frac{\partial}{\partial \xi} \left[\frac{1}{|J|\det(\mathbf{M})} \begin{pmatrix} x_\xi \\ y_\xi \end{pmatrix}^T \mathbf{M} \begin{pmatrix} x_\eta \\ y_\eta \end{pmatrix} \right] \right.$$
$$\left. + \frac{\partial}{\partial \eta} \left[\frac{1}{|J|\det(\mathbf{M})} \begin{pmatrix} x_\xi \\ y_\xi \end{pmatrix}^T \mathbf{M} \begin{pmatrix} x_\xi \\ y_\xi \end{pmatrix} \right] \right\}. \qquad (3.31)$$

This MMPDE can be easily discretized to move the mesh. For details of this derivation, and a series of computations using it, see Huang (2001a). Most of these variational methods can be straightforwardly extended from two dimensions to n dimensions.

A disadvantage of all of the above-mentioned methods, such as the MMPDEs given in (3.31), is that the computations have to be done on a highly nonlinear system. Ceniceros and Hou (2001) also consider a variational principle using a scalar monitor function, but this time in the *computational domain*. After certain further simplifications this leads (in two dimensions) to the equations

$$\nabla_{\boldsymbol{\xi}} \cdot (M \nabla_{\boldsymbol{\xi}} x) = 0, \quad \nabla_{\boldsymbol{\xi}} \cdot (M \nabla_{\boldsymbol{\xi}} y) = 0. \qquad (3.32)$$

Now *all derivatives are expressed in terms of the computational variables* so that $\nabla_{\boldsymbol{\xi}} = (\partial_\xi, \partial_\eta)^T$, and the monitor function is considered as a function of the computational coordinates, *i.e.*, $M = M(\xi, \eta)$. A relaxation method is proposed by Ceniceros and Hou (2001) to solve (3.32), which leads to a set of moving mesh PDEs of the form

$$x_\tau = \nabla_\xi \cdot (M \nabla_\xi x), \quad y_\tau = \nabla_\xi \cdot (M \nabla_\xi y). \tag{3.33}$$

This system is significantly simpler than (3.31) and can be easily discretized. The above equation, when discretized, is rather stiff and can also benefit from a degree of mesh smoothing. A low-pass filter smoothing is applied to the monitor function by Ceniceros and Hou (2001). Smoothing can also be applied directly to the mesh itself, *e.g.*,

$$(1 - \gamma \Delta_\xi) x_\tau = \nabla_\xi \cdot (M \nabla_\xi x), \quad (1 - \gamma \Delta_\xi) y_\tau = \nabla_\xi \cdot (M \nabla_\xi y), \tag{3.34}$$

where $\gamma > 0$ is related to M (typically, if a time step of Δt is used then $\gamma = \Delta t \max(M)$). We note that in one dimension this system is exactly that given by (3.9). The MMPDE method (3.34) in discretized form has been used with success in a number of different applications, including Tang (2005), Zegeling (2007), Ceniceros (2002) and some of the examples in Section 5.

The harmonic map method of Dvinsky (1991) given in (2.34) uses the functional

$$I(\xi) = \frac{1}{2} \int_{\Omega_P} \sqrt{\det(\mathbf{M})} \sum_i (\nabla \xi_i)^T \mathbf{M}^{-1} \nabla \xi_i \, d\mathbf{x}, \tag{3.35}$$

while the method of Brackbill and Saltzman (1982) (*cf.* (2.35)) takes the form

$$I(\xi) = \theta_a \int_{\Omega_P} w |J| \, d\mathbf{x} + \theta_s \int_{\Omega_P} \sum_i (\nabla \xi_i)^T \nabla \xi_i \, d\mathbf{x} \tag{3.36}$$

$$+ \theta_o \int_{\Omega_P} \sum_{i \neq j} ((\nabla \xi_i)^T \nabla \xi_j)^2 \, d\mathbf{x}.$$

Following Winslow (1967), Thompson *et al.* (1985) use a system of elliptic differential equations for generating body-fitted, adaptive meshes. They propose using the Poisson equations

$$\nabla^2 \xi_i = P_i(\mathbf{x})$$

to control the mesh concentration and direction, where P_i, $1 \leq i \leq d$, are control functions. The system can be interpreted as the Euler–Lagrange equation of the quadratic functional

$$I(\xi) = \int_{\Omega_P} \sum_i (|\nabla \xi_i|^2 - P_i \xi_i) \, d\mathbf{x}. \tag{3.37}$$

Knupp and his co-workers (Knupp 1995, Knupp 1996, Knupp and Robidoux 2000, Knupp *et al.* 2002) determine the coordinate transformation such that its Jacobian matrix is as close as possible to a reference Jacobian matrix in the least-squares sense. One of the functionals they use is

$$I(\xi) = \int_{\Omega_P} \left\| \frac{\partial \xi}{\partial x} - K \right\|_F^2 ds, \qquad (3.38)$$

where $\| \cdot \|_F$ is the Frobenius norm and $K = K(\mathbf{x})$ is the user-prescribed, reference Jacobian matrix. A detailed discussion on how to choose the matrix K is given in Knupp (1996); see also Knupp and Robidoux (2000) for a broader discussion on algebraic properties of the Jacobian matrix.

The method of Huang (2001b) given in Section 2 (cf. (2.40)) augments the above variational principles with an additional contribution based on mesh quality control of equidistribution alignment, and orientation. The choice of the monitor function $\mathbf{M}$ in this method, based on interpolation error estimates, was extensively studied in Huang and Sun (2003) and Huang (2005a). The idea of mesh quality control was also used by Branets and Carey (2003) in developing their grid-smoothing variational method.

3.2.2. Examples of meshes generated

We now consider using certain of these methods described above to generate a series of meshes. In Section 3.3 we compare these results to the meshes generated by using an optimal transport algorithm.

Example 1. This example is to generate adaptive meshes for a given weight function,

$$w(x,y) = 1 + 10\exp\left(-50\left(y - \frac{1}{2} - \frac{1}{4}\sin(2\pi x)\right)^2\right), \quad \text{in } \Omega \equiv (0,1) \times (0,1).$$

The monitor function is chosen to be $\mathbf{M} = wI$. Adaptive meshes are shown in Figure 3.1 using the harmonic mapping method, Winslow's method and the variational method with alignment control given in (2.40) with $\theta = 0.1$).

Example 2. In this example we generate adaptive moving meshes for the weight function

$$w(x,y,t) = 1 + 10\exp\left(-50\left|\left(x - \frac{1}{2} - \frac{1}{4}\cos(2\pi t)\right)^2\right.\right.$$
$$\left.\left. + \left(y - \frac{1}{2} - \frac{1}{4}\sin(2\pi t)\right)^2 - \left(\frac{1}{10}\right)^2\right|\right).$$

The monitor function is chosen as $\mathbf{M} = wI$. Adaptive meshes obtained using MMPDEs based on the methods in Example 1 are shown in Figures 3.2, 3.3, and 3.4.

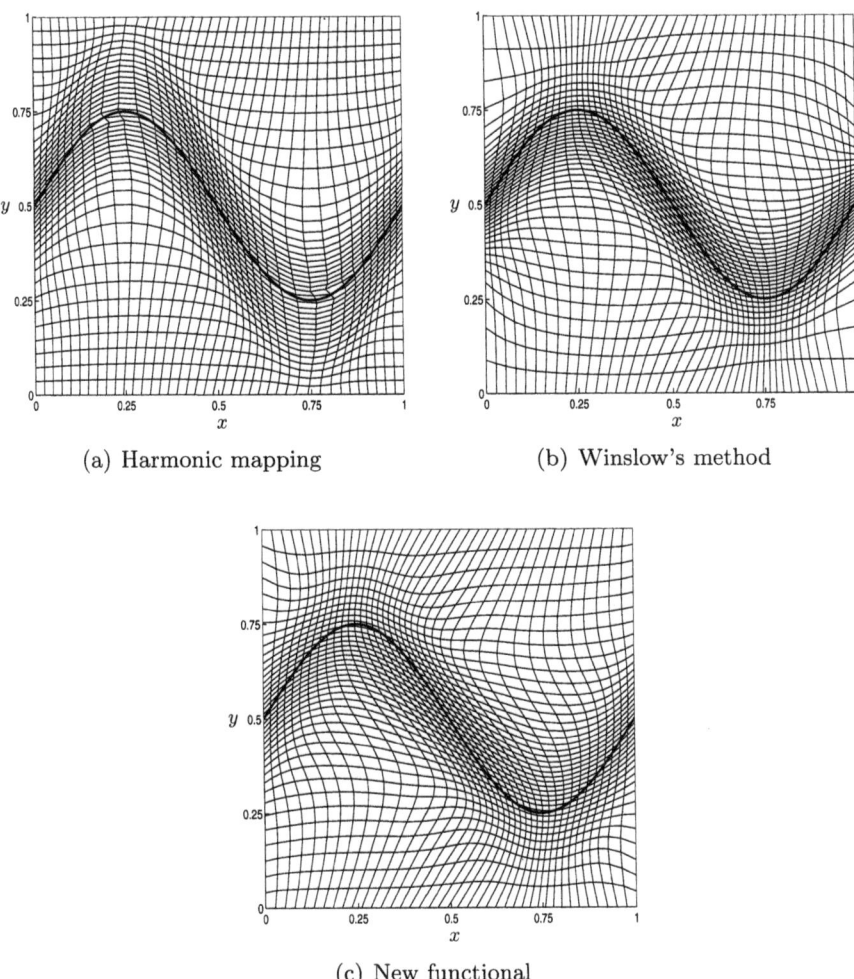

Figure 3.1. *Example 1.* Adaptive moving meshes obtained by the harmonic mapping method, Winslow's method, and the method based on equidistribution and alignment control (2.40) ($\theta = 0.1$).

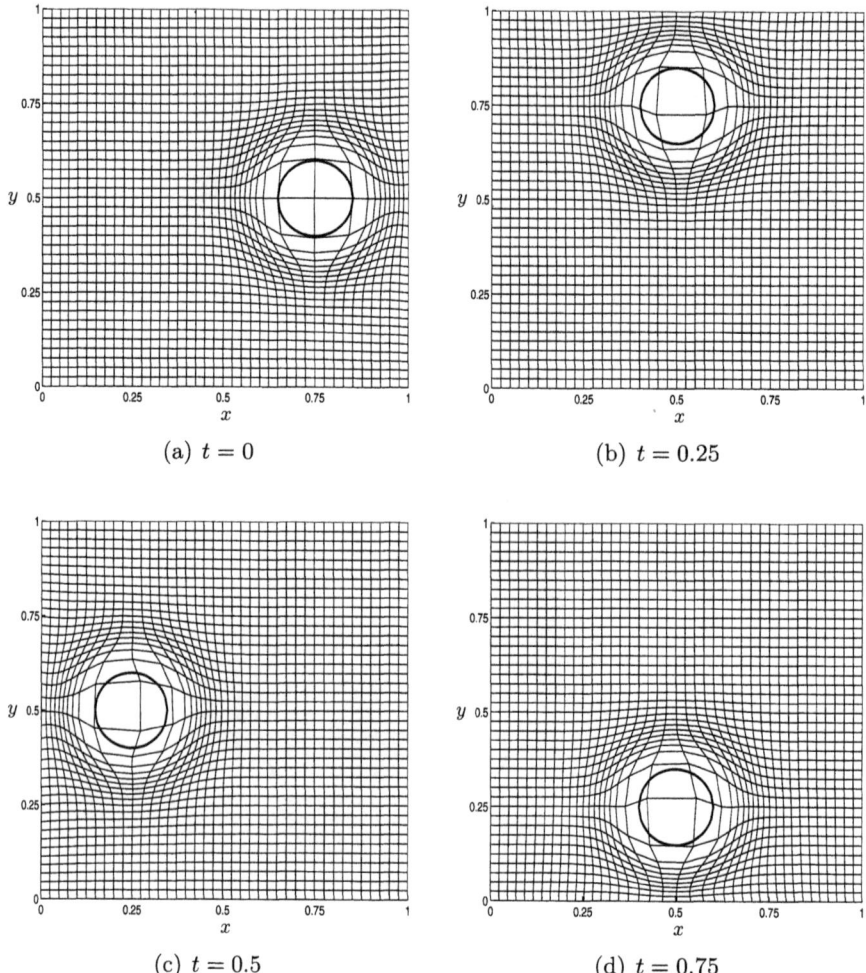

Figure 3.2. *Example 2.* Adaptive moving meshes obtained by MMPDEs based on the harmonic mapping method. The circles shown in this and the following figures indicate the locations where the mesh concentration is anticipated.

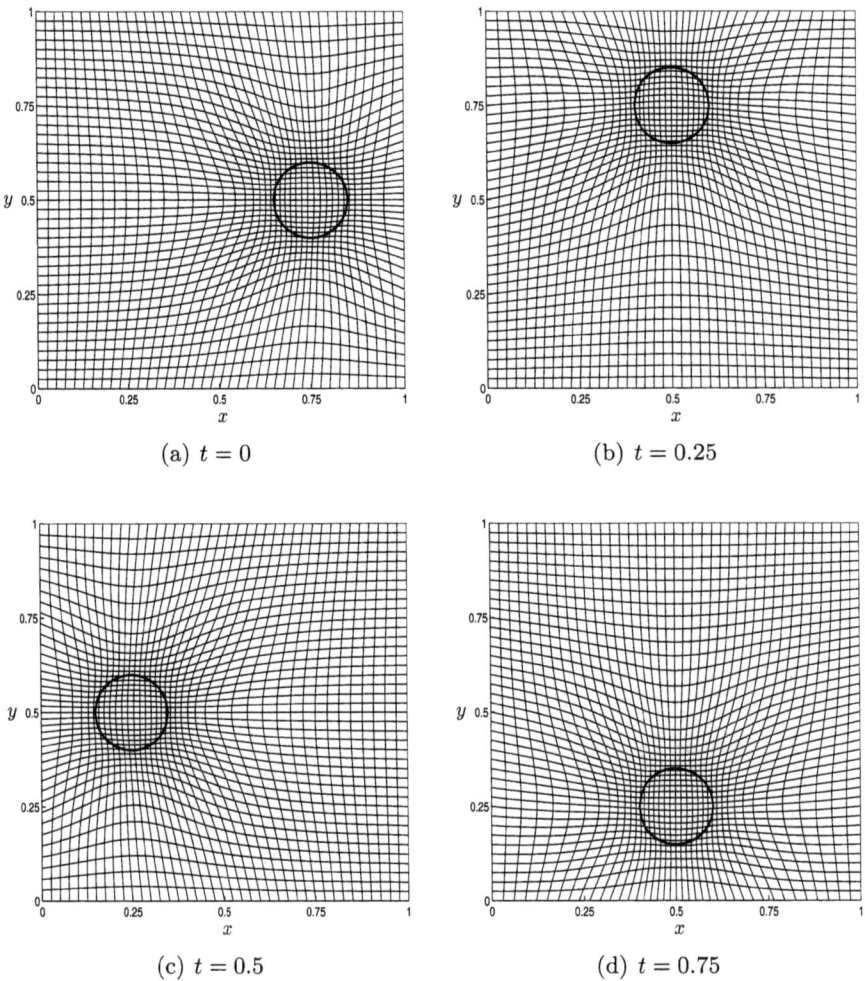

Figure 3.3. *Example 2.* Adaptive moving meshes obtained by MMPDEs based on Winslow's method.

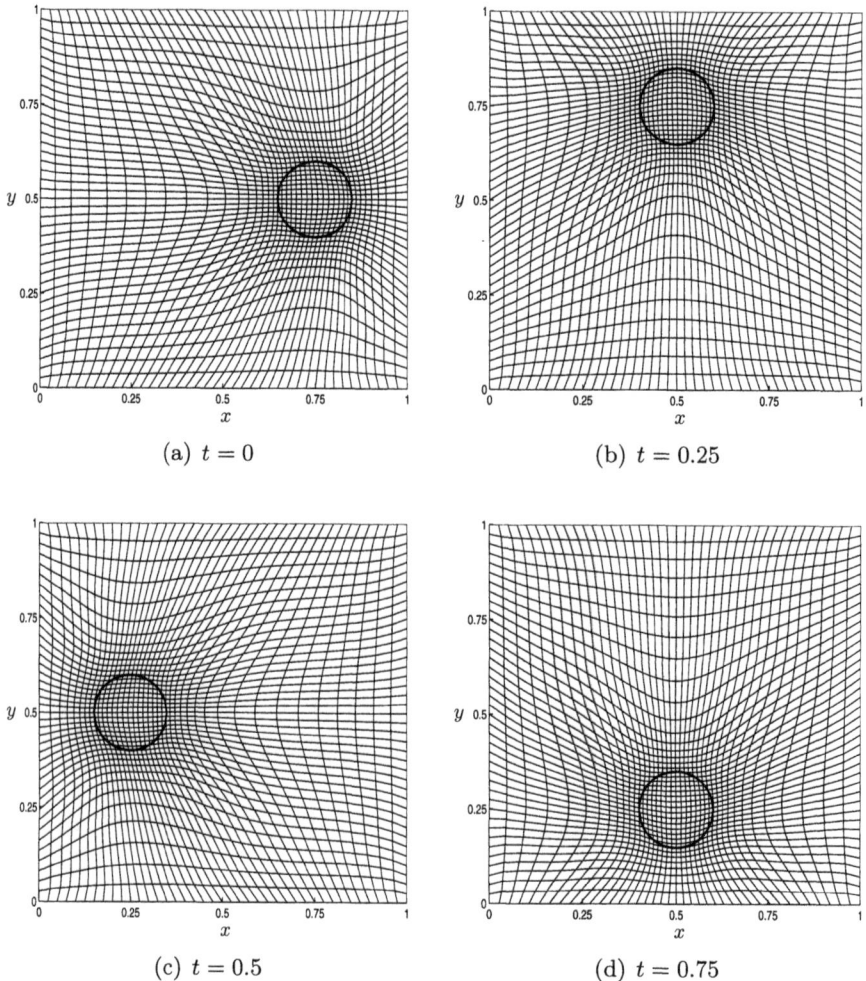

Figure 3.4. *Example 2.* Adaptive moving meshes obtained by MMPDEs based on the method with equidistribution and alignment control (2.40) ($\theta = 0.1$).

Example 3. This example generates an adaptive mesh for a given analytical solution

$$u(x,y) = \tanh\left(30\left(x^2 + y^2 - \frac{1}{8}\right)\right)$$
$$+ \tanh\left(30\left((x - 0.5)^2 + (y - 0.5)^2 - \frac{1}{8}\right)\right)$$
$$+ \tanh\left(30\left((x - 0.5)^2 + (y + 0.5)^2 - \frac{1}{8}\right)\right)$$
$$+ \tanh\left(30\left((x + 0.5)^2 + (y - 0.5)^2 - \frac{1}{8}\right)\right)$$
$$+ \tanh\left(30\left((x + 0.5)^2 + (y + 0.5)^2 - \frac{1}{8}\right)\right)$$

defined in $[-2, 2] \times [-2, 2]$. An adaptive mesh with a monitor function is based on isotropic and anisotropic estimates error in interpolating this function. This is expected to concentrate around five circles. Results are shown in Figure 3.5.

Example 4. This example generates a three-dimensional adaptive mesh for

$$u(x, y, z) = \tanh(100((x - 0.5)^2 + (y - 0.5)^2 + (z - 0.5)^2) - 0.0625)$$

defined in the unit cube. An adaptive mesh with a monitor function is based on the error in interpolating this function. This is expected to concentrate near the sphere centred at $(0, 0, 0)$ with radius 0.25. Results are shown in Figure 3.6.

3.3. Optimal transport methods

3.3.1. Derivation of the optimal transport equations

Optimal transport methods are a very natural generalization of MMPDE methods in one dimension, that retain much of the simplicity of the one-dimensional approach (such as always solving scalar equations and automatic calculation of the mesh on a boundary) whilst being general enough to deliver meshes of provable mesh quality (with many of the proofs following directly from the one-dimensional case). They have the disadvantage of being less flexible than some of the moving mesh methods described above. However, in practice they can give very regular meshes for a wide range of possible monitor functions. The key idea behind an optimal mesh is that it should be one which is closest to a uniform mesh in a suitable norm, consistent with satisfying the equidistribution principle. The simplest such

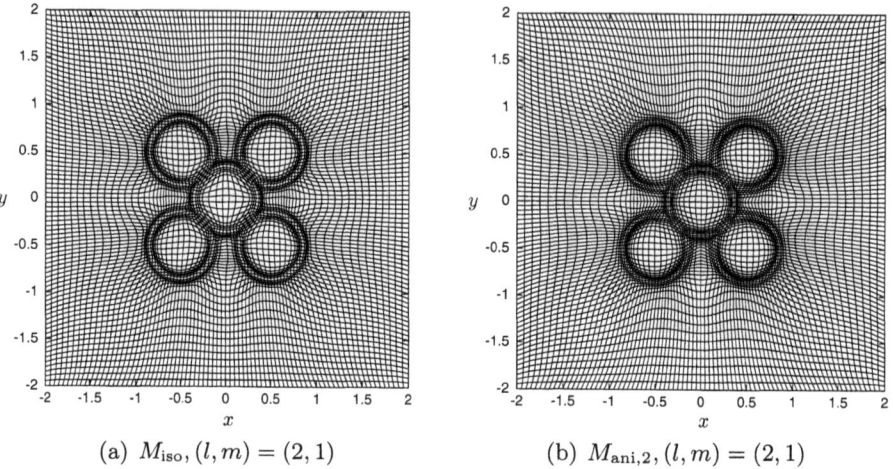

(a) M_{iso}, $(l, m) = (2, 1)$ (b) $M_{\text{ani},2}$, $(l, m) = (2, 1)$

Figure 3.5. *Example 3.* Adaptive meshes of size $N = 81 \times 81$ obtained using the variational method (2.40) ($\theta = 0.1$) for different monitor functions based on isotropic and anisotropic interpolation error estimates.

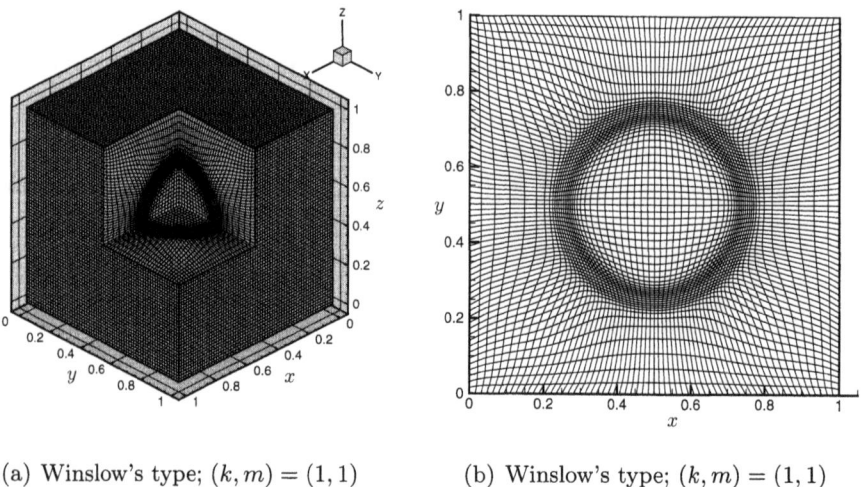

(a) Winslow's type; $(k, m) = (1, 1)$ (b) Winslow's type; $(k, m) = (1, 1)$

Figure 3.6. *Example 4.* An adaptive mesh of size $N = 65 \times 65 \times 65$ obtained using the variational method (2.40) ($\theta = 0.1$) for a monitor function based on interpolation error. (a) Cut-away plot of the mesh. (b) Plane projection of slice at $K_z = 32$ of the mesh in (a).

norm is the least-squares norm given by

$$I = \int_{\Omega_C} |\mathbf{F}(\boldsymbol{\xi}, t) - \boldsymbol{\xi}|^2 \, d\boldsymbol{\xi}. \tag{3.39}$$

Minimizing I subject to the equidistribution principle is in fact the celebrated Monge–Kantorovich problem from differential geometry. This principle is often called optimum transport, as it leads to a transformation, the creation of which takes a minimum amount of work as a deviation from the identity. Intuitively, this is likely to deliver a regular mesh, as this mesh will be as close (in an averaged sense) to the most regular possible mesh, i.e., a completely uniform one. Remarkably, this minimization problem has a unique solution with a very elegant expression for the transformation.

Theorem 3.3. There exists a unique optimal mapping $\mathbf{F}(\boldsymbol{\xi}, t)$ satisfying the equidistribution equation. This map has the same regularity as M. Furthermore, $\mathbf{F}(\boldsymbol{\xi}, t)$ is the unique mapping from this class which can be written as the gradient (with respect to $\boldsymbol{\xi}$) of a convex (mesh) potential $P(\boldsymbol{\xi}, t)$, so that

$$\mathbf{F}(\boldsymbol{\xi}, t) = \nabla_{\boldsymbol{\xi}} P(\boldsymbol{\xi}, t), \quad \Delta_{\boldsymbol{\xi}} P(\boldsymbol{\xi}, t) > 0. \tag{3.40}$$

Proof. See Brenier (1991) or Caffarelli (1992, 1996) for an abstract proof and Delzanno et al. (2008) for a proof in the context of adaptive mesh generation. □

The following is then immediate.

Lemma 3.4. The map $\mathbf{F}$ is *irrotational* so that $\nabla_{\boldsymbol{\xi}} \times \mathbf{F} = \mathbf{0}$, and the Jacobian of $\mathbf{F}$ is *symmetric*.

Significantly, the transformation above is an example of a *Legendre transformation* (Sewell 2002). Such transformations include translations and linear maps by positive definite symmetric matrices. We now show how to calculate such a transformation.

3.3.2. Properties of optimally transported meshes

It is immediate that if $\mathbf{x} = \nabla_{\boldsymbol{\xi}} P$ then

$$\frac{\partial \mathbf{x}}{\partial \boldsymbol{\xi}} = H(P),$$

where $H(P)$ is the *Hessian* of P. Additionally, if the measure $M \in W^2(\Omega_P)$ is strictly positive on its supports (assumed to be convex), then the potential $P \in W^2_{\text{loc}}(\Omega_C)$, and satisfies, in the classical sense, the Monge–Ampère equation

$$M(\nabla_{\boldsymbol{\xi}} P, t) H(P) = \theta(t). \tag{3.41}$$

Here $H(P)$ denotes the determinant of the Hessian matrix of P. This is a famous equation in differential geometry. Solving it defines the map **F** *uniquely*. In order to solve it we must specify boundary conditions for the solution. In general applications of mesh generation we consider bounded domains mapping to bounded domains. We can then prescribe a boundary condition where equation (3.41) is supplemented with the condition that the boundary must map to the boundary. Suppose that the points on the boundary of Ω_C satisfy the implicit equation $G_C(\boldsymbol{\xi}) = 0$ and those on the boundary of Ω_P satisfy the implicit equation $G_P(\mathbf{x}) = 0$, we then have the following (nonlinear) Neumann boundary condition for (3.41):

$$G_P(\nabla_\xi P, t) = 0 \quad \text{if} \quad G_C(\boldsymbol{\xi}, t) = 0. \tag{3.42}$$

The existence, uniqueness and regularity of the solutions of (3.41) and (3.42) has been well studied (Brenier 1991, Caffarelli 1992, Caffarelli 1996). From the point of view of grid generation this puts us into the nice situation of being able to infer properties of the mesh from those of a function P with known regularity. We have the following very important result.

Theorem 3.5. If both Ω_C and Ω_P are *smooth, convex* domains then (3.41) and (3.42) have a unique solution (up to an additive constant), which is as regular as the monitor function M. This in turn leads to a unique regular mesh.

Proof. This follows immediately from the results of Brenier (1991) and Caffarelli (1992, 1996). □

If Ω_C or Ω_P are not smooth then there is a possible loss of mesh regularity. In the case of solutions which are logical rectangles, this is not severe and applies only at the corners, where it can be shown (Cullen 1989) that $P \in C^3$ and $F \in C^2$. This is sufficient regularity for most applications.

An immediate consequence of this result is that the solution of the Monge–Ampère equation defines a map not only from Ω_C to Ω_P, but also between the boundaries of the respective domains. This is a desirable property from the perspective of mesh generation as we do not have to consider separate equations for the mesh on the boundary. This is in contrast to the variational methods described in Section 3.2.1, which require separate equations to describe the mesh on the boundaries. As an example, we can consider meshes on logical cubes, taking

$$\Omega_C = [0, 1]^d = \Omega_P. \tag{3.43}$$

The boundary conditions above then reduce to

$$P_\xi = 0, 1 \quad \text{if} \quad \xi = 0, 1, \quad P_\eta = 0, 1 \quad \text{if} \quad \eta = 0, 1. \tag{3.44}$$

The boundary condition (3.44) applied in two dimensions implies the orthogonality of the grid lines at the boundaries of the square. To see this, consider the bottom boundary of the square in the physical domain, for which $Y = 0$. A grid line Γ which intersects this side is given by

$$\Gamma = \{(x, y) : 0 \leq \eta \leq 1, \quad \xi \text{ fixed}\}.$$

The tangent to this line in the physical domain is given by $\boldsymbol{\tau} = (x_\eta, y_\eta)^T$, and, trivially, the tangent to the bottom boundary of the square by the vector $\mathbf{t} = (1, 0)^T$. However, on the bottom boundary, the function P satisfies the identity $P_\eta(\xi, 0) = 0$ and by definition $x = P_\xi$. It follows immediately that when $\eta = 0$,

$$x_\eta(\xi, 0) = P_{\eta\xi}(\xi, 0) = 0,$$

so that on the lower boundary $\boldsymbol{\tau} \cdot \mathbf{t} = 0$. Thus the mesh is orthogonal to the lower boundary. Similar results apply to the other sides as well. The condition of mesh orthogonality at the boundary of the square is often desirable in certain circumstances (Thompson et al. 1985). However, it may cause problems when resolving features, such as fronts, which intersect boundaries at an angle. In such cases it may be useful to refine the mesh close to the boundary, either by introducing a finer computational mesh there, or by locally increasing the value of the monitor function M at mesh points adjacent to the boundary.

Mesh symmetries. Some very desirable properties of the mesh follow immediately from the properties of the Monge–Ampère equation. It is trivial to see that the equation is invariant under translations in (ξ, η). It is also easy to see that (in a similar manner to the Laplacian operator) the Monge–Ampère equation is also invariant under any *orthogonal* map such as a rotation or a reflection. This is because the Monge–Ampère equation is the determinant of the Hessian, which transforms covariantly under such maps. This simple observation implies that (away from boundaries) the meshes generated by solving the Monge–Ampère equation should have no difficulty aligning themselves to structures, such as shocks, which may occur anywhere in a domain and at any orientation. It is immediate that the Monge–Ampère equation is also invariant under *scaling transformations* of the form $\xi \to \lambda\xi$, $\eta \to \mu\eta$, $P \to \nu P$, provided that M is chosen carefully.

Mesh skewness. The regularity of the mesh generated by this approach gives a useful feature in seeing control in the variation of the element size across the domain. In general the meshes generated by the Monge–Ampère equation have good regularity properties and are effective in interpolating functions (Delzanno et al. 2008). It is possible to make some estimates for the resulting skewness of the mesh in terms of the properties of the function P. Consider a two-dimensional problem for which the map from Ω_C to Ω_P has the Jacobian J. A measure for the skewness s of the mesh in Ω_P is given by

$$s = \frac{\lambda_1}{\lambda_2} + \frac{\lambda_2}{\lambda_1} = \frac{(\lambda_1 + \lambda_2)^2}{\lambda_1 \lambda_2} - 2 = \frac{\text{trace}(J)}{\det(J)} - 2 = \frac{\Delta(P)^2}{H(P)} - 2, \quad (3.45)$$

where λ_1 and λ_2 are the (real and positive) eigenvalues of J. The skewness can be estimated in certain cases. One example of this arises in the scale-invariant meshes for the local singularities blow-up problems studied in Section 5, in which a sequence of meshes are calculated which, close to the singularity, take the form $P(\xi, t) = \Lambda(t)\hat{P}(\xi)$ for an appropriate scaling function $\Lambda(t)$. It is immediate that

$$\frac{\Delta(P)^2}{H(P)} - 2 = \frac{\Delta(\hat{P})^2}{H(\hat{P})} - 2,$$

so that the skewness of the rescaled mesh is the same as the original. Hence, if an initially uniform mesh is used then the mesh close to the singularity will retain local uniformity.

3.3.3. Solution of the Monge–Ampère equation

The equation (3.41) can be solved either directly (Delzanno *et al.* 2008) or by a relaxation method.

The direct method. The Monge–Ampère equation (Evans 1999, Gutiérrez 2001) belongs to the class of fully nonlinear second-order equations and has two sources of nonlinearity. Firstly, the Hessian $H(P)$ is nonlinear in the second derivatives (except in the one-dimensional case). Secondly, the monitor function in general depends nonlinearly on the first derivatives of P (either directly or through the solution of the original PDE). For any suitably smooth positive monitor function the equation has a unique solution, which is a convex function. Linearization of the equation shows that it is elliptic in the space of convex functions. The Monge–Ampère equation arises from prescribing the product of the eigenvalues (the determinant) of the Jacobian matrix of a gradient mapping. If we prescribe the sum of the eigenvalues (the trace) then we obtain a standard Poisson equation. For the Poisson equation, multigrid methods can find the solution of a discretization on a grid with $\mathcal{O}(N)$ unknowns using $\mathcal{O}(N)$ operations. Methods for solving the Monge–Ampère equation aim to obtain the same computational complexity.

Oliker and Prussner (1988) propose a specially designed discretization and iterative method which explicitly preserve the convexity of the iterates. Benamou and Brenier (2000) transform the Monge–Ampère equation into a time-dependent fluid mechanics problem, which is solved using an iterative method based on an augmented Lagrangian approach. Dean and Glowinski (2003, 2004) propose finite element discretizations based on an augmented Lagrangian approach and a least-squares formulation respectively. Feng and Neilan (2009) consider the nonlinear second-order equation as the limiting equation of a singularly perturbed fourth-order quasilinear equation. Chartrand, Vixie, Wohlberg and Bollt (2007) show that the Monge–Ampère equation can be reformulated as an unconstrained optimization problem, which can be solved by a gradient descent method. A special property of the mappings generated by the Monge–Ampère equation is that they are irrotational, *i.e.*, the curl is zero. Haker and Tannenbaum (2003) propose a gradient descent method that uses a Poisson solve in each step to 'remove the curl'. The nonlinear multigrid method developed in Fulton (1989) for the semigeostrophic equation should be easily adapted to our problem. A Newton–Krylov-multigrid method is proposed in Delzanno *et al.* (2008). The above methods all try to solve the fully nonlinear Monge–Ampère equation directly. It was a remarkable achievement of Kantorovich to show that the problem can be relaxed to a linear one by considering not a transport map $\boldsymbol{\xi} \to \mathbf{x} = \mathbf{F}(\boldsymbol{\xi})$, but a transport plan $G(\boldsymbol{\xi}, \mathbf{x})$ indicating the amount of material to be transported from $\boldsymbol{\xi}$ to x (Rachev and Rüschendorf 1998, Evans 1999). Robust methods exist for solving the corresponding linear programming problem, but to the best of our knowledge these methods typically require $\mathcal{O}(N^2)$ operations (Kaijser 1998, Balinski 1986), which is unacceptable except for small problems.

The parabolic Monge–Ampère (PMA) *method.* An alternative approach motivated by the discussion of the MMPDEs given earlier, is to introduce a parabolic regularization to (3.41) so that the gradient of solutions of this evolve toward the gradient of the solutions of (3.41) over a (relatively) short time scale. This method also couples naturally to the solution of a time-dependent PDE. Accordingly we consider using *relaxation* to generate an approximate solution of (3.41), which evolves together with the solution of the underlying PDE. Accordingly we consider a time-evolving function $Q(\boldsymbol{\xi}, t)$ with associated mesh $\mathbf{x}(\boldsymbol{\xi}, t) = \boldsymbol{\nabla}_{\boldsymbol{\xi}} Q(\boldsymbol{\xi}, t)$, with the property that this mesh should be close to that determined by the solution of the Monge–Ampère equation. To do this we consider a relaxed form of (3.41) taking the form of a parabolic Monge–Ampère equation (PMA) of the form

$$\epsilon(I - \gamma \Delta_{\boldsymbol{\xi}}) Q_t = \bigl(H(Q) M(\boldsymbol{\nabla}_{\boldsymbol{\xi}} Q)\bigr)^{1/n}. \tag{3.46}$$

To find a moving mesh, we start with an initially uniform mesh for which

$$Q(\boldsymbol{\xi}, 0) = \frac{1}{2} |\boldsymbol{\xi}|^2.$$

The function Q then evolves according to (3.46). In (3.46) the scaling power $1/n$ is necessary for global existence of the solution. This is because if Q is scaled by a factor $L(t)$ then the Hessian term $H(Q)$ scales as $L(t)^n$.

If M is constant, then equation (3.46) without the power law scaling admits a variables-separable solution for which $L_t = CL^n$. If $n > 1$ then this equation has solutions which blow up in a finite time. The rescaling prevents this possibility. The operator on the left of this system is a smoothing operator, similar to the operator used in (2.14), (3.9), which reduces the stiffness of this system when it is discretized. Observe that the term $\theta(t)$ has not been included in (3.46). Indeed this term arises naturally as a constant of integration. The PMA equation (3.46) has many properties in common with the moving mesh equation (3.9). In particular, if M is independent of time then the solution of (3.41) corresponds to a stable solution of (3.46). Indeed, we have the following results.

Lemma 3.6.

(a) Suppose that $M_t = 0$ and the Monge–Ampère equation (3.41) admits a (steady) convex solution $P(\xi)$ with associated map $\mathbf{x}(\xi)$ for which $H(P) > 0$, so that P satisfies the equation

$$M(\nabla P)H(P) = \theta.$$

Then:

(i) the PMA equation (3.46) admits a time-dependent solution

$$Q(\xi, t) = \frac{\theta^{1/n}t}{\epsilon} + P(\xi) \qquad (3.47)$$

for which $\nabla_\xi Q = \nabla_\xi P = \mathbf{x}(\xi)$;

(ii) the resulting mesh is locally stable.

(b) If M is slowly varying, then the solution of (3.46) remains ϵ-close to a solution of (3.41) for all time.

Proof. This is given in Budd and Williams (2009) and is very similar to the corresponding proof given for the stability of (3.9) □

It is also possible to show (Budd and Williams 2009) that, throughout the evolution of the mesh, if $H(Q)$ and ΔQ are initially positive then they stay positive for all time. This application of the maximum principle guarantees that the map generating the mesh is locally invertible for all time, and hence no mesh tangling can occur. This is a very useful feature of such r-adaptive methods, and we will see numerical examples of this in the following calculations.

3.3.4. Discretizing the parabolic Monge–Ampère equation (3.46)
To discretize (3.46) in space we can impose a uniform grid of mesh size $\Delta\xi$ on the computational space and assume that Q and Q_t take point values

$Q_{i,j}(t)$, $i,j = 0 \ldots N$, etc., on this grid. The Hessian operator $H(Q)$ can be discretized with central differencing using a nine-point stencil interior to the domain Ω_C to evaluate all second derivatives in $H(Q)$. For the boundary points, the central differences are replaced by Taylor series expansions, using the respective conditions $Q_\xi = 0$ or $Q_\xi = 1$, and so on. The gradient ∇Q is calculated similarly, so that the right-hand side of (3.46) can be determined. To determine the values of Q_t at the mesh points we then invert the operator $(I - \gamma \Delta_\xi)$. As the system is posed on a uniform (rectangular) mesh in the computational domain, this inversion can be done very rapidly by using a fast spectral solver based upon the discrete cosine transform (invoking the Neumann boundary conditions for Q_t). Knowing Q_t we may then determine $\mathbf{x}_t$ by taking the gradient.

3.3.5. Examples of meshes generated using the PMA method

To give some flavour of the behaviour of moving mesh methods, we now consider some meshes generated using the PMA method for a known monitor function $M(x, y, t)$, choosing examples which can be compared with other methods presented both in the literature and in other sections of this article. In all of these following calculations we take $\epsilon = 0.01$, and the ODEs obtained by the discretization described above are solved in MATLAB using the routine **ode45**. All runs took under 5 minutes on a standard desktop computer.

Example 1. We first take a case motivated by Cao et al. (2002) (see also Section 4) with a monitor function localized over a moving circle of the form

$$M_1(x, y, t) = 1 + 5 \exp(-50|(x - 1/2 - 1/4\cos(2\pi t))^2 \\ + (y - 1/2 - \sin(2\pi t)/2)^2 - 0.01|).$$

This can be a severe test of a moving mesh method, and the meshes calculated from this using the (velocity-based) geometric conservation law method (Cao et al. 2002) have a high degree of skewness. To compute a corresponding moving mesh using the parabolic Monge–Ampère method (and also the solutions in all of the examples in this subsection) we use a uniform computational mesh to solve (3.46) with $N = 30$ mesh points in each direction and mapping the unit square to the unit square. We see from this calculation that the resulting mesh closely follows the moving circle with no evidence of skewness or irregularity. Note the orthogonality and regularity of the mesh at the boundary of the domain. Notice further that in Figure 3.7 the solution partially exits the domain with no ill consequence, and that the grid at $t = 10$ is virtually indistinguishable from that at $t = 0$. It is clear from these figures that the mesh generated has excellent regularity. Observe the high degree of mesh uniformity close to the solution maximum.

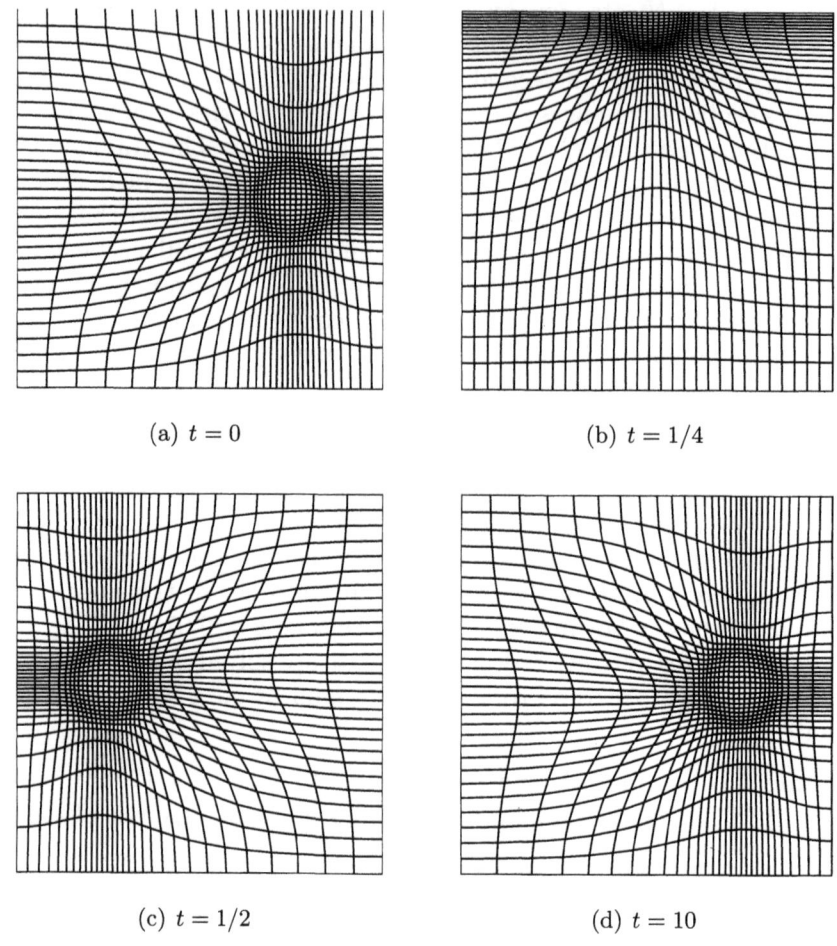

Figure 3.7. *Example 1.* Snapshots at $t = 0$ (a), $t = 1/4$ (b), $t = 1/2$ (c), $t = 10$ (d). As the solution is advected, the grid follows the maxima of the solution and does not ever fall behind or become distorted.

Example 2. In an example taken directly from Cao et al. (2002), we consider a monitor function localized on a sine wave of the form

$$M_2(x, y, t) = 1 + 5\exp(-50|(y - 1/2 - 1/4\sin(2\pi x)\sin(2\pi t)|),$$

and proceed to use exactly the same method as described in Example 1. See Figure 3.8.

In this example we again see that the PMA method has generated a very regular and periodic (in time) mesh.

Example 3. In this example we consider a monitor function localized on the support of two travelling linear fronts moving toward each other and show the resulting mesh in Figure 3.9. For this we take

$$x_0 = t, \qquad y_0 = 0.2 + t/2, \qquad u_0 = \gamma \operatorname{sech}(\lambda(x - x_0 + y - y_0)),$$
$$x_1 = 1 - t, \qquad y_1 = 0.8 - t/2, \qquad u_1 = \gamma \operatorname{sech}(\lambda(x - x_1 + y - y_1)),$$
$$M_3(x, y, t) = 1 + u_0 + u_1,$$

with $\gamma = 5$ and $\lambda = 100$. We note that the fronts pass through each other without generating spurious oscillations in the mesh. Note also that the solution points do not follow the front (moving in and out of the region of high mesh density), but that the *density* of the solution points does. Observe that the mesh automatically aligns itself along the front. A close-up of the mesh close to the front is given in Figure 3.11 (see p. 187). This shows very good resolution of the local structure close to the front and a smooth transition from a uniform mesh to one refined at the front. If γ is large, then close to the front we see a mesh compression proportional to γ orthogonal to the front, with a local skewness of $s = \gamma$. The PMA method in this case thus generates a smooth mesh, for which the degree of skewness can be controlled via the choice of γ. More details of this calculation are given in Walsh et al. (2009). We can also see the effects of mesh orthogonality close to the boundaries of the domain.

Example 4. In this example we look at a monitor function localized on the support of two fronts meeting at an angle: see Figure 3.10. The behaviour of the mesh close to the front is similar to that in the last example. Significantly, there is no mesh tangling as the two fronts intersect. In Figure 3.12 we show a close-up of the intersection region, showing the high degree of mesh regularity. For this we take

$$x_0 = t, \qquad y_0 = 0.2 + t/2, \qquad u_0 = \gamma \operatorname{sech}(\lambda(x - x_0 + y - y_0)),$$
$$x_1 = 1 - t, \qquad y_1 = 0.8 - t, \qquad u_1 = \gamma \operatorname{sech}(\lambda(x - x_1 + (y - y_1)/2)),$$
$$M_4(x, y, t) = 1 + u_0 + u_1.$$

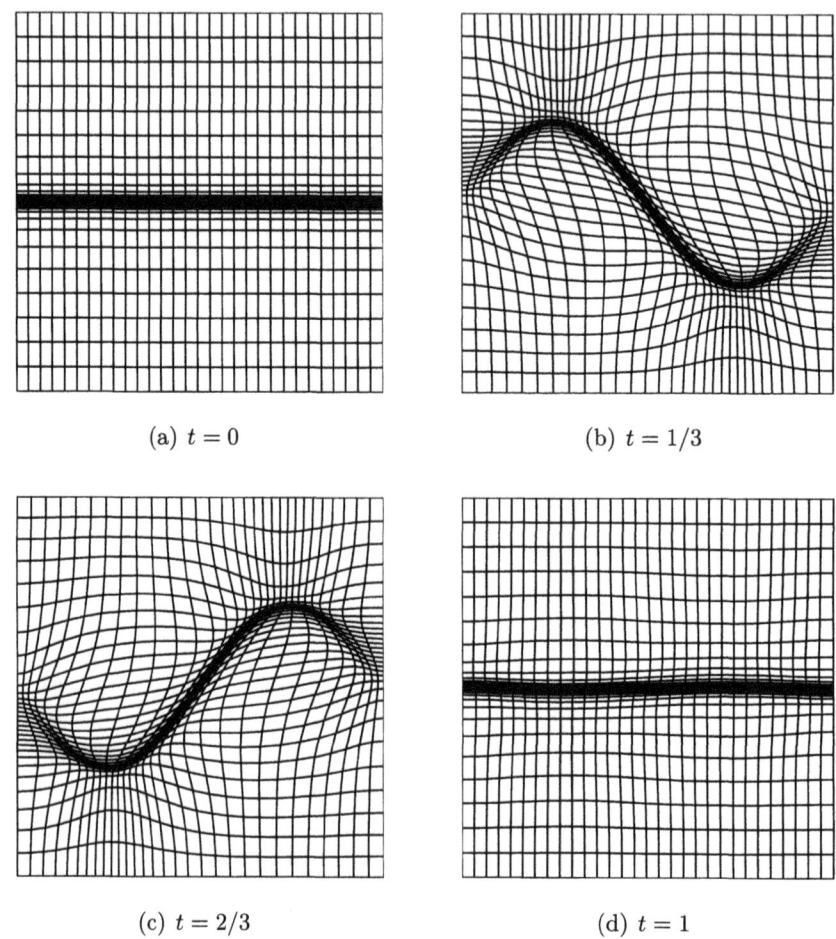

Figure 3.8. *Example 2.* Snapshots at $t = 0$ (a), $t = 1/3$ (b), $t = 2/3$ (c), $t = 1$ (d). In this figure we see both the smoothness and the periodic (in time) form of the mesh.

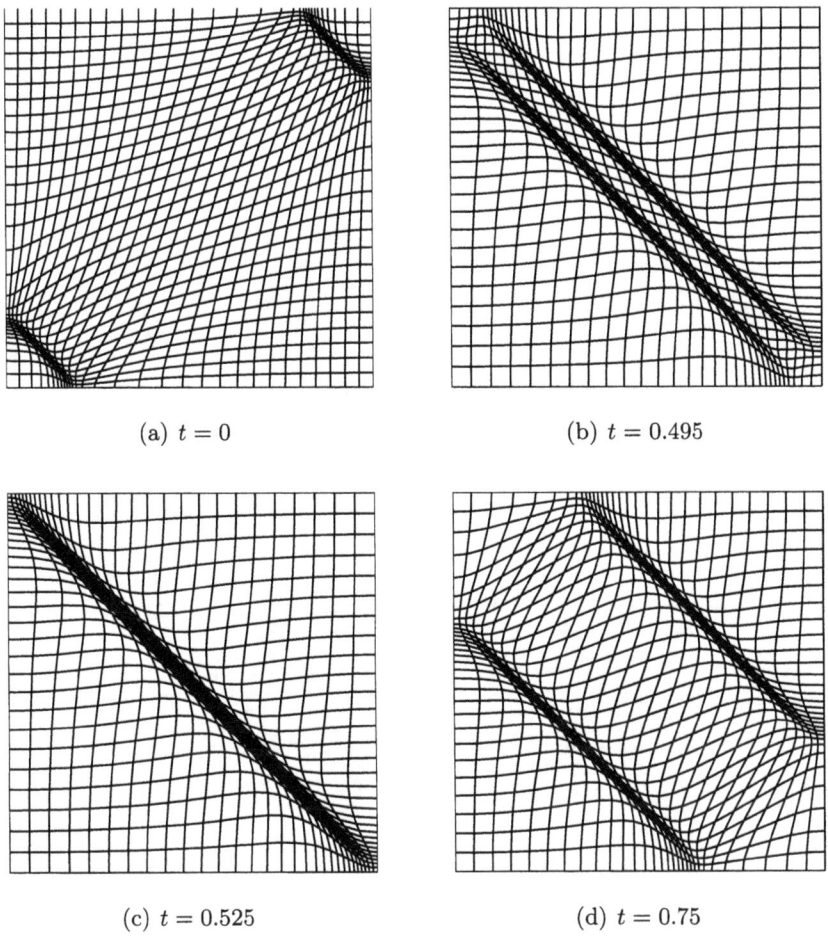

Figure 3.9. *Example 3*. Snapshots at $t = 0$ (a), $t = 0.495$ (b), $t = 0.525$ (c), $t = 0.75$ (d). Here we have simulated two fronts passing through each other in parallel, and see no difficulties with the resulting mesh. Note the way that the mesh automatically aligns itself parallel to the front.

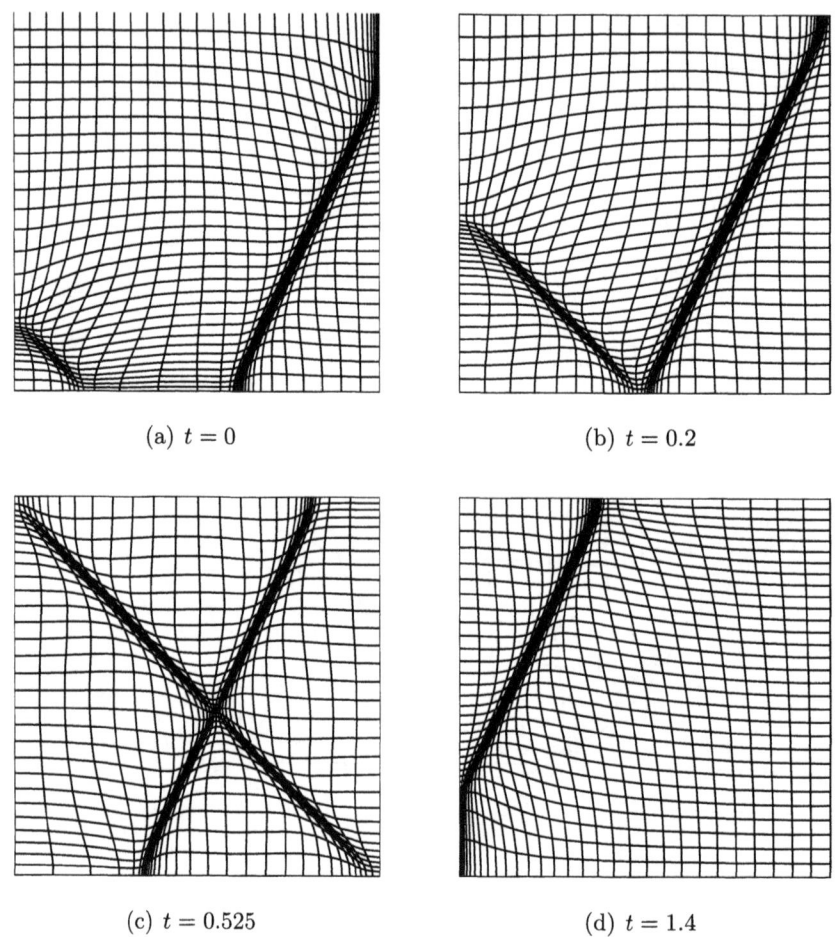

Figure 3.10. *Example 4.* Snapshots at $t = 0$ (a), $t = 0.2$ (b), $t = 0.525$ (c), $t = 1.4$ (d). Here we have simulated two linear travelling fronts passing through each other at an angle and see no difficulties in generating and moving the mesh.

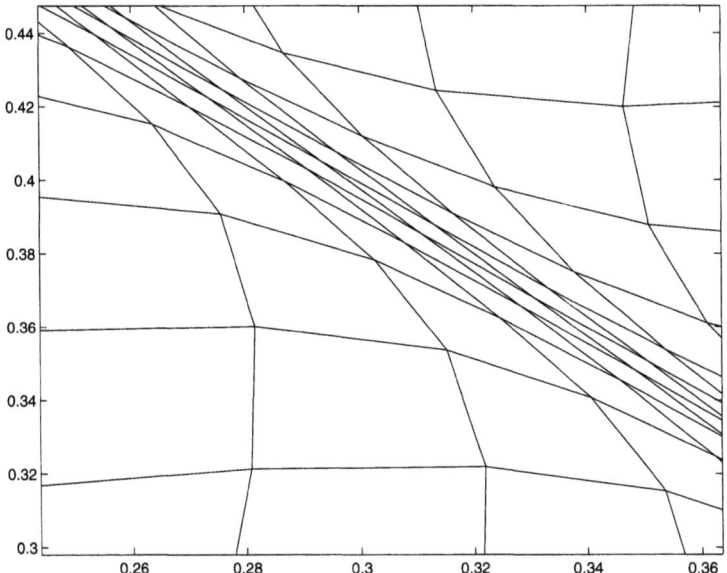

Figure 3.11. *Example 3*. A close-up of the mesh close to the front, showing the transition from a uniform mesh to one compressed by a factor γ orthogonal to the front.

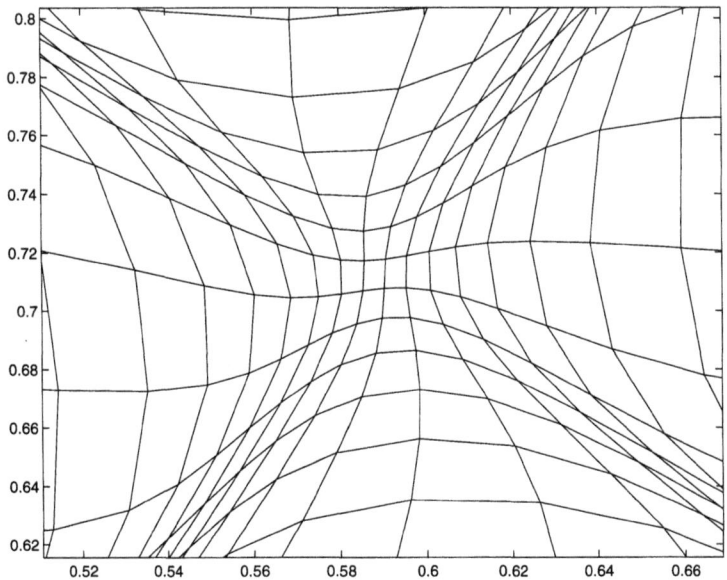

Figure 3.12. *Example 4*. A close-up of the region where the two fronts intersect.

Further examples in which we couple the PMA algorithm to generate moving meshes for certain partial differential equations are presented in Section 5.

3.4. Adaptive discretization of PDEs in higher spatial dimensions

We now extend the discussion earlier, in Sections 3.1.1–3.1.4, in which we looked at coupling an MMPDE to a one-dimensional PDE, to consider the harder problem of coupling a moving mesh method (derived either from a variational approach or from a Monge–Ampère-based approach) to a partial differential equation in several spatial dimensions, which we assume has the form

$$\mathbf{u}_t = \mathbf{f}(t, \mathbf{x}, \mathbf{u}, \nabla \mathbf{u}, \Delta \mathbf{u}). \tag{3.48}$$

Obviously, in this case the errors in using a non-uniform mesh are more pronounced, and the additional convective terms introduced by the moving mesh are potentially destabilizing. A moving mesh method becomes an adaptive method when it is coupled to a discretization of a partial differential equation. There is no unique way to perform this coupling, and it depends upon whether the PDE is discretized in the *computational domain* (typically with a finite difference method) or in the *physical domain* (typically with a finite element or a finite volume method). The nature of the coupling also depends upon whether or not the adaptive method is going to be coupled to existing software (such as a standard CFD method). The former usually involves some form of interpolation of the solution to map it onto a mesh suitable for the existing solver to use. The coupling of the mesh to the underlying PDE should also preserve important structures of the PDE; in particular, any conservation laws or solution symmetries should ideally be preserved in the coupled system. This can be done in various ways: see Tang (2005), Huang (2007) for reviews of these. The *quasi-Lagrangian* approaches (which avoid interpolation) allow directly for mesh movement, and express the PDE in Lagrangian variables, generalizing the expression (3.16):

$$\dot{\mathbf{u}} = f(t, \mathbf{x}, \mathbf{u}, \nabla_\mathbf{x} \mathbf{u}, \Delta_x u) + \nabla_\mathbf{x} \mathbf{u} \cdot \dot{\mathbf{x}}, \tag{3.49}$$

where $\dot{\mathbf{u}}, \dot{\mathbf{x}}$ denote derivatives with respect to time, with the computational variable $\boldsymbol{\xi}$ fixed. The MMPDEs and (3.49) can then both be discretized on the *computational mesh* and solved *simultaneously*. As described earlier, this procedure is generally the method of choice when used in calculations in one spatial dimension (Huang and Russell 1996). However, it is much harder to use in higher dimensions as the coupled system can be very stiff and the equations are very nonlinear. This method has the advantages that there is no need to interpolate a solution from one mesh to the next as the mesh evolves, and also that the mesh can inherit useful dynamical properties of the solution such as scaling structures (Budd *et al.* 1996, Budd and Williams 2006, Baines *et al.* 2006). Alternatively, the moving mesh equations and (3.49) can be solved *alternately* (Huang 2007, Huang and Russell 1999, Ceniceros and Hou 2001). This reduces the stiffness problems but can lead to a lag in the mesh movement.

Alternatively, a *rezoning* method can be used (Tang 2005). In such methods, the MMPDEs are solved to advance the mesh by one time step. The current solution **u** is then interpolated onto this new mesh in the physical domain, and the original PDE (3.48) solved on the new mesh in this domain (often using a finite element or finite volume method). The advantage of this method is that standard software can be used to solve the PDE, but at the expense of an interpolation step. We now describe these methods in more detail.

3.4.1. Simultaneous solution in the computational domain

We first briefly detail the implementation of the *simultaneous solution method* solving the MMPDEs and (3.49) by using finite differences in the *computational* domain. To do this we associate each mesh point in the fixed computational mesh with a solution point $U_{i,j}(t)$. To discretize (3.49) we transform the derivatives in the physical domain to ones in the computational domain. For example, in two dimensions, if we have general transformation F with Jacobian J then the derivatives of u can be expressed in terms of the computational variables in the following manner:

$$u_x = \frac{1}{J}(y_\eta u_\xi - y_\xi u_\eta), \tag{3.50}$$

$$u_y = \frac{1}{J}(-x_\eta u_\xi + x_\xi u_\eta),$$

$$u_{xx} = \frac{1}{J}\left(y_\eta (J^{-1} y_\eta u_\xi)_\xi - y_\eta (J^{-1} y_\xi u_\eta)_\xi - y_\xi (J^{-1} y_\eta u_\xi)_\eta + y_\xi (J^{-1} y_\xi u_\eta)_\eta\right),$$

$$u_{yy} = \frac{1}{J}\left(x_\eta (J^{-1} x_\eta u_\xi)_\xi - x_\eta (J^{-1} x_\xi u_\eta)_\xi - x_\xi (J^{-1} x_\eta u_\xi)_\eta + x_\xi (J^{-1} x_\xi u_\eta)_\eta\right).$$

As the computational domain has a uniform mesh, and the Jacobian J may be determined directly from the mesh mapping F, it follows that (3.50) can be discretized to high accuracy in space using a standard finite difference or finite element discretization. The solution u is then determined by solving (3.46) and (3.49) together using an appropriate ODE solver such as SDIRK or a predictor–corrector method. Examples of the use of this method are given in Section 5.

3.4.2. Alternating solution in the computational domain

The alternating solution method is often used in higher-dimensional calculations to alternatively solve the system and to move the mesh. This method avoids the highly nonlinear coupling of the mesh and the physical solution and preserves many structures such as ellipticity and sparsity in each of the mesh and physical PDEs. This can lead to significant efficiency gains, but at the disadvantage of a possible lag in the movement of the mesh and possible mesh instabilities.

Suppose that the physical solution $\mathbf{u}^n$, the mesh $\mathbf{x}^n$, and a solution time step Δt^n are known at a time t^n. The alternating solution procedure is typically implemented as follows.

(1) The monitor function $M^n(\mathbf{x}) = M(t^n, \mathbf{u}^n, \mathbf{x}^n)$ is calculated using $\mathbf{u}^n$ and $\mathbf{x}^n$.

(2) The MMPDE is discretized in space and then integrated over the time $[t^n, t^n + \Delta t^n]$ to give a new mesh $\mathbf{x}^{n+1}$ at the time $t^n + \Delta t_n]$. The underlying solution u^n is not changed during this calculation. This calculation can be done by using the (fixed) monitor function M^n, or by updating it during the integration by using linear interpolation (Huang 2007).

(3) The physical PDE (3.49) is then discretized in (computational) space and integrated using an SDIRK or similar method. In this calculation, the convective terms involving $\mathbf{x}_t$ use the approximation

$$\dot{\mathbf{x}} = \frac{\mathbf{x}^{n+1} - \mathbf{x}^n}{\Delta t^n}.$$

The mesh used in the discretization is calculated by using linear interpolation so that

$$\mathbf{x}(t) = \mathbf{x}^n + (t - t^n)\dot{\mathbf{x}}.$$

(4) It may be necessary to use the new solution $\mathbf{u}^{n+1}$ to update the monitor function M^n and to iterate from step (2) above in order to gain better control of the grid. In this case repeat these steps until the new mesh does not change.

(5) This procedure is then repeated from step (1) above.

Ceniceros and Hou (2001) use this method together with the MMPDE (3.34) (without updating the monitor function between time steps) to solve problems involving vortex singularities in the Boussinesq equations.

3.4.3. Rezoning in the physical domain

There are various problems associated with solving the Lagrangian form of the PDE (3.49) in the computational domain. A significant one of these is that certain properties of the original PDE may be lost when it is put into the Lagrangian form and coupled to a moving mesh equation. Two examples of this occurring are the loss of a conservation form of the equation (for example when adaptive methods are used to solve the Euler equations) or the loss of a Hamiltonian structure when solving problems such as the KdV or the NLS equations. These problems can be avoided by completely decoupling the solution of the PDE and the moving mesh equations over each time step, so that when solving the PDE the mesh is regarded as being static. The PDE can then be solved over that time step using a method (for example the finite volume method for problems with a conservation law) that preserves the significant features of the solution. The rezoning method can be briefly described as follows.

(1) At time step t^n let the solution be $\mathbf{u}^n$ and the mesh $\mathbf{x}^n$. Calculate the corresponding monitor function M^n.

(2) Using this monitor function, solve the moving mesh PDE over the time interval $[t^n, t^n + \Delta t^n]$ to give a new mesh $\mathbf{x}^{n+1}$.

(3) *Interpolate* the solution $\mathbf{u}^n$ onto the *new mesh* $\mathbf{x}^{n+1}$, to give an interpolated solution $\hat{\mathbf{u}}^n$.

(4) If necessary, repeat steps (2) and (3), updating M until the new mesh does not change.

(5) Starting from $\hat{\mathbf{u}}^n$ solve the original PDE (3.48). Typically, in the *physical domain* using high-resolution standard software such as the finite volume method) over the time interval $[t^n, t^n + \Delta t^n]$, doing all the calculations on the new mesh $\mathbf{x}^{n+1}$.

(6) Repeat this from step (1).

This method is of course closely related to the quasi-Lagrangian approach. To see this in a one-dimensional example, suppose that the mesh velocity is $\dot{x}$ and the underlying solution is u with $U_i^n \approx u(X_i^n, t^n)$; then in step (3) we have

$$\hat{U}_i^n \approx u(X_i^{n+1}, t^n) = u(X_i^n + \delta t_n \dot{x}) \approx U_i^n + \Delta t_n u_x \dot{x}.$$

Thus, if we use a simple forward Euler method to approximate the solution of (3.48) we obtain

$$U_i^{n+1} = \hat{U}_i^n + \Delta t_n f_i = U_i^n + \Delta t_n u_x \dot{x} + \Delta t_n f_i = U_i^n + \delta t_n (f_i + u_x \dot{x}),$$

which is of course the result of applying the forward Euler method to the Lagrangian form of the PDE given by (3.49).

The key to the success of this approach is the interpolation step (3), and an interpolation scheme that preserves some quantities of the solution is often necessary. We summarize some rezoning methods based on this approach for solving conservation laws in one and two dimensions, using the finite volume method, which are described in Tang (2005) and Tang and Tang (2003).

A one-dimensional calculation. Suppose that the ith new grid point at time t^{n+1} is X_i^{n+1}. In a finite volume method the key quantities are the *cell averages* given by

$$U_{j+1/2} = \frac{1}{X_{j+1} - X_j} \int_{X_j}^{X_{j+1}} u \, dx, \quad X_{j+1/2} = \frac{1}{2}(X_j + X_{j+1}).$$

If these are known on the mesh x^n then they can be interpolated onto the new mesh x^{n+1}. Suppose that the new mesh points satisfy $X_{j+1/2}^{n+1} \in [X_{k-1/2}^n, X_{k+1/2}^n]$; then, naively, this can be done via the formula

$$\hat{U}_{j+1/2}^n = U_{k+1/2}^n + \frac{U_{k+1/2}^n - U_{k-1/2}^n}{X_{k+1/2}^n - X_{k-1/2}^n}(X_{j+1/2}^{n+1} - X_{k+1/2}^{n+1}).$$

Unfortunately, this simple linear interpolation *does not conserve discrete solution mass*, in the sense that

$$\sum \hat{U}_{j+1/2}^n (X_{j+1}^{n+1} - X_j^{n+1}) \neq \sum U_{j+1/2}^n (X_{j+1}^n - X_j^n),$$

and consequently leads to unsatisfactory results (Tang and Tang 2003) when used to solve hyperbolic conservation laws. An improved method in Tang and Tang (2003) mimics in part the discussion above for the relation between the quasi-Lagrangian method and the rezoning method. Suppose that $x^{n+1} = x^n + \delta t \, \dot{x}$; then it follows from an application of the Reynolds transport theorem that, to leading order,

$$(X_{j+1}^{n+1} - X_j^{n+1})\hat{U}_{j+1/2}^n \approx \int_{X_j^{n+1}}^{X_{j+1}^{n+1}} \hat{u}^n \, dx$$

$$\approx (X_{j+1}^n - X_j^n)U_{j+1/2}^n + \Delta t((\dot{x}U^n)_{j+1} - (\dot{x}U^n)_j).$$

This prompts the use of the conservative (to leading order) interpolation formula given by

$$(X_{j+1}^{n+1} - X_j^{n+1})\hat{U}_{j+1/2}^n = (X_{j+1}^n - X_j^n)u_{j+1/2}^n + \Delta t((\dot{x}U^n)_{j+1} - (\dot{x}U^n)_j), \tag{3.51}$$

which automatically conserves discrete mass. Tang and Tang (2003) use this method with success to solve conservation laws of the form

$$u_t + f(u)_x = 0$$

with the PDE being integrated using a MUSCL method (LeVeque 1990) and using the MMPDE (3.9) to advance the mesh.

Two-dimensional calculations. In two dimensions, Tang and Tang (2003) use the moving mesh method described in Ceniceros and Hou (2001) to advance the mesh (X, Y) from time t^n to time t^{n+1} using the MMPDE (3.34).

In this calculation we suppose that
$$(x^{n+1}, y^{n+1}) = (x^n, y^n) + \Delta t(\dot{x}, \dot{y}).$$

Cell averages at time t^n over a (time-evolving mesh cell) of area $A^n_{j+1/2,k+1/2}$ are now given by $U^n_{j+1/2,k+1/2}$, etc., and the normal mesh speed relative to the surfaces of the mesh cell is given by

$$v = \dot{x} n_x + \dot{y} n_y, \quad \text{where the unit normal is given by } (n_x, n_y).$$

The conservative interpolation scheme proposed is

$$A^{n+1}_{j+1/2,k+1/2} \hat{U}^n_{j+1/2,k+1/2} = A^n_{j+1/2,k+1/2} U^n_{j+1/2,k+1/2} \qquad (3.52)$$
$$+ \Delta t \big([(vU^n)_{j+1,k+1/2} + (vU^n)_{j,k+1/2}]$$
$$+ [(vU^n)_{j+1/2,k+1} + (vU^n)_{j+1/2,k}] \big).$$

This scheme preserves discrete mass to leading order. Again a MUSCL method can be used to advance the solution of the PDE. This method is then used to solve the double-Mach reflection problem and various other problems with contact discontinuities arising in the solution of the Euler equations.

4. Velocity-based moving mesh methods

In this section we will look in some more detail at *velocity-based methods* for moving meshes. These are also called *Lagrangian methods*, and they rely on calculating the *mesh point velocities* and from this the mesh point locations. In some ways these methods are very natural, since in (say) fluid mechanics calculations, natural solution features are often convected with the flow, and it is natural to evolve the mesh points to follow the flow itself. (Note the huge popularity of the semi-Lagrangian and the characteristic Galerkin methods.) However, velocity-based methods can easily have severe implementation problems, and overcoming them remains a challenging issue. These include significant mesh tangling, with associated skewness, and also a tendency to create meshes which lag behind the solution. Indeed, it is very possible for such meshes to have unstable movement, to move well away from equidistributed solutions and to lead to permanently distorted skewed and even frozen meshes, even after the significant solution structures have long gone. Some examples of this type of behaviour will be presented in our discussion of the GCL method. For these reasons the overall performance of these methods is, in general, not as good as that of the position-based methods described in the last section, and hence we will spend less time discussing them. We now describe three velocity-based methods: the moving finite element method (MFE), the geometric conservation law method (GCL), and the deformation map method.

4.1. Moving mesh finite element methods

The moving finite element method was originally developed by Miller and Miller (1981) and Miller (1981), and represents a very important class of velocity-based moving mesh methods, with much underlying theory and many applications. See, for example, Adjerid and Flaherty (1986), Baines et al. (2005), Beckett et al. (2001a), Cao, Huang and Russell (1999a), Carlson and Miller (1998a, 1998b), Di et al. (2005), Lang et al. (2003), Li et al. (2002) and Wathen and Baines (1985). A very complete survey of these and related methods is given in Baines (1994) and we will only describe them briefly here. The MFE method determines a mesh velocity $\dot{\mathbf{x}} = \mathbf{v}$ through a variational principle coupled to the solution of a PDE by using a finite element method. Specifically, we consider the time-dependent PDE

$$\frac{\partial u}{\partial t} = \mathcal{L}u, \qquad (4.1)$$

with $\mathcal{L}$ a spatial differential operator. The continuous version of the MFE determines the *solution and the mesh together* by minimizing the residual given by

$$\min_{v, \frac{Du}{Dt}} I\left[\mathbf{v}, \frac{Du}{Dt}\right] \equiv \int_{\Omega_P} \left(\frac{Du}{Dt} - \nabla u \cdot \mathbf{v} - \mathcal{L}u\right)^2 W \, d\mathbf{x}. \qquad (4.2)$$

Here the function W is a weight function for which

$$W = 1$$

in the usual version of MFE described in Miller and Miller (1981) and Miller (1981); alternatively, we can take

$$W = \frac{1}{1 + |\nabla u|^2},$$

for the weighted form of MFE described in Carlson and Miller (1998a, 1998b). Observe that MFE is naturally trying to advect the mesh along with the solution flow. Of course, in practice this equation is discretized using a Galerkin method.

This method is elegant, and when tuned correctly works well (Baines 1994). However, it does have significant disadvantages. One of these is that the functional derivative of I with respect to $\mathbf{v}$ can become singular, and regularization is needed in practice.

4.2. Geometric conservation law (GCL) methods

The geometric conservation law (GCL) methods are based upon a direct differentiation of the equidistribution equation with respect to time, to derive an equation for the mesh velocity. In its simplest form the method assumes that the monitor function M is normalized so that the integral of M over

Ω_P is constant (typically unity). If A is an arbitrary measurable set in Ω_C it follows that

$$I = \int_A d\xi = \int_B M\, dx,$$

where $B = F(A)$. Now, even if A is fixed, then, as the mesh is moving the set B will typically change with time, with the points on the boundary of B moving with velocity $\mathbf{v}$. An application of the Reynolds transport theorem implies that

$$\frac{d}{dt}\int_B M\, d\mathbf{x} = \int_B M_t\, d\mathbf{x} + \int_{\partial B} M\mathbf{v} \cdot dS = \int_B (M_t + \nabla \cdot (M\mathbf{v}))\, d\mathbf{x}.$$

However, as A is fixed, it follows that $dI/dt = 0$. Furthermore, the set A is arbitrary. It follows that M and hence $\mathbf{v}$ must satisfy the (geometric) conservation law

$$M_t + \nabla \cdot (M\mathbf{v}) = 0. \tag{4.3}$$

If M is known, then (4.3) gives an equation for the (mesh) velocity $\mathbf{v}$. This equation must be augmented with the boundary condition

$$\mathbf{v} \cdot \mathbf{n} \quad \text{on } \partial \Omega_P. \tag{4.4}$$

The equations (4.3) and (4.4) have a unique solution in one dimension but many solutions in higher dimensions. To determine $\mathbf{v}$ uniquely, additional conditions must be imposed. In the derivation of the optimal transport methods we saw the use of an additional condition on the curl of the solution in the computational space. In the various forms of the geometric conservation law (GCL) methods the curl of the velocity is imposed in the *physical space*. In particular, for a suitable weight function w and a background velocity field $\mathbf{u}$, the condition

$$\nabla \times w(\mathbf{v} - \mathbf{u}) \tag{4.5}$$

is imposed, so that for an appropriate potential function ϕ we have

$$\mathbf{v} = \mathbf{u} + \frac{1}{w}\nabla \phi.$$

Here ϕ is unknown, and we presume that $\mathbf{u}$ is specified in advance. Substituting into the conservation law, it then follows that ϕ satisfies the *elliptic partial differential equation*

$$\nabla \cdot \left(\frac{M}{w}\nabla\phi\right) = -M_t - \nabla \cdot (M\mathbf{u}), \tag{4.6}$$

with the boundary condition on $\partial\Omega_P$ given by

$$\frac{\partial \phi}{\partial n} = -w\mathbf{u} \cdot \mathbf{n}.$$

This equation is a *scalar linear equation* for ϕ when posed (and solved) in the computational space. Unlike the MFE methods, this partial differential equation is well-defined for all time. In a similar manner to the optimal mesh methods, we need only solve a scalar equation; however, unlike the optimal mesh methods this equation is linear in the physical coordinates. Observe, further, that unlike certain of the MMPDE-based and variational-based methods described in Section 3, this system automatically deals with the mesh location on the boundaries. Having determined ϕ from this equation, the mesh velocity $\mathbf{v}$ can be determined directly, and the mesh then found from integrating the equation $\mathbf{x}_t = \mathbf{v}$ with respect to time using, for example, an SDIRK method. This time integration to give $\mathbf{x}$ has the disadvantage of possibly introducing mesh tangling, and of mesh points moving out of the domain during the course of the integration, if an appropriately coarse discretization is used. This method is described in Cao et al. (2002)

An alternative formulation, also described in Cao et al. (2002), considers a variational formulation, where it is shown that the solution of (4.3) is also the minimizer of the functional

$$I[\mathbf{v}] = \frac{1}{2} \int_{\Omega_P} \left(|\nabla \cdot (M\mathbf{v}) + M_t|^2 + \left(\frac{M}{w}\right)^2 |\nabla \times w(\mathbf{v} - u)|^2 \right) d\mathbf{x}. \quad (4.7)$$

This equation can be discretized using a finite element method with basis functions defined over the mesh in the physical space, and $\mathbf{v}$ found using a simple Galerkin method. The mesh points can then also be found from $\mathbf{v}$ through a Galerkin calculation. Details of these calculations are given in Baines et al. (2005, 2006), where the GCL method is coupled to an ALE (arbitrary Lagrangian–Eulerian) method for discretizing the underlying PDE.

In the usual implementation of the GCL method the weight function

$$w = 1$$

is taken. This gives an irrotational mesh (in the physical coordinates) when the background velocity $\mathbf{u} = 0$. In many implementations of GCL, $\mathbf{u} = 0$. However, for certain applications, such as in computational fluid dynamics, the background velocity can be taken to be the flow velocity.

The appeal of the GCL methods is that to find ϕ (and hence $\mathbf{v}$) we need only solve a scalar linear elliptic partial differential equation. This seems to have the advantage over the optimal transport methods, which require the solution of a nonlinear equation. However, like other velocity-based methods it has the potential disadvantages of having problems with mesh tangling and mesh skewness as the mesh points follow the moving features in a monitor function. Furthermore, we require the solution of the mesh equations in the physical domain rather than the computational domain. This loses some of the speed advantages gained (by, for example, the use of spectral methods) when solving the mesh equations on a very uniform mesh in the computational domain.

We now consider two examples taken from Cao et al. (2002) of the application of the GCL method, which allow direct comparison with the optimal-transport-based methods described in Section 3.

Example 1. This first example looks at the mesh generated by a monitor function which concentrates points around a moving disc. This is considered to be a difficult test problem for a moving mesh method. We have

$$M_1(x,y,t) = \frac{d(x,y,t)}{\int_{\Omega_P} d(\tilde{x},\tilde{y},t)\,d\tilde{x}\,d\tilde{y}},$$

where

$$d(x,y,t) = 1 + 5\exp\left(-50\left|\left(x - \frac{1}{2} - \frac{1}{4}\cos(2\pi t)\right)^2 \right.\right.$$
$$\left.\left. + \left(y - \frac{1}{2} - \frac{1}{4}\sin(2\pi t)\right)^2 - 0.01\right|\right).$$

Results are shown in Figure 4.1.

If we compare the calculated mesh to that of the identical Example 1 in the application of the PMA method in Section 3 (see p. 181), we see that the GCL method has not performed as well in this case. In particular, we see some significant effects of mesh distortion, and of mesh points lagging behind, with the Lagrangian-based method leading to mesh skewness and eventually to tangling and singular behaviour. This is in contrast to the much more regular mesh generated by the position-based PMA method.

Example 2. In this second example we consider a monitor function which concentrates mesh points around an oscillating front:

$$M_1(x,y,t) = \frac{d(x,y,t)}{\int_{\Omega_P} d(\tilde{x},\tilde{y},t)\,d\tilde{x}\,d\tilde{y}},$$

where

$$d(x,y,t) = 1 + 5\exp\left(-50\left|y - \frac{1}{2} - \frac{1}{4}\sin(2\pi x)\sin(2\pi t)\right|\right).$$

The results of using the GCL method in this case are shown in Figure 4.2.

We can compare this mesh to that generated in Example 2 of the application of the PMA method (see p. 183). We can see that the GCL method has successfully followed the moving sine wave, but unlike the mesh generated by the PMA method, there is a small degree of instability, manifested by some oscillations of the mesh visible when $t = 1$ which are not present in the mesh when $t = 0$.

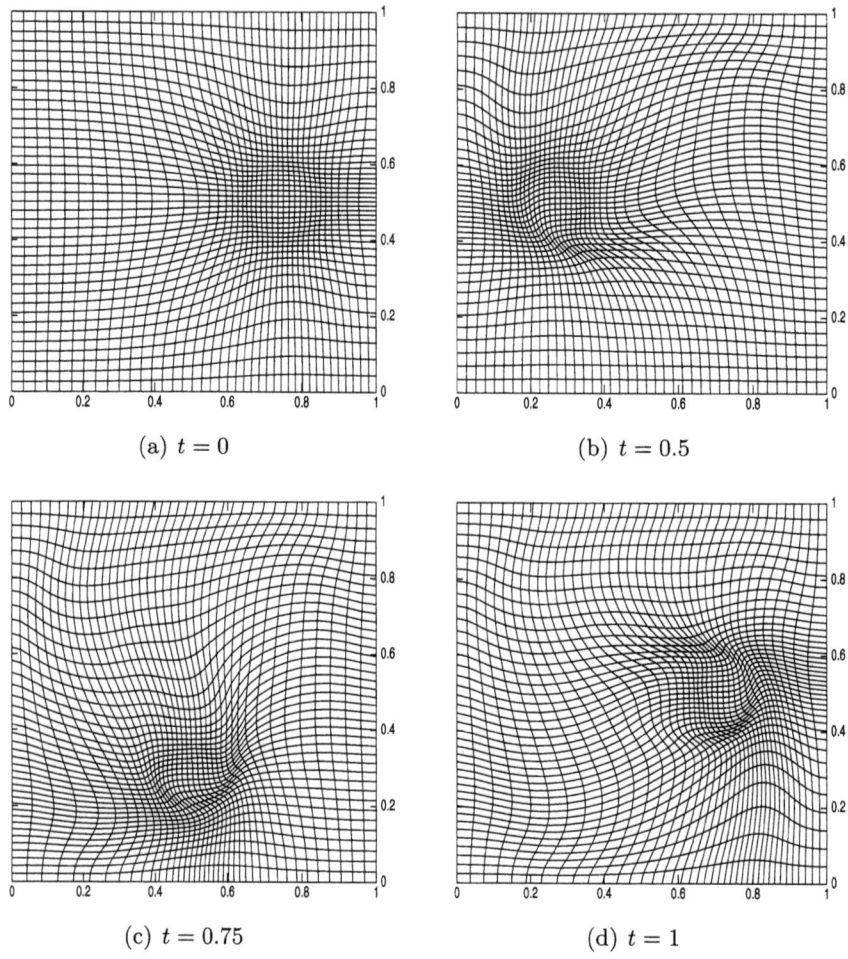

Figure 4.1. *Example 1.* Moving meshes at time $t = 0$ (a), $t = 0.5$ (b), $t = 0.75$ (c), $t = 1$ (d).

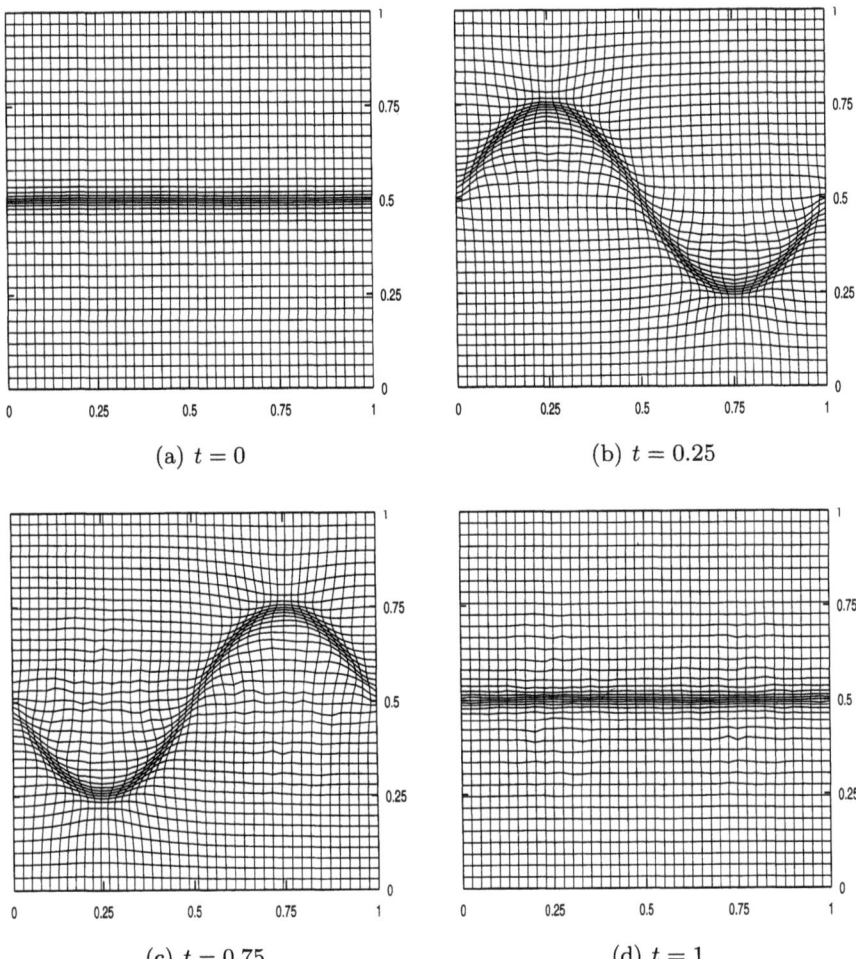

Figure 4.2. *Example 2*. Moving meshes generated by the GCL method at times $t = 0$ (a), $t = 0.25$ (b), $t = 0.75$ (c), $t = 0$ (d).

4.3. The deformation map method

Deformation map methods take a strongly differential geometric approach to moving mesh generation. Liao and his co-workers (Liao and Anderson 1992, Semper and Liao 1995, Liao and Xue 2006) have proposed a moving mesh method based on *deformation maps*. Such maps were introduced by Moser (1965) and Dacorogna and Moser (1990) in their study of volume elements of a compact Riemannian manifold, to prove the existence of a C^1 diffeomorphism with specified Jacobian. In this method, the mapping $\mathbf{x} = F(\boldsymbol{\xi}, t) : \Omega_C \to \Omega_P$ is determined from the system of equations

$$\mathbf{x}_t = \frac{1}{M(\mathbf{x}, t)} \nabla \phi(\mathbf{x}, t) \quad \text{in } \Omega_P,$$

$$\Delta \phi = -\frac{\partial M}{\partial t} \quad \text{in } \Omega_P, \quad (4.8)$$

$$\frac{\partial \phi}{\partial n} = 0 \quad \text{on } \partial \Omega_P.$$

It is easy to see that (4.8) corresponds to the system (4.6), the GCL method formulation which involves the potential function ϕ in the case where $\mathbf{u} = 0$ and $w = M$.

Dacorogna and Moser also use the alternative formulation in Bochev, Liao and Pena (1996), given by

$$\mathbf{x}_t = \frac{\boldsymbol{\nu}(\mathbf{x}, t)}{M(\mathbf{x}, t)} \quad \text{in } \Omega_P,$$

$$\nabla \cdot \boldsymbol{\nu} = -\frac{\partial M}{\partial t} \quad \text{in } \Omega_P, \quad (4.9)$$

$$\nabla \times \boldsymbol{\nu} = 0 \quad \text{on } \partial \Omega_P.$$

Note that letting $\boldsymbol{\nu} = M\mathbf{v}$, this becomes the div-curl system (4.3) and (4.5) with $\mathbf{u} = 0$ and $w = M$.

The equation for $\mathbf{x}_t$ in both cases is nonlinear, and for simplicity explicit time integration schemes are used by Liao and his co-workers. For (4.8), the solution of the potential equation for ϕ is straightforward. However, an implicit integration of (4.8) requires interpolation of the potential function ϕ for values at points other than grid nodes, and the flux-free boundary condition cannot generally be preserved. For (4.9), the div-curl system is solved using a least-squares approach.

4.4. Some other velocity-based methods

The natural idea of moving mesh points, so that the velocity of the points reflects the underlying dynamics of the solution, has led to many other velocity-based methods besides those described above. In the velocity-based methods of Anderson and Rai (1983), the mesh is moved according to attractive and repulsive pseudo-forces between nodes, motivated by a spring model in mechanics. Petzold (1987) obtains an equation for mesh velocity by minimizing the time variation of both the unknown variable and the spatial coordinate in computational coordinates and adding a diffusion-like term to the mesh equation. The method of Hyman and Larrouturou (1986, 1989) also simulates attraction and repulsion pseudo-forces.

In contrast, Dorodnitsyn (1991, 1993a, 1993b) and his co-workers (Dorodnitsyn and Kozlov 1997, Kozlov 2000) have derived a series of velocity-based methods in which the underlying symmetries of the PDE to be solved are used to guide the movement of the grid points. Such methods have the potential to preserve all of the symmetric invariants of the underlying PDE in the discretized system, and indeed precisely these invariants are used in the calculation of the mesh points. They are closely related to the (position-based) scale-invariant methods described in the next section. Such methods can also lead to conservation laws derived from a discrete version of Noether's theorem. They have been applied in Dorodnitsyn and Kozlov (1997) to solve a variety of nonlinear diffusion equations, and in Budd and Dorodnitsyn (2001) to integrate the nonlinear Schrödinger equation. The main disadvantage of these methods is that they tend to be highly nonlinear, and the equations for the mesh points hard to solve. More details of the implementation and application of these methods are given in Budd and Piggott (2005).

5. Applications of moving mesh methods

In this final section we look at some applications of moving mesh methods to a variety of problems related to the solution of partial differential equations. As described in Section 1, there are a vast number of problems for which moving mesh methods have been used with great success, and we cannot hope to summarize all of them in this section. For example, one of the most important applications arises in fluid mechanics, and reviews of these, comparing many different methods and problems, are given in Baines (1994), Eisman (1985), Tang (2005), Tang and Tang (2003) and Yanenko *et al.* (1976).

Instead, we look in this section at some specific problems and classes of problems which aim to highlight some of the special advantages and disadvantages of the moving mesh method for solving partial differential equations. In particular we look at problems where moving meshes can exploit natural solution scaling structures (including self-similarity); blow-up problems in which solution singularities arise in finite time; and problems such as Burgers' equation, where moving meshes are used to capture the motion of moving fronts, and two physical problems, one in combustion and the other in meteorology. Two primary goals are to describe analytical techniques for solving nonlinear PDEs which can often be naturally modified for moving mesh algorithms, in such a way that the discrete solutions share common properties with the (unknown) analytical solutions; and to show that the sharing of the common structures leads to efficient adaptive methods.

5.1. Symmetry, self-similar solutions and scale invariance

Many naturally occurring physical systems have an invariance under symmetries including *translations*, *rotations* and *changes of scale*, and this is reflected in the partial differential equations that describe them. The solutions of these equations may then either be themselves invariant under (combinations of) these symmetries, the *self-similar solutions*, or they can be transformed into other solutions through the application of symmetry operators. Conservation laws can be linked through Noether's theorem (Olver 1986, Dorodnitsyn 1993b) to many of the continuous symmetries.

An example of such a system is Burgers' equation,

$$\mathbf{u}_t + \mathbf{u} \cdot \nabla u = \nu \Delta \mathbf{u},$$

which is invariant under *rotations* in space and *translations* in both space and time. It admits travelling-wave solutions which are self-similar solutions coupling spatial and temporal translation, with the wave speed giving the coupling. The waves can be at any orientation, and the action of a rotation is to map one wave to another. In other systems a rescaling of space, time and of the solution leaves the partial differential equations governing the system invariant. Such changes of scale were first observed by Kepler in his studies of the solar system and summarized in his famous *third law*, in which he observed that if a planetary orbit existed with a solution (in polar coordinates) given by $(r(t), \theta(t))$ then there was also a solution of the form $((\lambda^3 r(\lambda^2 t)), \mathcal{O}(\lambda^2 t))$ for any positive value of λ. Such scaling invariance also arises in the equations of fluid and gas dynamics, nonlinear optics and many biological systems. An excellent summary of scale-invariant systems with such properties is given in the book by Barenblatt (1996).

We can then ask the question of whether a numerical method can be constructed which is also invariant under symmetries. More generally, an adaptive method can be designed to exploit the symmetries in the problem. In a sense, the answer to this question is obvious as we can always perform an *a priori* scaling of the problem and then use a method in the scaled coordinates. However, this assumes more knowledge of the physical system than we may easily have available to us. Indeed, for some systems there are a number of different possible changes of scale, and it is not *a priori* obvious which one correctly describes the system evolution.

We now show how several of the moving mesh strategies described earlier, particularly those based on moving mesh partial differential equations, can be very effectively applied to such problems with symmetry. Such methods have several important properties. If carefully designed they can admit discrete self-similar solution, they can have (local truncation) errors which are invariant under changes in the scale of the problem, they can have discrete conservation laws, and they can cope with symmetric singular structures. Such problems have been studied by a number of authors: Budd *et al.* (1996), Baines *et al.* (2006), Budd and Piggott (2005), Ceniceros and Hou (2001), Dorodnitsyn (1991, 1993b) and Kozlov (2000).

The underlying problem that we will consider is the partial differential equation

$$u_t = f(u, \nabla u, \Delta u, \ldots). \tag{5.1}$$

Observe that this equation is invariant under translations in space and time and also rotations in space. We also assume it to be invariant under the action of the scaling symmetry (or symmetries)

$$t \to \lambda, \quad \mathbf{x} \to \lambda^\alpha \mathbf{x}, \quad u \to \lambda^\beta u. \tag{5.2}$$

The objective is now to derive an adaptive method which reflects the underlying symmetries. This can be achieved if, away from any finite boundaries, the equations describing the mesh (regardless of whether these are position- or velocity-based) are *invariant under the action of the same symmetry operations as the underlying PDE*, so that translations in space and time rotations and scalings of the form (5.1) leave the mesh equations invariant. This is in many ways a very natural question to ask of an *r*-adaptive method, for which the mesh may be regarded as a dynamic object, amenable to the action of symmetries involving space and time. It is much harder to see how *h*- and *p*-type methods can be considered in this manner.

As an example of such a problem (see also Baines *et al.* (2006) for the use of an adaptive ALE method in higher dimensions, and Dorodnitsyn (1993a) and Dorodnitsyn and Kozlov (1997) for a more abstract treatment), we consider the one-dimensional *porous medium equation* given by

$$u_t = (u^m)_{xx}, \tag{5.3}$$

where we consider this equation posed on the whole real line, with mild decay conditions on the solution u at ∞. This partial differential equation is invariant under two different, and independent, changes of scale, one of space and the other of the solution. These are given by

$$t \to \lambda t, \quad x \to \lambda^{1/2} x$$

and

$$t \to \mu t, \quad u \to \mu^{1/(m-1)} u.$$

Any (linear) combination of these two changes of scale will also leave the equation (5.3) invariant.

Whilst the PDE may be invariant under the action of a symmetry group, not all of the solutions have this property, although any one solution can be mapped into another by the action of the symmetry operator. Those solutions which are themselves invariant under the action of the symmetry operator are termed the *self-similar solutions*. A self-similar solution satisfies the functional equation

$$u(\lambda t, \lambda^\alpha \mathbf{x}) = \lambda^\beta u(t, \mathbf{x}). \qquad (5.4)$$

Such a solution can be described in terms of a new set of coordinates, typically of the form

$$u(t, \mathbf{x}) = t^\beta v(\mathbf{z}), \quad \mathbf{z} = \mathbf{x}/t^\beta. \qquad (5.5)$$

In the case of the porous medium equation, the self-similar solution with *constant mass* takes the form

$$u(t, x) = t^{-1/3} v(x/t^{1/3}).$$

The function $v(\mathbf{z})$ generally satisfies an ordinary differential equation, which is much simpler than the original PDE. The solutions of this ODE which correspond to solutions of the PDE are generally those with certain decay or boundedness conditions as $|\mathbf{z}| \to \infty$. This allows the construction of exact solutions to PDEs in many cases. In the example of the porous medium equation we have the famous Barenblatt–Pattle solution (Barenblatt 1996), given by

$$u(x, t) = t^{-1/3}(a - x^2/t^{2/3})_+,$$

where a is a constant. Significantly such solutions are *globally attracting* (in the sense of L_1-convergence. An extensive description of such problems is given in Barenblatt (1996) and Olver (1986).

More generally, solutions of PDEs with symmetries may also take the form

$$u(t, \mathbf{x}) = U(t)v(\mathbf{z}), \quad \text{where } \mathbf{z} = \mathbf{x}/L(t). \qquad (5.6)$$

Here $U(t)$ and $L(t)$ are appropriate solution and length scales. In the case of self-similar solutions these are pure powers of t (or a translation of t) but they can take more general forms. For example, in the case of the blow-up equation,

$$u_t = u_{xx} + u^3,$$

with $u \to \infty$ as $t \to T$, we have

$$U(t) = (T - t)^{-1/2} \quad \text{and} \quad L(t) = (T - t)^{1/2} |\log(T - t)|.$$

Such solutions are called *approximately self-similar solutions* (Samarskii, Galaktionov, Kurdyumov and Mikhailov 1973). Examples of this type of behaviour also arise in the nonlinear Schrödinger equation in dimension two, and in the chemotaxis equations of mathematical biology.

The self-similar (and approximately self-similar) solutions of PDEs can play an important role in the description of the solution (beyond the fact that they lead to exact solutions). This is because they often describe very well the *intermediate asymptotics* of the solution, which is the behaviour of the solution after the transient effects of any initial conditions and before boundary terms become important. They are also often effective in describing certain singular types of behaviour such as the peaks in the blow-up and related problems, and also the interfaces in various problems in gas dynamics (Barenblatt 1996). It is therefore useful to have numerical methods which can accurately reproduce self-similar behaviour when it arises in applications. One method that has been used is to make an *a priori* choice of (self-similar) variables, so that the PDE can be reduced to an ODE and to then solve this ODE numerically. This method, however, has a number of disadvantages. Firstly, it cannot deal with general initial and boundary conditions satisfied by the PDE. Secondly, there may often be several symmetry groups acting on a partial differential equation and it may often not be at all clear which (if any) leads to a self-similar solution. Indeed, there are many problems (for example the heat equation posed on a finite interval with an initially highly localized solution) which may have one form of self-similar behaviour for part of the evolution and another over longer times (for example when the solution of the heat equation interacts with the boundary). Problems of this form (called type II self-similar solutions in Barenblatt (1996)) are extremely hard to analyse in advance of any PDE calculation.

An alternative approach, which makes considerable use of the r-adaptive methods, is to use a numerical method which *admits the same scaling invariances as the original PDE away from any boundaries*. Such methods are called *scale-invariant* (Budd and Piggott 2005, Baines *et al.* 2006) as they perform identically under the scaling transformations.

The key to the design and implementation of such methods lies in the use of the moving mesh partial differential equations to describe the location of the mesh points. By an appropriate choice of monitor function it is often possible to construct such MMPDEs to be invariant under the action of the scaling transformations. The advantages of such methods are as follows.

- They usually have *discrete self-similar solutions* which inherit many of the properties of the underlying self-similar solutions.

- If designed carefully they may work for several types of scaling symmetry and thus can be used in the case of type II self-similarity when the exact scaling group is not known in advance.

- They can be applied to problems with arbitrary initial and boundary conditions

- They can have relative truncation errors which are *independent of the scale of the solution* (Budd, Leimkuhler and Piggott 2001, Baines et al. 2006)

We give a partial proof of these results presently. A further, but less general, advantage of such methods is that they also often preserve the asymptotic properties of the (approximately) self-similar solutions. This is seen both in the global convergence towards the self-similar solutions of the porous medium equation, and the local convergence towards the singular profile described by the approximately self-similar solution of the blow-up equation.

We start this calculation by looking at a moving mesh method in one dimension for which the moving mesh PDE is given by MMPDE1 and for which the monitor function is a function of u and u_x, so that

$$(M(u, u_x) x_\xi)_\xi = 0.$$

If this is to be invariant under the action of the scaling symmetry $t \to \lambda t, x \to \lambda^\beta x, u \to \lambda^\gamma u$, we require that

$$(M(\lambda^\gamma u, \lambda^{\gamma-\beta} u_x) \lambda^\beta x_\xi)_\xi = 0.$$

This is satisfied (for all β) provided that the monitor function satisfies the functional equation

$$M(\lambda^\gamma u, \lambda^{\gamma-\beta} u_x) = \lambda^\theta M(u, u_x), \qquad (5.7)$$

where θ is arbitrary. Many monitor functions do not satisfy this functional equation, for example the simple arclength monitor $M = \sqrt{1 + u_x^2}$ (although it approximately satisfies it when $|u_x|$ is large). However, it is certainly possible to find functions that do, and a simple example is given by

$$M(u, u_x) = u^\delta$$

for some choice of δ. Observe that this monitor function is invariant under a very arbitrary set of scaling symmetries and using it poses no explicit *a priori* scaling on the solution. It is thus very useful when considering self-similar solutions of type II.

As an example we consider the porous medium equation in the form

$$u_t = (u u_x)_x \quad |u| \to 0 \quad \text{as} \quad |x| \to \infty. \qquad (5.8)$$

It is easy to see that the first integral of this solution is constant, and we may scale the solution so that, for all t, we have

$$\int_{-\infty}^{\infty} u \, dx = 1.$$

It is then natural to choose $M = u$ so that the monitor function has unit integral over the physical domain. It follows immediately that

$$Mx_\xi = 1. \tag{5.9}$$

Furthermore, from the geometric conservation law given by

$$M_t + (M\dot{x})_x = 0,$$

we have

$$u_t + (u\dot{x})_x = 0.$$

Substituting for u_t and integrating gives

$$u(u_x + \dot{x}) = 0,$$

so that the Lagrangian equation for the mesh points is given by

$$\dot{x} = -u_x. \tag{5.10}$$

This equation is used in Dorodnitsyn (1993a) and Dorodnitsyn and Kozlov (1997) as the equation of motion of all of the mesh points. Note that this is also the equation for the movement of the leading edge of the front of those solutions of the porous medium equation which have compact support. The same monitor function is used in Baines et al. (2006) to compute solutions of the porous medium equation using the scale-invariant ALE method.

In the usual manner, either of the equations (5.9) and (5.10) can be discretized and solved simultaneously with the porous medium equation (5.8) (see Budd, Collins, Huang and Russell (1999b) for more details), so that the discrete solution and mesh points are given by

$$U_i \approx u(X_i, t), \quad X_i = x(i\Delta\xi).$$

A similar procedure can also be used with a variational formulation in higher dimensions (Baines et al. 2006)

It is *immediate* (see Budd et al. (1999b)) that any such discretization admits a *discrete self-similar solution* of the form

$$U_i = t^{-1/3} V_i, \quad X_i = t^{1/3} Z_i. \tag{5.11}$$

It can also be shown (Budd and Piggott 2005) that such self-similar solutions are not only locally stable, but are also *global attractors*, so that they correctly organize the qualitative long-term dynamics of the solution and even obey a discrete maximum principle.

5.1.1. Construction of scale-invariant MMPDEs

The porous medium equation has relatively benign dynamics, which allows us to use a relatively simple moving mesh equation to evolve the mesh. In the case of systems with more extreme forms of dynamics, such as that arising in shocks or localized singularities, it is usually necessary to use a more sophisticated moving mesh equation to avoid mesh instabilities.

In one dimension two possible equations for the mesh are the moving mesh PDEs MMPDE5 and MMPDE6, given by

$$\epsilon x_t = (M x_\xi)_\xi, \quad -\epsilon x_{\xi\xi t} = (M x_\xi)_\xi. \qquad (5.12)$$

Suppose that these are used to solve a system which has intrinsic solution, length and time scales given respectively by U, L and T and a derived scale Λ for the monitor function M. As the computational variable ξ is *independent of scale*, the left-hand side of each of these two MMPDEs scales as L/T and the right-hand side as ΛL. These two balance (so that the mesh evolves at the same rate as the underlying solution) provided that

$$\frac{1}{T} = \Lambda. \qquad (5.13)$$

Observe that this condition is independent of the spatial length scale L. The implication of this is that if the monitor function $M(u, x)$ depends upon u and x, then this satisfies (5.13) provided that

$$M(Uu, Lx) = \frac{1}{T} M(u, x). \qquad (5.14)$$

This is a more severe condition than the condition (5.14) for MMPDE1, but is necessary to ensure that the mesh calculation does not destabilize the calculation of the solution of the PDE. It has the useful property that it often gives a precise characterization of the necessary form of the monitor function appropriate to one scaling transformation. A disadvantage of this approach is, however, that it may not always (or indeed generally) be possible to capture *all* possible scaling transformations in *a single monitor function* satisfying (5.14). Thus some *a priori* knowledge of the expected solution behaviour might be necessary in this case.

As an example we will consider the radially symmetric solutions of the cubic nonlinear Schrödinger equation. This equation has the form

$$i u_t + u_{rr} + \frac{n-1}{r} u_r + u|u|^2 = 0, \qquad (5.15)$$

where $r = |x|$ and n is the spatial dimension. If $n \geq 2$ this can have solutions which develop singularities (in amplitude and phase) in a finite time. The equation (5.15) is invariant under the action of the scaling group

$$t \to \lambda t, \quad r \to \lambda^{1/2} r \quad u \to \lambda^{-1/2} u,$$

as well as the unitary multiplicative group

$$t \to t, \quad r \to r, \quad u \to e^{i\phi} u, \quad \phi \in R.$$

This system develops singularities in a finite time T with a natural time scale of $(T-t)$, a length scale of $L = (T-t)^{1/2}$, and a solution scale of $U = (T-t)^{-1/2}$. For the moving mesh PDE based on a simple monitor function of the form $M(u)$ to be invariant under the action of the unitary multiplicative group, we require that

$$M(u) = M(|u|).$$

The condition (5.14) then implies that

$$M((T-t)^{-1/2}u) = \frac{1}{(T-t)}M(u).$$

Both conditions are satisfied if

$$M(u) = |u|^2. \tag{5.16}$$

Calculations using a regularized form of this monitor function are described in Section 5.2.2.

Scale-invariant moving mesh methods for a general class of scale-invariant PDEs can also be constructed in higher dimensions, say $\mathbf{x} \in \mathbb{R}^n$. We consider two cases, firstly the method of Ceniceros and Hou (2001), and then the optimal transport method. The first of these describes a two-dimensional moving mesh generated by the PDE system

$$x_t = \nabla_\xi \cdot (M \nabla_\xi x), \quad y_t = \nabla_\xi \cdot (M \nabla_\xi y).$$

This system scales in an identical manner to both MMPDE5 and MMPDE6, and consequently is scale-invariant provided that the monitor function M satisfies the condition (5.14). In the case of optimal transport in an n-dimensional system, the PMA equation gives a mesh from $\nabla_\xi Q$, where Q satisfies the PDE

$$(I - \gamma \Delta)Q_t = (M(u)H(Q))^{1/n}. \tag{5.17}$$

Now, if the underlying problem has the natural scaling symmetry $\mathbf{x} \to \lambda \mathbf{x}$, then this is equivalent to the scaling symmetry $Q \to LQ$. It is immediate that

$$H(LQ) = L^n H(Q).$$

It then follows immediately that (5.17) is invariant under the action of the scaling symmetries provided that the monitor function satisfies the functional equation

$$M(Uu, L\mathbf{x})^{1/n} = \frac{1}{T}M(u, \mathbf{x}). \tag{5.18}$$

We note that, in two dimensions, the scaling structure of the function Q also implies that the mesh skewness s, as defined in Section 3 by the relation

$$s = \frac{\Delta(Q)^2}{H(Q)} - 2,$$

is *invariant under the scale change* $Q \to LQ$. This implies that, if the mesh is generated by a scale-invariant PMA-type method, then any mesh regularity is *preserved under scaling*.

5.1.2. Discrete self-similar and approximately self-similar solutions

As mentioned above, a significant benefit of using a scale-invariant adaptive scheme is that it admits discrete self-similar solutions. One reason for this is the almost trivial, yet very important, observation that:

> The actions of discretization and of rescaling commute,

where here a discretization can be any of a finite difference, collocation, finite element or a finite volume method. (This means that a discretization of a rescaled solution will be identical to a rescaling of a discrete solution.) More formally, if the PDE is invariant under the action of the scaling group

$$t \to \lambda t, \quad \mathbf{x} \to \lambda^\beta \mathbf{x}, u \to \lambda^\alpha u,$$

then a *continuous* self-similar solution takes the form

$$u(\mathbf{x}, t) = t^\alpha v(\mathbf{y}) \quad \mathbf{y} = \mathbf{x}/t^\beta. \tag{5.19}$$

In terms of the computational variables, this becomes

$$u(\mathbf{x}(\boldsymbol{\xi}, t)) = t^\alpha v(\boldsymbol{\xi}), \quad \mathbf{x}(\boldsymbol{\xi}, t) = t^\beta \mathbf{y}(\boldsymbol{\xi}),$$

for appropriate functions v and y. This leads immediately to a *discrete self-similar solution* of the form

$$U_i(t) = t^\alpha V_i, \quad X_i(t) = t^\beta Y_i, \tag{5.20}$$

For convenience we now study this discrete self-similar solution in the context of a prototypical example, a one-dimensional system governed by a semilinear second-order PDE of the form

$$u_t = u_{xx} + f(u), \quad \text{or in Lagrangian form} \quad u_t = u_{xx} + f(u) + \dot{x} u_x.$$

This PDE is invariant under the action of the scaling group

$$t \to \lambda t, \quad x \to \lambda^{1/2} x, \quad u \to \lambda^\alpha u$$

(so that $\beta = 1/2$), provided that the function $f(u)$ satisfies the functional equation

$$f(\lambda^\alpha u) = \lambda^{\alpha-1} f(u).$$

Substituting the expression for the self-similar solution (5.19), and setting $\beta = 1/2$, we see that the function $v(y)$ must satisfy the *ordinary differential equation*

$$\alpha v - \frac{y}{2} v_y = v_{yy} + f(v), \quad \text{so that} \quad \alpha v = v_{yy} + f(v) + \frac{y}{2} v_y. \tag{5.21}$$

Now consider a simple centred finite difference discretization of the Lagrangian form of the underlying PDE, which takes the form

$$\dot{U}_i = \frac{\frac{U_{i+1}-U_i}{X_{i+1}-X_i} - \frac{U_i-U_{i-1}}{X_i-X_{i-1}}}{\frac{1}{2}(X_{i+1} - X_{i-1})} + f(U_i) + \dot{X}_i \frac{U_{i+1} - U_{i-1}}{(X_{i+1} - X_{i-1})}.$$

We can substitute (5.20) directly into this expression to give

$$\alpha V_i = \frac{\frac{V_{i+1}-V_i}{Y_{i+1}-Y_i} - \frac{V_i-V_{i-1}}{Y_i-Y_{i-1}}}{\frac{1}{2}(Y_{i+1} - Y_{i-1})} + f(V_i) + \frac{Y_i}{2} \frac{V_{i+1} - V_{i-1}}{(Y_{i+1} - Y_{i-1})}. \quad (5.22)$$

It is immediately obvious that (5.22) is a consistent discretization of the ordinary differential equation (5.21) so that the function $v(y)$ will be approximated by V_i at the (time-independent computational) mesh point Y_i. Note further that the discretization error in approximating v by V_i is *independent of the solution scale*. In the case of systems such as the blow-up problems of Section 5.2, this implies that the asymptotic form of the singularity at the peak will be approximated with uniform accuracy for peaks with very small spatial scales (and correspondingly large solution scales). We note, however, that other errors do arise at points where the (rapidly) moving mesh following the evolving peak matches a nearly stationary mesh in the regions closer to the boundary of the domain.

To determine the location of the points Y_i in terms of the computational variables, we must apply the MMPDE used to evolve the mesh. Again, to give an example we consider MMPDE5. Substituting (5.20) into a standard discretization of MMPDE5 gives the following discrete equation for Y_i:

$$\frac{1}{2}Y_i = \frac{1}{2\Delta\xi^2}\left(M_{i+1/2}(Y_{i+1} - Y_i) - M_{i-1/2}(Y_i - Y_{i-1})\right). \quad (5.23)$$

Crucially, we note that the functional equation (5.14) satisfied by the monitor function M allows this rescaling to be made. Similar discrete equations arise for other choices of MMPDE, provided that M satisfies (5.14).

We note that exactly the same rescalings are possible in higher-dimensional problems using moving meshes generated by any other methods described in this subsection, and for any other form of discretization (provided that the processes of discretization and rescaling commute).

It is also possible to construct approximate discrete self-similar solutions in cases where the underlying solution is better described by approximate self-similar variables. Details of the calculations in this case are given in Budd and Williams (2006).

As an example of this we return to the porous medium equation in one dimension. For this problem, for all constants $C > 0$ there is a self-similar solution $u_s(x,t)$ of the form

$$u_s(x,t) = (t+C)^{-1/3}v(y), \quad x = (t+C)^{1/3}y,$$

where the function $v(y)$ is given by the Barenblatt–Pattle profile (Barenblatt 1996). It is also well known that any positive initial data lead to a solution $u(x,t)$, which converges towards to a self-similar solution in the sense that

$$t^{1/3} u(t^{1/3} y, t) \to v(y).$$

Similarly, it is shown in Budd et al. (1999b) that if the monitor function is chosen to give a scale-invariant scheme (for example $M = u$), then this scheme admits a set of *discrete self-similar solutions* on a moving mesh which take the form

$$U_i(t) = (t+C)^{-1/3} V_i, \quad X_i(t) = (t+C)^{1/3} Y_i.$$

Note that the product

$$W_i \equiv U_i X_i = V_i Y_i$$

is invariant in time. In Figure 5.1 we show the results of a computation presented in the computational domain, using a scale-invariant moving mesh method with $M = u$ and a centred finite difference discretization. In this calculation the initial data at $t = 0$ were taken to be an irregular function, and results are plotted at times $t = 0$ and $t = 10$. We also show (dashed) two discrete self-similar solutions with values of C chosen so that they initially lie above and below the solution. Note that the solution calculated is sandwiched between these two functions. We also present the corresponding mesh $X_i(t)$ and the scaled mesh $Y_i(t) = X_i(t)/t^{1/3}$. It is clear from these figures that the computed solution converges towards the discrete self-similar solutions (in correspondence with the continuous theory) and that the moving mesh tends towards the one for which Y_i is constant in time so that $X_i(t)$ scales asymptotically as $t^{1/3}$. Similar figures for solutions of the porous medium equation computed using a scale-invariant ALE method in two dimensions are given in Baines et al. (2006).

5.2. Blow-up and related problems

5.2.1. Parabolic blow-up

As mentioned in Section 5.1, a significant success in the application of moving mesh methods occurs in the study of parabolic partial differential equations (and also of systems of PDEs) which have solutions that blow up, so that the solution, or some derivative of it, becomes infinite in a finite time T. We now consider such problems in a little more detail. Blow-up in the solution often represents an important change in the properties of the model that the equation represents (such as the ignition of a heated gas mixture), and it is important that it is reproduced accurately in a numerical computation. A survey of many different types of blow-up problem is presented in Samarskii et al. (1973). Blow-up typically occurs on increasingly small time and length scales, and hence it is usually essential to use both

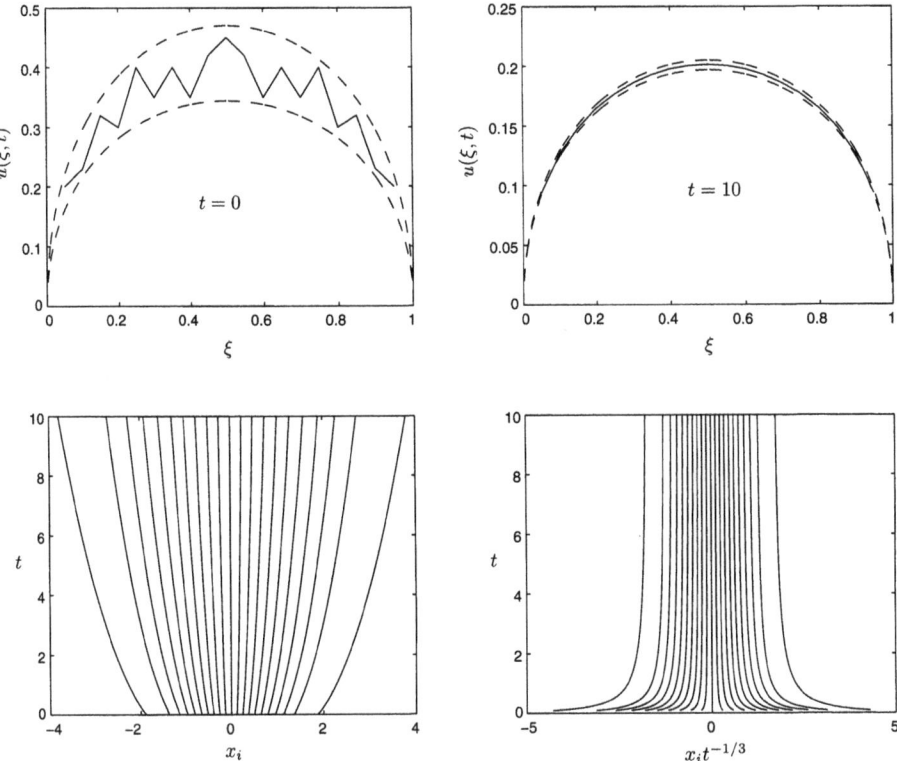

Figure 5.1. Convergence of the solution in the computational domain to the discrete self-similar solution. Note the invariance of the solution profile in this domain. We also show the evolution of the mesh and note that this scales as $t^{1/3}$.

temporarily and spatially adaptive methods for any such computation. A survey of methods and underlying numerical theory for blow-up-type problems is given in Budd *et al.* (1996), with more recent references in Budd and Williams (2006), Ceniceros and Hou (2001) and Huang, Ma and Russell (2008). Mesh refinement (h-adaptive) methods have been used in some numerical studies of blow-up, including the dynamic rescaling methods, such as those described in Berger and Kohn (1988), but moving mesh methods have proved to be more effective in both one and two dimensions. This is due, in part, to the strong scaling symmetry properties of the solution close to blow-up (where the effects of boundary and initial conditions become progressively less important), and the way that this can be exploited by using the scale-invariant methods described in Section 5.1.

A prototype example of a blow-up system is the parabolic partial differential equation

$$u_t = \Delta u + u^p, \quad x \in \Omega, \quad p > 1, \quad u|_{\partial\Omega} = 0. \tag{5.24}$$

This has the property of *single-point blow-up* in that for certain types of sufficiently large initial data, there is a point $\mathbf{x}^*$ and a finite time T so that

$$u(x^*, t) \to \infty \quad \text{as} \quad t \to T.$$

Close to the point $\mathbf{x}^*$ the solution develops a narrow peak which evolves in an approximately self-similar manner. It is not uncommon to consider solutions which change by ten orders of magnitude, with a reduction in the solution length scale by a similar amount. It is essential to use an adaptive method to capture such behaviour accurately.

Example 1. The first example that we consider is given by

$$\begin{aligned} u_t &= \Delta u + u^3, & \mathbf{x} &= (x, y) \in \Omega = (0, 1)^2, \\ u(\mathbf{x}, t) &= 0, & \mathbf{x} &\in \partial\Omega, \\ u(\mathbf{x}, 0) &= 5\exp(-25(x-0.45)^2 - 25(y-0.35)^2). \end{aligned} \tag{5.25}$$

This problem has the natural scaling symmetry

$$t \to \lambda t, \quad \mathbf{x} \to \lambda^{1/2}\mathbf{x}, \quad u \to \lambda^{-1/2}u,$$

it has a natural time scale of $(T-t)$, and it is shown in Samarskii et al. (1973) that the natural space and solution scales are given by the approximately self-similar variables

$$L = (T-t)^{1/2}|\log(T-t)|^{1/2} \quad U = (T-t)^{-1/2}.$$

To compute a solution, we augment this problem with the PMA equation (5.17) to determine the moving mesh with a monitor function of the form $M \equiv M(u)$. To obtain scale invariance for this system we require that M must satisfy the function equation (5.18) so that

$$M((T-t)^{-1/2}u)^{1/2} = \frac{1}{(T-t)} M(u).$$

A simple solution of this is $M(u) = u^4$. In practice this monitor function can lead to instabilities due to placing too many points in the singular region and not sufficiently closer to the boundary of the domain, and to overcome this we apply a McKenzie regularization to give

$$M(u) = u(\mathbf{x}, t)^4 + \int_{\Omega_P} u(\mathbf{x}', t)^4 \, d\mathbf{x}'.$$

This choice of monitor leads to a mesh which automatically inherits the correct dynamic length scale of the underlying solution in the singular regions

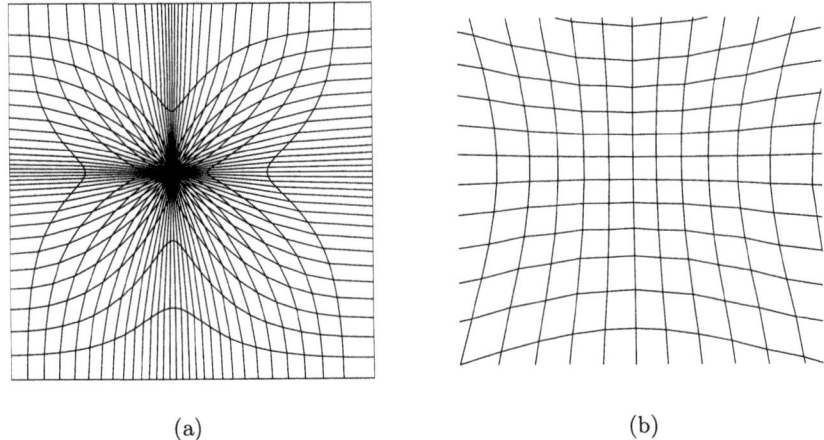

(a) (b)

Figure 5.2. *Example 1.* Final grid for the blow-up example.
(a) Entire grid. (b) Detail near the blow-up point. Note that
the grid is quite regular in the vicinity of the singularity, with
no evidence of any skewness or tangling.

where u is large. We can then find a solution of the blow-up problem in the computational domain Ω_C, using a finite difference method with a uniform $N = 30 \times 30$ mesh in the computational domain to discretize both the PMA equation and the Lagrangian form of the underlying PDE. The resulting system of ODEs was then solved *simultaneously* using a BDF method. In this calculation, as the blow-up time was approached an adaptive time step was used by applying the Sundman transformation, as described in Budd et al. (2001). For this we take

$$\Delta t = \frac{1}{(\max u)^2}.$$

In Figure 5.2 we show the final grid both over the whole domain and close to the peak near the centre, as well as the initial and final solutions. The integration was performed until $|u|_\infty = 10^{15}$, for which the peak has an approximate length scale of 10^{-15}.

Over the course of the evolution we see a mesh compression (and a solution amplification) by a factor of 10^{12} in the physical domain. Note that this has been achieved with a very modest number of mesh points. The final mesh shows a similar gradation of mesh elements from size $\mathcal{O}(10^{-2})$ to size $\mathcal{O}(10^{-14})$. However, if we study the mesh close to the solution peak, as illustrated in Figure 5.2, we see that this shows a strong degree of local regularity, with no evidence of long thin elements or skewness in the region where the solution gradients are very large. This is exactly as predicted by the previous theory. (Away from the solution peak the mesh is less smooth; however, in this region the solution gradients are much smaller than in the peak, and the local truncation errors are thus lower.)

Example 2. In order to consider these results in the context of some of the error analysis presented in Sections 2 and 3, it is instructive to look at a second one-dimensional example, in which we consider the blow-up problem

$$u_t = u_{xx} + u^3, \quad u_x(0) = u(1) = 0.$$

It is known (Samarskii et al. 1973) that this equation has solutions which blow up at the origin, and in the peak the asymptotic blow-up profile of this solution takes the form

$$u(x,t) = \frac{1}{\sqrt{T-t}} \frac{1}{\sqrt{1 + ax^2/L^2}}, \quad (5.26)$$

where a is a constant (depending weakly on the initial conditions) which we may take equal to unity, and $L(t)$ is the natural length scale given by

$$L(t) = \sqrt{(T-t)|\log(T-t)|}.$$

As this is now a problem in one dimension, we initially take $M = u^2$ so that, in the peak,

$$M = \frac{1}{(T-t)} \frac{1}{1 + x^2/L^2}.$$

The integral of M is overwhelmingly dominated by the contribution in the peak, so that

$$\theta = \int_0^1 M \, dx = \frac{L}{(T-t)} \tan^{-1}(1/L) \approx \frac{\pi L}{2(T-t)}.$$

We can then take a regularized monitor function of the form

$$\bar{M} = M + \theta, \quad \text{with} \quad \int_0^1 \bar{M} = 2\theta,$$

to ensure that we have a 50:50 mesh. In the peak, $\bar{M} \approx M$ and an equidistributed mesh satisfies the equation $M x_\xi = 2\theta$ so that

$$x_\xi = \pi L(1 + x^2/L^2).$$

It follows immediately that in the peak we have

$$x(\xi, t) = L(t) \tan(\pi \xi). \quad (5.27)$$

Observe that, if $\xi < 1/2$, then $x = \mathcal{O}(L)$, and that this mesh matches naturally to one for which $x = \mathcal{O}(1)$ as $\xi \to 1/2$, so that we have (as required) half of the mesh points in the peak and half outside the peak.

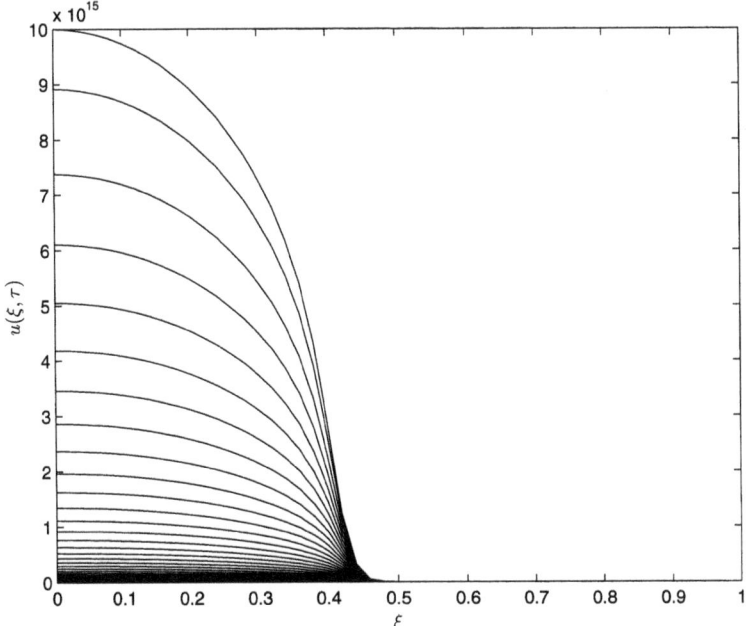

Figure 5.3. *Example 2.* Solution profile for the blow-up problem computed in the computational space up to a peak value of 10^{15}.

In Figure 5.3 we show the solution computed in the computational plane using a simple centred discretization of the PDE using a uniform mesh with $N = 30$ points in the computational space. Observe that the peak in the region $\xi < 1/2$ has a regular form with no evidence of instability.

In Figure 5.4 we present the mesh evolving in time and also a plot of $\tau \equiv \log(T - t)$ as a function of $\log(x)$. Observe that the mesh in the first of these figures has concentrated in the peak, but that there is still good resolution of the solution away from the peak. In the second figure we see that all of the grid cells move at the same rate close to the origin, demonstrating the self-similar form there.

The expression (5.27) allows a direct evaluation of the leading-order terms in the truncation error Tr. Recall from (2.49) that the leading term in the truncation error in the discretization of u_{xx} is given by

$$\text{Tr} = \Delta \xi^2 \left(\frac{1}{3} x_{\xi\xi} u_{xxx} + \frac{1}{12} x_\xi^2 u_{xxxx} \right).$$

From (5.27) we have

$$x = L \tan(\pi \xi), \quad x_\xi = \pi L \sec^2(\pi \xi), \quad x_{\xi\xi} = 2\pi^2 L \tan(\pi \xi) \sec^2(\pi \xi).$$

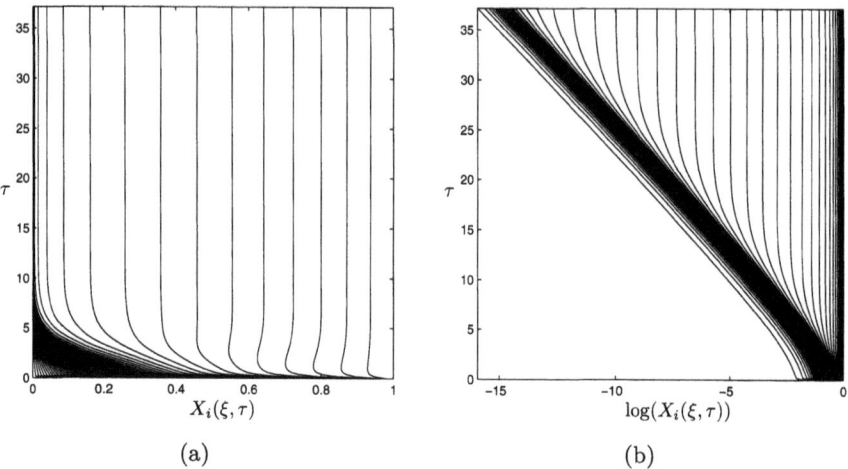

Figure 5.4. *Example 2.* Blow-up mesh. (a) Grid in the natural coordinates. (b) Plotting $\tau = |\log(T-t)|$ as a function of $\log(X_i)$ at equally spaced values of ξ, showing the contraction rate and approximately self-similar behaviour close to the origin.

A straightforward calculation based on the scaling structures of u shows that, in the peak region,

$$u = \mathcal{O}\left(\frac{1}{\sqrt{T-t}}\right), \qquad u_{xx} = \mathcal{O}\left(\frac{1}{L^2\sqrt{T-t}}\right),$$

$$u_{xxx} = \mathcal{O}\left(\frac{1}{L^3\sqrt{T-t}}\right), \qquad u_{xxxx} = \mathcal{O}\left(\frac{1}{L^4\sqrt{T-t}}\right).$$

If $\xi < 1/2$ we can also see that x and its derivatives are all of $\mathcal{O}(L)$. As a consequence, substituting into the expression for T, we have that in this region

$$\mathrm{Tr} = \mathcal{O}\left(\frac{\Delta\xi^2}{L^2\sqrt{T-t}}\right), \quad \text{so that} \quad \frac{\mathrm{Tr}}{u_{xx}} = \mathcal{O}(\Delta\xi^2).$$

We can see from this expression that the *relative truncation error* Tr/u_{xx} in discretizing u_{xx} depends only on $\Delta\xi$ and not on the intrinsic solution scale. This analysis starts to break down as $\xi \to 1/2$, and the terms $x_{\xi\xi}$, etc., become large as the mesh adapted to the peak evolves into a uniform mesh close to the boundary. However, the results shown above do not indicate instability in the mesh in this limit.

5.2.2. Focusing solutions of the nonlinear Schrödinger equation

The nonlinear Schrödinger equation described in Section 5.1.1,

$$iu_t + \Delta u + u|u|^2 = 0, \quad \mathbf{x} \in \mathbb{R}^n, \tag{5.28}$$

is a model for the modulational instability of water waves and plasma waves, and is important in studies of nonlinear optics where the refractive index of a material depends on the intensity of a laser beam. In all dimensions it has the conserved quantities

$$\int |u|^2 \, d\mathbf{x} \quad \text{and} \quad \int \left(|\nabla u|^2 - \frac{1}{2}|u|^4 \right) d\mathbf{x},$$

corresponding to mass and energy respectively. In one dimension (5.28) is integrable and can give rise to soliton-type solutions. More generally, it is an example of a Hamiltonian PDE. Many numerical (usually non-adaptive) methods have been derived to take advantage of this integrability (Budd and Piggott 2005, McLachlan 1994). If posed in n dimensions, where $n \geq 2$, then the PDE (5.28) is no longer integrable and it may admit singular (focusing) solutions for certain initial data. A review of these is given by Sulem and Sulem (1999), who also describe some numerical computations using a moving mesh method based on Winslow's algorithm. Numerically this is a very difficult problem, as high resolution in time and space is required to capture the strong self-focusing of the solution, to deal with the highly oscillatory nature of the solution 'tail', and to compute over a large domain to avoid boundary effects (Budd et al. 1999a, Ceniceros 2002). In such singular solutions both the maximum value of the solution modulus, and its phase, blow up in a finite time T. The precise form of this blow-up and the initial conditions that lead to blow-up are the subject of much investigation (see the review in Sulem and Sulem (1999)), and much remains unresolved. Two significant open questions are: (1) What is the exact nature of the blow-up profile in two dimensions (where it is known to be approximately self-similar but the precise form is still unclear), and (2) Do there exist radially symmetric self-similar solutions in three dimensions? In the latter case these are conjectured to take the form

$$u(r,t) = \frac{1}{\sqrt{(T-t)}} e^{ia \log(T-t)} Q(y), \quad r = \sqrt{T-t}\, y, \quad r = |\mathbf{x}|,$$

where a is an unknown constant. The existence of such a solution can be addressed by a scale-invariant method applied to the (one-dimensional) class of radially symmetric solutions. In this we take the monitor function

$$M = |u(r,t)|^2,$$

derived in Section 5.1.1, and use a collocation-based discretization with $N = 81$ points. In such a calculation we would expect to observe a discrete

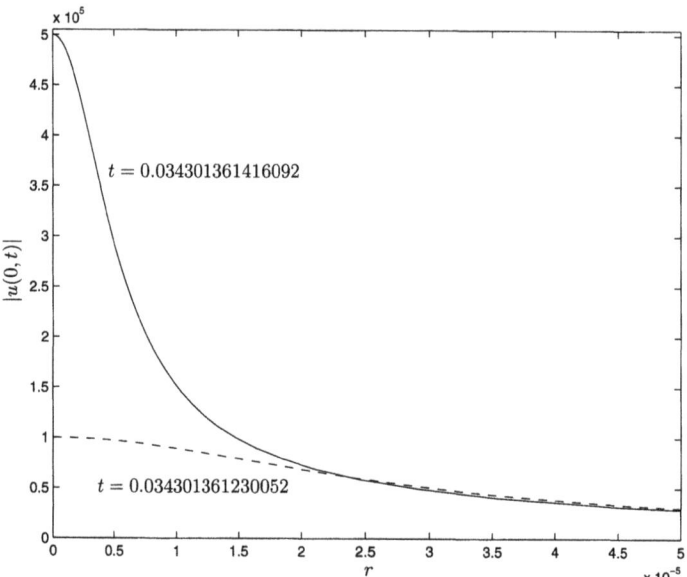

Figure 5.5. The solution when $|u(0,t)| = 100000$ and 500000. Note the narrow width of the peak.

self-similar solution for which the mesh points X_i should take the form $X_i = \sqrt{T-t}\,Y_i$, so that

$$|u(0,t)|X_i = Q(0)Y_i$$

is constant in time. In Figure 5.5 we plot two solutions starting from initial data $u(r,0) = 6\sqrt{2}\exp(-r^2)$, which have a computed blow-up time of $T = 0.0343013614215$. Observe the excellent resolution of the peak even when the peak amplitude is around 10^5 and the peak width is around 10^{-5}. In Figure 5.6 we show the computed values of $W_i \equiv |u(0,t)|X_i$ as functions of $\tau = \log(T-t)$ for a range in which $|u|$ varies from 100 to 500000. These are clearly tending towards constants, indicating that both the solution and the mesh are evolving in a self-similar manner.

It is interesting to note that the scale-invariant methods are easy to use and give rather better results in this case than symplectic methods described in McLachlan (1994), despite the Hamiltonian structure of the problem. This is because the adaptive methods give much better resolution of the peak.

In contrast, further computations of focusing solutions of the nonlinear Schrödinger equation are given by Ceniceros (2002). In this case the following highly dispersive problem was studied:

$$i\epsilon u_t + \frac{1}{2}\epsilon^2 u_{xx} + u|u|^2 = 0, \quad u(x,0) \equiv u_0 = A_0(x)e^{iS_0(x)/\epsilon}, \qquad (5.29)$$

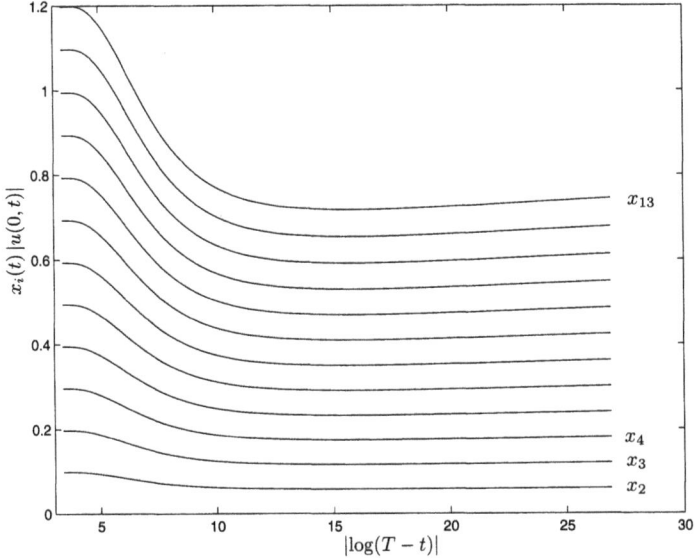

Figure 5.6. The evolution of the mesh as the blow-up time T is approached. Observe that the product $X_i|u(0,t)|$ converges to a constant, indicating that both solution and mesh are evolving in a self-similar manner.

with $\epsilon \ll 1$. In this calculation a moving mesh method with smoothing was used, as described in Section 3. This took the form

$$(1 - \gamma \partial_{\xi\xi}^2)\dot{x} = (Mx_\xi)_\xi, \tag{5.30}$$

with $\gamma = \max(M)$. In Ceniceros (2002) the Lagrangian form of the PDE (5.29) was discretized in the computational domain and solved alternatively with the MMPDE (5.30) using the methods described in Section 3. In particular, a semi-implicit BDF method was used for the time integration of both the moving mesh equation and the underlying PDE. A monitor function of the form

$$M = \sqrt{1 + \beta^2|u_\xi|^2 + \alpha^2|u|^4}$$

was employed, with

$$\beta = (2L)^2 \frac{\|u_{0,x}\|_\infty}{\|u_0\|}$$

and L a measure of the size of the computational domain. In this calculation, particular care had to be taken with the spatial discretization of the mesh movement terms of the form $\dot{x}u_x$, arising in the Lagrangian form of (5.29) as the dispersive nature of (5.29) meant that there was no natural damping to the high-frequency terms in the errors associated with a central difference discretization This is in contrast to the highly dissipative nature of the discretization of (5.30), which took the form

$$\frac{X_j^{n+1} - X_j^n}{\Delta t} = \gamma \frac{X_{j+1}^{n+1} - 2X_j^{n+1} + X_{j-1}^{n+1}}{\Delta \xi^2} - \gamma \frac{X_{j+1}^n - 2X_j^n + X_{j-1}^n}{\Delta \xi^2}$$

$$+ \frac{1}{\Delta \xi^2}\left[M_{j+1/2}^n(X_{j+1}^n - X_j^n) - M_{j-1/2}^n(X_j^n - X_{j-1}^n)\right].$$

To avoid some of the resulting instabilities reported in Li et al. (1998) and discussed in Section 5.3 in the context of the solution of Burgers' equation, a high-order (fourth-order central difference) expression was used to discretize the Lagrangian advective term $\dot{x}u_x$. The resulting semi-implicit BDF discretization then took the form

$$\frac{1}{2\Delta t}[3U_j^{n+1} - 4U_j^n + U_j^{n-1}] =$$

$$\mathrm{i}\frac{\epsilon \sigma_j^{n+1}}{2\Delta \xi^2}\left[\sigma_{j+1/2}^{n+1}(U_j^{n+1} - U_j^{n+1}) - \sigma_{j-1/2}^{n+1}(U_j^{n+1} - U_{j-1}^{n+1})\right]$$

$$+ 2\left[\frac{\mathrm{i}}{\epsilon}|U_j^n|^2 U_j^n + \dot{X}_j^n \frac{U_{j-2}^n - 8U_{j-1}^n + 8U_{j+1}^n - U_{j+2}^n}{X_{j-2}^n - 8X_{j-1}^n + 8X_{j+1}^n - X_{j+2}^n}\right]$$

$$- \left[\frac{\mathrm{i}}{\epsilon}|U_j^{n-1}|^2 U_j^{n-1} + \dot{X}_j^{n-1} \frac{U_{j-2}^{n-1} - 8U_{j-1}^{n-1} + 8U_{j+1}^{n-1} - U_{j+2}^{n-1}}{X_{j-2}^{n-1} - 8X_{j-1}^{n-1} + 8X_{j+1}^{n-1} - X_{j+2}^{n-1}}\right],$$

with $\sigma = 1/x_\xi$. The resulting moving mesh method was then found to be highly effective in computing the focusing solutions.

Remark. It bears mentioning that finding self-similar or approximately self-similar structures for analytical solutions to PDEs can be extremely demanding, and moving mesh methods with built-in scale invariance can often be used to gain insight into the form of such structures. A case in point is the paper by Budd, Galaktionov and Williams (2004), which analysed an unexpected form of blow-up for a fourth-order PDE, entirely motivated by the results of numerical calculations with a moving mesh method.

5.3. Some problems with moving fronts

A classical problem leading to the formation of sharp fronts is Burgers' equation given (in two dimensions) by

$$u_t + \frac{1}{2}(u^2)_x + \frac{1}{2}(u^2)_y = \nu \Delta u, \quad \nu \ll 1. \tag{5.31}$$

This equation has been used as a benchmark for a number of different moving mesh algorithms (Huang and Russell 1997a, Mackenzie and Mekwi 2007a, Zhang and Tang 2002, Tang and Xu 2007, Li et al. 1998).

Example 1. As a first calculation we consider a problem with the initial and boundary values chosen over the unit square, so that (5.31) has the exact solution given by

$$u(x, y, t) = \left(1 + e^{(x+y-t)/2\nu}\right)^{-1}. \tag{5.32}$$

This solution has a sharp moving front. For the purposes of this example we consider coupling this system to the PMA algorithm described in Section 3, to generate a moving mesh which can both compute an approximation to this solution and follow the front as it evolves over the time interval $t \in [1/4, 2]$. To do this we discretize (5.31) in the Lagrangian form

$$u_t = \nu \Delta u - \left(\frac{1}{2}(u^2)_x + \frac{1}{2}(u^2)_y\right) + \dot{x}u_x + \dot{y}u_y, \quad \nu \ll 1. \tag{5.33}$$

with all discretizations made in the computational variables. The conservation form of the equation above is used for this calculation so that, for example, the advective term u_x^2 is rescaled as

$$u_x^2 = \frac{1}{J}\left[y_\eta u_\xi^2 - y_\xi u_\eta^2\right],$$

which is then discretized using a central difference scheme. For this calculation we also take similar discretizations for the other advective terms. The equation posed in the computational variables is then coupled to the discretized form of PMA equation (5.17) with the values of the approximation $U_{i,j}$ to $u(X_i, Y_j)$ and of Q given on the mesh vertices. In these computations we take an $N \times N$ computational mesh (typically $N = 40$), a viscosity of $\nu = 0.005$, and use the arclength monitor function

$$M = \sqrt{1 + \alpha\left(u_x^2 + u_y^2\right)},$$

with $\alpha = 1$. A number of different strategies could be used to evolve this coupled system forward in time, but in practice a simple scheme which solved the PMA equation and (5.33) *simultaneously* using a simple forward Euler method is effective with a time step Δt, as given in Zhang and Tang (2002), determined by the CFL condition. In the PMA equation we used $\epsilon = 1, \gamma = 0.335$ to find the initial mesh (before evolving the solution PDE) when $t = 1/4$, and then $\epsilon = 0.01, \gamma = \sqrt{\max(M)}$ to follow the front up to $t = 2$. More details are given in Walsh et al. (2009). We note that solving the PMA equation coupled to (5.33), using an alternating solution strategy coupled with an up-winding discretization, has also been shown to be effective (Sulman 2008). In Figure 5.7 we present the results of a series of computations using this method, when $N = 40$, showing both the solution for (5.31) and the corresponding mesh. In this figure we see excellent resolution of the solution at the front.

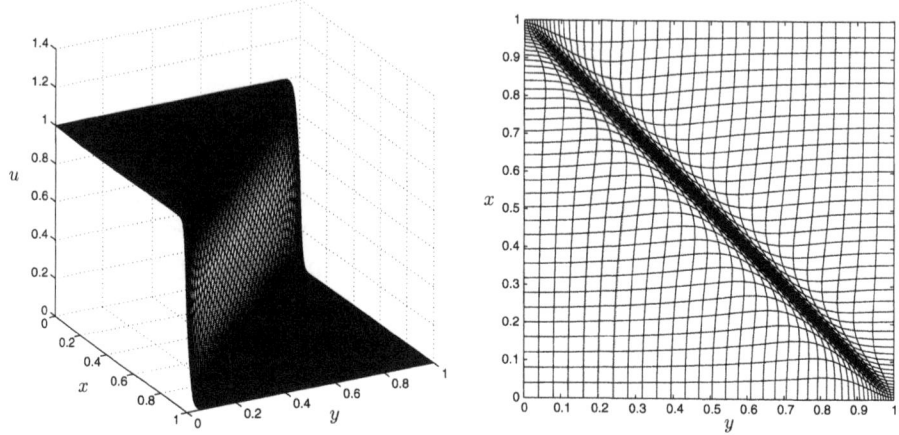

Figure 5.7. *Example 1.* The solution of Burgers' equation and the corresponding mesh at the time $t = 1$.

Table 5.1.

N	L_2-error on a uniform mesh	L_2-error on the moving mesh
20	6e-2	8e-3
40	2e-2	2e-3
80	6e-3	1e-3

A quantitative measure of the error in the computed solution at $t = 1$ can be given by determining the L_2-norm of the difference between it and the exact solution. We consider this error both for a uniform and a moving mesh for various values of N: see Table 5.1.

Sulman (2008) obtained a similar table by using a PMA method in which the convective terms are calculated by using an upwind method and the PMA system, and solved the underlying PDE alternately. It is also interesting to compare this calculation with the results presented in Zhang and Tang (2002). In this paper a harmonic mapping method of the form described in Section 4 was used to generate the mesh, and this was then coupled to a finite volume discretization of (5.31.) We see from Table 5.1 that a useful reduction in the error for calculating the solution to (5.31) resulted from the PMA method. This reduction is similar to that observed in Zhang and Tang (2002) for the harmonic map method.

Example 2. In closely related calculations we can compare with a similar moving mesh calculation of the solution of Burgers' equation, over the interval $x \in [0,1]$, when $\nu = 1e-4$ and with initial data $u_0 = \sin(2\pi x) + (1/2)\sin(\pi x)$. In this case we compute the mesh using a one-dimensional form of PMA with the arclength monitor function and $N = 30$ mesh points, central differencing in the computational domain for all spatial derivatives, and solve the resulting ODEs using the MATLAB routine ode15s. We start with an initially uniform mesh, and evolve the mesh and solution together using PMA with $\epsilon = 1$. In this calculation, and with this choice of ϵ, the mesh evolves from being uniform to one which is equidistributed, over a time $t \approx 0.05$. In this time period the underlying solution remains fairly smooth. At the time $t \approx 0.2$ the solution develops a sharp front, which is well approximated, and then followed, by the evolving mesh. The solution is presented in the physical domain at a series of different times in Figure 5.8, and the resulting mesh trajectories are presented in Figure 5.9. Observe the manner in which the mesh points resolve the front with no oscillations in this case (due in part to the regularity of the initial data).

It is interesting to compare this calculation with a very similar calculation made by Li and Petzold (1997), who looked at the solution of Burgers' equation when $\nu = 1e-4$ with the less regular initial data given by

$$u(x,0) \equiv u_0 = 0.2, \quad x \leq 0.1, \quad u_0 = 8x - 0.6, \quad 0.1 \leq x \leq 0.2$$
$$u_0 = 1, \quad 0.2 \leq x \leq 0.5, \quad u_0 = -10x + 6, \quad 0.5 \leq x \leq 0.6,$$
$$u_0 = 0, \quad 0.6 \leq x \leq 1.$$

This is a more severe test of the moving mesh method than the previous example, as the initial data are much less smooth. This calculation employed an arclength-based monitor function together with a regularized differential algebraic formulation of the moving mesh equations, very similar to using MMPDE6, and based on a method proposed in Adjerid and Flaherty (1986), with the mesh-smoothing algorithm proposed by Dorfi and Drury (1987), described in Section 2. Li and Petzold (1997) considered three different discretizations: a central difference discretization, an ENO (Roe) method, and a piecewise hyberbolic method (PHM) due to Marquina (1994). It was found that in this case the central difference method tended to lead to spurious oscillations due (as commented on in the example of the nonlinear Schrödinger equation) to the destabilizing effect of the anti-diffusive terms arising in the discretization of the advective terms describing the mesh motion. Li and Petzold (1997) showed that these oscillations were reduced with the ENO scheme and eliminated using the PHM method.

The related paper by Li et al. (1998) also considered applying the same moving mesh method to a number of other reaction–diffusion problems,

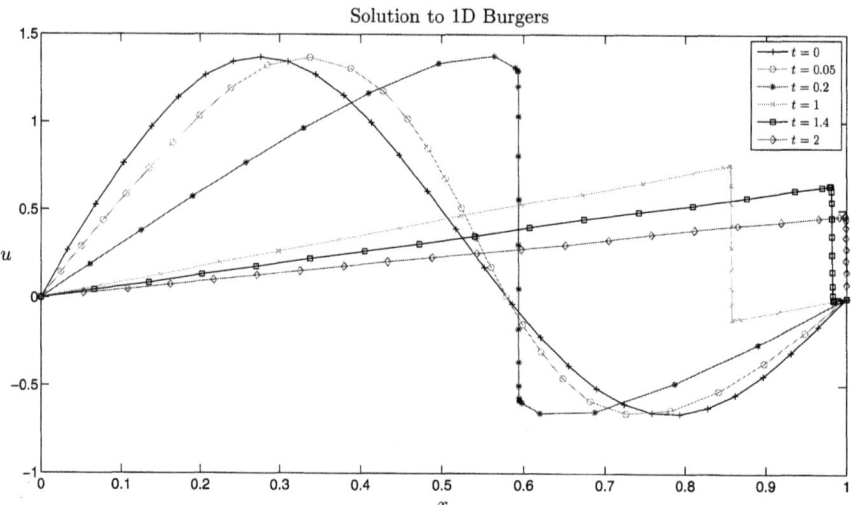

Figure 5.8. The computed solution of Burgers' equation in one dimension at times $t = 0, 0.05, 0.2, 1, 1.4, 2$.

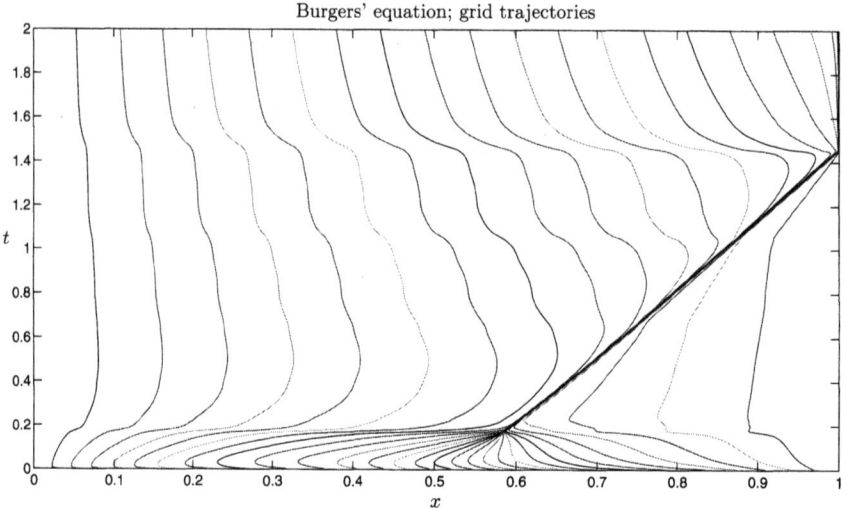

Figure 5.9. The motion of the mesh points when solving Burgers' equation in one dimension.

including the Fisher equation,

$$u_t = \beta u_{xx} + \alpha u(1-u). \qquad (5.34)$$

This equation has an exact solution with a moving front given by

$$u = (1 + \exp(\sqrt{\alpha/6}x - 5\alpha t/6))^{-2}.$$

Whilst it was found that the wave front was well resolved, the moving mesh method appeared to give worse results than for the fixed grid, with the front moving at the wrong speed (too fast). Li et al. (1998) proposed that this error was largely due to the effects of discretizing the additional convective terms due to the moving mesh, as discussed in Section 3, which can lead both to a high truncation error and instabilities in the solution. A possible solution is to use higher-order methods or a curvature-dependent monitor function; e.g., see Qiu and Sloan (1998). However, there is an interesting alternative reason for the error. In problems such as (5.34), the speed of motion of the front is not so much determined by the shape of the front itself but by the nature of the exponential terms in the tails of the solution and how they interact with the boundaries. Similar behaviour arises in the Cahn–Hilliard equation, and also in similar systems such as the Gray–Scott equation in chemistry. Somewhat paradoxically, the solution needs to be well resolved in the boundary regions where it appears to be behaving smoothly, in order for the front speed to be accurately resolved. Thus a uniform mesh may well perform better than a moving mesh in these cases, as it will probably be placing more mesh points close to the boundary.

5.4. Problems involving a change of phase and/or combustion

A very rich set of examples of the use of moving mesh methods is given by Stefan-type problems, involving phase changes or combustion, and examples of these can be found in Beckett et al. (2001a), Mackenzie and Mekwi (2007a), Mackenzie and Robertson (2002), Miller, Gleyzer and Imhoff (1998), Huang and Zhan (2004), Zegeling (2005), Tan (2007) and Tan et al. (2007).

As an example we consider the combustion problem described in Cao et al. (1999a) and Moore and Flaherty (1992). The mathematical model is a system of coupled nonlinear reaction–diffusion equations,

$$\frac{\partial u}{\partial t} - \nabla^2 u = -\frac{R}{\alpha\delta} u\, e^{\delta(1-1/T)},$$

$$\frac{\partial T}{\partial t} - \frac{1}{Le}\nabla^2 T = \frac{R}{\delta Le} u\, e^{\delta(1-1/T)},$$

where u and T represent the dimensionless concentration and temperature of a chemical which is undertaking a one-step reaction. We consider the J-shape solution domain shown in Figure 5.10. The initial and boundary

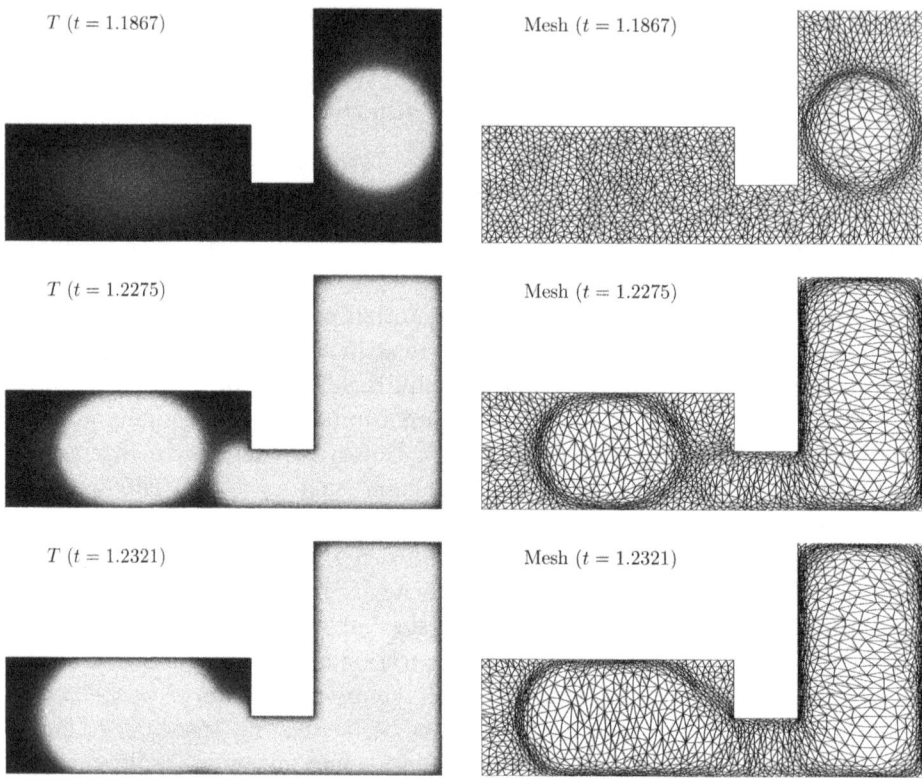

Figure 5.10. The contour plot of the temperature T (where white represents 2.2 and black represents 1) and the moving meshes are shown at various times.

conditions are
$$u|_{t=0} = T|_{t=0} = 1, \quad \text{in } \Omega,$$
$$u|_{\partial\Omega} = T|_{\partial\Omega} = 1, \quad \text{for } t > 0$$
and the physical parameters are set to be $Le = 0.9, \alpha = 1, \delta = 20$, and $R = 5$. A resulting calculation obtained with an MMPDE finite element method as described in Section 3 (with details in Cao et al. (1999a)) is shown in Figure 5.10.

5.5. Convection-driven problems in meteorology

A system of interest to meteorologists is the *Eady problem*, which describes the evolution of (localized) extra-tropical (mid-latitude) cyclones. The Eady problem is a two-dimensional reduction of the Euler equations describing the (incompressible) air velocity (u, w) and pressure P in the (x, z)-plane, where $x \in [-L, L]$ is a horizontal coordinate along lines of constant latitude, and $z \in [0, H]$ represents height. (A shallow atmosphere model is used with $H \ll L$, together with an f-plane approximation which uses a locally flat approximation to the Earth's curvature.) This model also includes both the potential temperature θ (relative to a reference temperature θ_0) and Coriolis effects. It is described in detail in Cullen (2006) and takes the form

$$\frac{Du}{Dt} - fv + P_x = 0,$$

$$\frac{Dv}{Dt} + fu - \frac{Cg}{\theta_0}(z - H/2) = 0,$$

$$\frac{D\theta}{Dt} - Cv = 0,$$

$$\frac{Dw}{Dt} + P_z - \frac{g\theta}{\theta_0} = 0,$$

$$u_x + w_z = 0.$$

Here D is the advective (total) derivative, f is the Coriolis parameter (assumed constant), g is the gravitational constant and $C = -\theta_y$ is assumed constant. All variables are periodic in x, with w and P_x vanishing at $z = 0$ and $z = H$. From certain initially smooth data (as described by Nakamura (1994)), it is possible (Cullen 2006) for the solutions of the Eady problem to develop severe fronts in a small number of days, and some sort of adaptive mesh is needed to resolve the fine structure of the solution close to the fronts. In Figure 5.11 we present the solution to the Eady problem close to the formation of a severe tropical storm, obtained by using a pressure-correction method on a semi-staggered grid, looking at the contours in the horizontal and vertical coordinates (x, z) of the longitudinal wind speed v and potential temperature θ.

We consider two different monitor functions coupled to PMA to find adaptive meshes for this problem. In the first case we take the arclength monitor function

$$M_1 = \sqrt{1 + |\nabla v|^2}.$$

In the second case we take the monitor function M_2 to be an estimate of the *potential vorticity* q of the solution, so that M_2 is taken to be the maximum eigenvalue of the matrix

$$Q = \begin{pmatrix} v_x + f & v_z \\ \theta_x & \theta_z \end{pmatrix},$$

for which $q = \det(Q)$. The resulting meshes are shown in Figure 5.12. In both cases we see well-structured and regular meshes with good resolution at the boundaries and of the front, and with no evidence of mesh tangling. However, use of the potential vorticity monitor function M_2 leads to a mesh which more precisely follows the physical solution. Further details are given in Walsh et al. (2009).

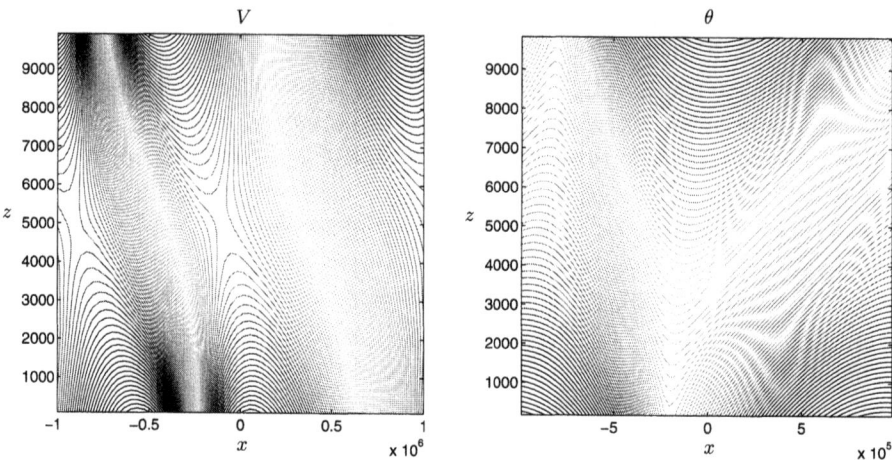

Figure 5.11. The (x, z) contours of the longitudinal velocity and potential temperature of the Eady problem close to the formation of a tropical storm.

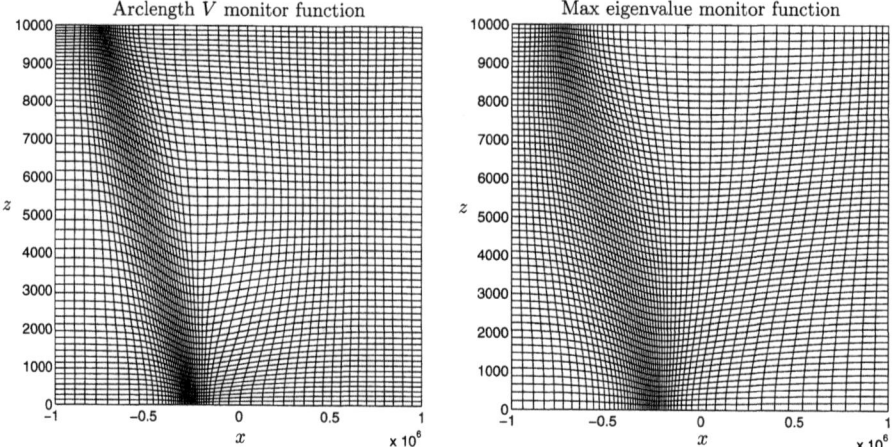

Figure 5.12. Two meshes adapted to the solution of the Eady problem presented above. In the first figure we use the arclength monitor function M_1 and in the second the potential vorticity monitor function M_2.

Acknowledgements

It is a pleasure to thank Emily Walsh for Figures 5.7, 5.8, 5.9, 5.11 and 5.12 and J. F. Williams for Figures 3.7, 3.8, 3.9, 3.11, 5.2 and 5.4. Both also helped by reading this text and through many useful discussions. This work was supported in part by the EPSRC Critical Mass Grant GR/586525/01.

REFERENCES

S. Adjerid and J. E. Flaherty (1986), 'A moving finite element method with error estimation and refinement for one-dimensional time dependent partial differential equations', *SIAM J. Numer. Anal.* **23**, 778–795.

M. Ainsworth and J. T. Oden (2000), *A Posteriori Error Estimation in Finite Element Analysis*, Pure and Applied Mathematics, Wiley-Interscience.

V. F. Almeida (1999), 'Domain deformation mapping: Application to variational mesh generation', *SIAM J. Sci Comput.* **20**, 1252–1275.

D. A. Anderson and M. M. Rai (1983), The use of solution adaptive grids in solving partial differential equations, in *Numerical Grid Generation* (J. H. Thompson, ed.), pp. 317–338.

V. B. Andreev and N. B. Kopteva (1998), 'On the convergence, uniform with respect to a small parameter, of monotone three-point difference approximations', *Diff. Urav.* **34**, 921–929.

U. Ascher, J. Christiansen and R. D. Russell (1981), 'Collocation software for boundary value ODEs', *ACM Trans. Math. Software* **7**, 209–222.

I. Babuška and W. C. Rheinboldt (1979), 'Analysis of optimal finite element meshes in $\mathbb{R}^1$', *Math. Comput.* **33**, 435–463.

M. J. Baines (1994), *Moving Finite Elements*, Clarendon Press, Oxford.

M. J. Baines and S. L. Wakelin (1991), Equidistribution and the Legendre transformation. Numerical Analysis report 4/91, University of Reading.

M. J. Baines, M. E. Hubbard, and P. K. Jimack (2005), 'A moving mesh finite strategy for the adaptive solution of time-dependent partial differential equations with moving boundaries', *Appl. Numer. Math.* **54**, 450–469.

M. J. Baines, M. E. Hubbard, P. K. Jimack, and A. C. Jones (2006), 'Scale-invariant moving finite elements for nonlinear partial differential equations in two dimensions', *Appl. Numer. Math.* **56**, 230–252.

M. L. Balinski (1986), 'A competitive (dual) simplex method for the assignment problem', *Math. Program.* **34**, 125–141.

R. E. Bank and R. K. Smith (1997), 'Mesh smoothing using *a posteriori* error estimates', *SIAM J. Numer. Anal.* **34**, 979–997.

G. I. Barenblatt (1996), *Scaling, Self-Similarity, and Intermediate Asymptotics: Dimensional Analysis and Intermediate Asymptotics*, Cambridge Texts in Applied Mathematics, Cambridge University Press.

G. Beckett and J. A. Mackenzie (2000), 'Convergence analysis of finite-difference approximations on equidistributed grids to a singularly perturbed boundary value problem', *Appl. Numer. Math.* **35**, 87–109.

G. Beckett and J. A. Mackenzie (2001a), 'On a uniformly accurate finite difference approximation of a singularly perturbed reaction–diffusion problem using grid equidistribution', *J. Comput. Appl. Math.* **131**, 381–405.

G. Beckett and J. A. Mackenzie (2001b), 'Uniformly convergent high order finite element solutions of a singularly perturbed reaction–diffusion equation using mesh equidistribution', *Appl. Numer. Math.* **39**, 31–45.

G. Beckett, J. A. Mackenzie and M. L. Robertson (2001a), 'A moving mesh finite element method for the solution of two-dimensional Stefan problems', *J. Comput. Phys.* **186**, 500–518.

G. Beckett, J. A. Mackenzie, A. Ramage and D. M. Sloan (2001b), 'On the numerical solution of one-dimensional PDEs using adaptive methods based on equidistribution', *J. Comput. Phys.* **167**, 372–392.

J. D. Benamou and Y. Brenier (2000), 'A computational fluid mechanics solution to the Monge–Kantorovich mass transfer problem', *Numer. Math.* **84**, 375–393.

M. Berger and R. Kohn (1988), 'A rescaling algorithm for the numerical calculation of blowing-up solutions', *Comm. Pure. Appl. Math.* **41**, 841–863.

M. Berzins (1998), 'A solution-based triangular and tetrahedral mesh quality indicator', *SIAM J. Sci. Comput.* **19**, 2051–2060.

J. G. Blom and J. G. Verwer (1989), On the use of the arclength and curvature monitor in a moving grid method which is based on the method of lines. Technical Report NM-N8902, CWI, Amsterdam.

P. Bochev, G. Liao and G. d. Pena (1996), 'Analysis and computation of adaptive moving grids by deformation', *Numer. Methods PDEs* **12**, 489–506.

C. de Boor (1973), *Good Approximations by Splines with Variable Knots II*, Vol. 363 of *Lecture Notes in Mathematics*, Springer, Berlin.

J. U. Brackbill (1993), 'An adaptive grid with directional control', *J. Comput. Phys.* **108**, 38–50.

J. U. Brackbill and J. S. Saltzman (1982), 'Adaptive zoning for singular problems in two dimensions', *J. Comput. Phys.* **46**, 342–368.

L. Branets and G. F. Carey (2003), A local cell quality metric and variational grid smoothing algorithm, in *Proc. 12th International Meshing Roundtable*, Sandia National Laboratories, Albuquerque, NM.

Y. Brenier (1991), 'Polar factorization and monotone rearrangement of vector-valued functions', *Comm. Pure Appl. Math.* **44**, 375–417.

C. J. Budd and V. A. Dorodnitsyn (2001), 'Symmetry adapted moving mesh schemes for the nonlinear Schrödinger equation', *J. Phys. A* **34**, 103887–10400.

C. J. Budd and M. D. Piggott (2005), Geometric integration and its applications, in *Handbook of Numerical Analysis* (F. Cucker, ed.), pp. 35–139.

C. J. Budd and J. F. Williams (2006), 'Parabolic Monge–Ampère methods for blow-up problems in several spatial dimensions', *J. Phys. A* **39**, 5425–5444.

C. J. Budd and J. F. Williams (2009), Mesh generation using the parabolic Monge–Ampère method. Submitted.

C. J. Budd, W. Z. Huang, and R. D. Russell (1996), 'Moving mesh methods for problems with blow-up', *SIAM J. Sci. Comput.* **17**, 305–327.

C. J. Budd, S.-N. Chen and R. D. R. Russell (1999a), 'New self-similar solutions of the nonlinear Schrödinger equation, with moving mesh computations', *J. Comput. Phys.* **152**, 756–789.

C. J. Budd, G. J. Collins, W.-Z. Huang and R. D. Russell (1999b), 'Self-similar discrete solutions of the porous medium equation', *Philos. Trans. Roy. Soc. London A* **357**, 1047–1078.

C. J. Budd, B. Leimkuhler, and M. D. Piggott (2001), 'Scaling invariance and adaptivity', *Appl. Numer. Math.* **39**, 261–288.

C. J. Budd, V. A. Galaktionov and J. F. Williams (2004), 'Self-similar blow-up in higher-order semilinear parabolic equations', *SIAM J. Appl. Math.* **64**, 1775–1809.

C. J. Budd, R. Carretero-Gonzalez, and R. D. Russell (2005), 'Precise computations of chemotactic collapse using moving mesh methods', *J. Comput. Phys.* **202**, 462–487.

C. J. Budd, M. D. Piggott, and J. F. Williams (2009), Adaptive numerical methods and the geostrophic coordinate transformation. Submitted to *Monthly Weather Review*.

L. A. Caffarelli (1992), 'The regularity of mappings with a convex potential', *J. Amer. Math. Soc.* **5**, 99–104.

L. A. Caffarelli (1996), 'Boundary regularity of maps with convex potentials', *Ann. of Math.* **3**, 453–496.

X. Cai, D. Fleitas, B. Jiang and G. Liao (2004), 'Adaptive grid generation based on the least squares finite element method', *Comput. Math. Appl.* **48**, 1077–1085.

D. A. Calhoun, C. Helzel and R. J. LeVeque (2008), 'Logically rectangular grids and finite volume methods for PDEs in circular and spherical domains', *SIAM Review* **50**, 723–752.

W. Cao (2005), 'On the error of linear interpolation and the orientation, aspect ratio and internal angles of a triangle', *SIAM J. Numer. Anal.* **43**, 19–40.

W. Cao (2007a), 'An interpolation error estimate on anisotropic meshes in $\mathbb{R}^n$ and optimal metrics for mesh refinement', *SIAM J. Numer. Anal.* **45**, 2368–2391

W. Cao (2007b), 'Anisotropic measures of third order derivatives and the quadratic interpolation error on triangular elements', *SIAM J. Sci. Comput.* **29**, 756–781.

W. Cao (2008), 'An interpolation error estimate in $\mathbb{R}^2$ based on the anisotropic measures of higher order derivatives', *Math. Comput.* **77**, 265–286.

W. Cao, W. Huang, and R. D. Russell (1999a), 'An r-adaptive finite element method based upon moving mesh PDEs', *J. Comput. Phys.* **149**, 221–244.

W. Cao, W. Huang, and R. D. Russell (1999b), 'A study of monitor functions for two-dimensional adaptive mesh generation', *SIAM J. Sci. Comput.* **20**, 1978–1994.

W. Cao, W. Huang, and R. D. Russell (2002), 'A moving mesh method based on the geometric conservation law', *SIAM J. Sci. Comput.* **24**, 118–142.

W. Cao, W. Huang, and R. D. Russell (2003), 'Approaches for generating moving adaptive meshes: Location versus velocity', *Appl. Numer. Math.* **47**, 121–138.

M. Capiński and E. Kopp (2004), *Measure, Integral and Probability*, Springer Undergraduate Mathematics Series, Springer.

G. Carey (1997), *Computational Grids: Generation, Adaptation and Solution Strategies*, Taylor and Francis.

N. Carlson and K. Miller (1998a), 'Design and application of a gradient-weighted moving finite element code I: In 1-D', *SIAM J. Sci. Comput.* **19**, 728–765.

N. Carlson and K. Miller (1998b), 'Design and application of a gradient-weighted moving finite element code II: In 2-D', *SIAM J. Sci. Comput.* **19**, 766–798.

H. D. Ceniceros (2002), 'A semi-implicit moving mesh method for the focusing nonlinear Schrödinger equation', *Comm. Pure Appl. Anal.* **4**, 1–14.

H. D. Ceniceros and T. Y. Hou (2001), 'An efficient dynamically adaptive mesh for potentially singular solutions', *J. Comput. Phys.* **172**, 609–639.

L. Chacón and G. Lapenta (2006), 'A fully implicit, nonlinear adaptive grid strategy', *J. Comput. Phys.* **212**, 703–717.

R. Chartrand, K. R. Vixie, B. Wohlberg and E. M. Bollt (2007), A gradient descent solution to the Monge–Kantorovich problem.
math.lanl.gov/Research/Publications/Docs/chartrand-2007-gradient.pdf.

K. Chen (1994), 'Error equidistribution and mesh adaptation', *SIAM J. Sci. Comput.* **15**, 798–818.

L. Chen, P. Sun and J. Xu (2007), 'Optimal anisotropic meshes for minimizing interpolation errors in the L^p-norm', *Math. Comput.* **76**, 179–204.

S. Chynoweth and M. J. Baines (1989), Legendre transform solutions to semi-geostrophic frontogenesis, in *Finite Element Analysis in Fluids* (T. J. Chung and G. R. Kerr, eds), pp. 697–703.

S. Chynoweth and M. J. Sewell (1989), 'Dual variables in semigeostrophic theory', *Proc. R. Soc. London A* **424**, 155–186.

M. J. P. Cullen (1989), 'Implicit finite difference methods for modelling discontinuous atmospheric flows', *J. Comput. Phys.* **81**, 319–348.

M. J. P. Cullen (2006), *A Mathematical Theory of Large-Scale Atmosphere/Ocean Flow*, Imperial College Press.

M. J. P. Cullen and R. J. Purser (1984), 'An extended Lagrangian theory of semigeostrophic frontogenesis', *J. Atmos. Sci.* **41**, 1477–1497.

M. J. P. Cullen, J. Norbury, and R. J. Purser (1991), 'Generalised Lagrangian solutions for atmospheric and oceanic flows', *SIAM J. Appl. Math.* **51**, 20–31.

B. Dacorogna and J. Moser (1990), 'On a partial differential equation involving the Jacobian determinant', *Ann. Inst. Henri Poincaré Analyse non linéaire* **7**, 1–26.

E. Dean and R. Glowinski (2003), 'Numerical solution of the two-dimensional elliptic Monge–Ampère equation with Dirichlet boundary conditions: An augmented Lagrangian approach', *Comptes rendus Mathématique* **336**, 779–784.

E. Dean and R. Glowinski (2004), 'Numerical solution of the two-dimensional elliptic Monge–Ampère equation with Dirichlet boundary conditions: A least-squares approach', *Comptes rendus Mathématique* **339**, 887–892.

G. Delzanno, L. Chacón, J. Finn, Y. Chung and G. Lapenta (2008), 'An optimal robust equidistribution method for two-dimensional grid adaptation based on Monge–Kantorovich optimization', *J. Comput. Phys.* **227**, 9841–9864.

Y. Di, R. Li, T. Tang and P. Zhang (2005), 'Moving mesh finite element methods for the incompressible Navier–Stokes equations', *SIAM J. Sci. Comput.* **26**, 1036–1056.

E. A. Dorfi and L. O'C. Drury (1987), 'Simple adaptive grids for 1-D initial value problems', *J. Comput. Phys.* **69**, 175–195.

V. A. Dorodnitsyn (1991), 'Transformation groups in mesh spaces', *J. Sov. Math.* **55**, 1490–1517.

V. A. Dorodnitsyn (1993a), Finite-difference models exactly inheriting symmetry of original differential equations, in *Modern Group Analysis: Advanced Analytical and Computational Methods in Mathematical Physics* (N. Ibragimov et al., eds), Kluwer, Dordrecht, pp. 191–201.

V. A. Dorodnitsyn (1993b), 'Finite difference analog of the Noether theorem', *Dokl. Akad. Nauk* **328**, 678–690.

V. A. Dorodnitsyn and R. Kozlov (1997), The whole set of symmetry preserving discrete versions of a heat transfer equation with a source. Preprint 4/1997, NTNU, Trondheim.

A. S. Dvinsky (1991), 'Adaptive grid generation from harmonic maps on Riemannian manifolds', *J. Comput. Phys.* **95**, 450–476.

P. R. Eisman (1985), 'Grid generation for fluid mechanics computation', *Ann. Rev. Fluid Mech.* **17**, 487–522.

P. R. Eisman (1987), 'Adaptive grid generation', *Comput. Meth. Appl. Mech. Engrg* **64**, 321–376.

L. C. Evans (1999), Partial differential equations and Monge–Kantorovich mass transfer, in *Current Developments in Mathematics, 1997* (Cambridge, MA), International Press, Boston, MA, pp. 65–126.

X. Feng and M. Neilan (2009), 'Vanishing moment method and moment solutions for fully nonlinear second order partial differential equations', *J. Sci. Comput.*, to appear.

W. M. Feng, P. Yu, S. Y. Hu, Z. K. Liu, Q. Du and L. Q. Chen (2006), 'Spectral implementation of an adaptive moving mesh method for phase-field equations', *J. Comput. Phys.* **220**, 498–510.

S. Fulton (1989), 'Multigrid solution of the semigeostrophic invertibility relation', *Monthly Weather Review* **117**, 2059–2066.

W. Gangbo and R. J. McCann (1996), 'The geometry of optimal transport', *Acta Math.* **177**, 113–161.

C. E. Gutiérrez (2001), *The Monge–Ampère Equation*, Vol. 44 of *Progress in Nonlinear Differential Equations and their Applications*, Birkhäuser, Boston, MA.

S. Haker and A. Tannenbaum (2003), On the Monge–Kantorovich problem and image warping, in *Mathematical Methods in Computer Vision*, Vol. 133 of *IMA Vol. Math. Appl.*, Springer, New York, pp. 65–85.

D. F. Hawken, J. J. Gottlieb and J. S. Hansen (1991), 'Review of some adaptive node-movement techniques in finite element and finite difference solutions of PDEs', *J. Comput. Phys.* **95**, 254–302.

Y. He and W. Huang (2009), *A posteriori* error analysis for finite element solution of elliptic differential equations using equidistributing meshes. Submitted.

W. Huang (2001a), 'Practical aspects of formulation and solution of moving mesh partial differential equations', *J. Comput. Phys.* **171**, 753–775.

W. Huang (2001b), 'Variational mesh adaption: Isotropy and equidistribution', *J. Comput. Phys.* **174**, 903–924.

W. Huang (2005a), 'Measuring mesh qualities and application to variational mesh adaption', *SIAM J. Sci. Comput.* **26**, 1643–1666.

W. Huang (2005b), 'Metric tensors for anisotropic mesh generation', *J. Comput. Phys.* **204**, 663–665.

W. Huang (2005c), 'Convergence analysis of finite element solution of one-dimensional singularly perturbed differential equations on equidistributing meshes', *Internat. J. Numer. Anal. Model.* **2**, 57–74.

W. Huang (2007), Anisotropic mesh adaption and movement, in *Adaptive Computations: Theory and Algorithms* (T. Tang and J. Xu, eds), Science Press, Beijing, pp. 68–158.

W. Huang and B. Leimkuhler (1997), 'The adaptive Verlet method', *SIAM J. Sci. Comput.* **18**, 239–256.

W. Huang and X. P. Li (2009), 'An anisotropic mesh adaptation method for the finite element solution of variational problems', *Finite Elements in Analysis and Design*, to appear.

W. Huang and R. D. Russell (1996), 'A moving collocation method for solving time dependent partial differential equations', *Appl. Numer. Math.* **20**, 101–116.

W. Huang and R. D. Russell (1997a), 'Analysis of moving mesh partial differential equations with spatial smoothing', *SIAM J. Numer. Anal.* **34**, 1106–1126.

W. Huang and R. D. Russell (1997b), 'A high dimensional moving mesh strategy', *Appl. Numer. Math.* **26**, 63–76.

W. Huang and R. D. Russell (1999), 'A moving mesh strategy based on a gradient flow equation for two-dimensional problems', *SIAM J. Sci. Comput.* **20**, 998–1015.

W. Huang and R. D. Russell (2001) 'Adaptive mesh movement: The MMPDE approach and its applications', *J. Comput. Appl. Math.* **128**, 383–398.

W. Huang and D. Sloan (1994), 'A simple adaptive grid method in two dimensions', *SIAM J. Sci. Comput.* **15**, 776–797.

W. Huang and W. Sun (2003), 'Variational mesh adaption II: Error estimates and monitor functions', *J. Comput. Phys.* **184**, 619–648.

W. Huang and X. Zhan (2004), Adaptive moving mesh modeling for two dimensional groundwater flow and transport, in *Recent Advances in Adaptive Computation*, Vol. 383 of *Contemporary Mathematics*, AMS, pp. 283–296.

W. Huang, Y. Ren, and R. D. Russell (1994), 'Moving mesh partial differential equations (MMPDEs) based on the equidistribution principle', *SIAM J. Numer. Anal.* **31**, 709–730.

W. Huang, L. Zheng and X. Zhan (2002), 'Adaptive moving mesh methods for simulating one-dimensional groundwater problems with sharp moving fronts', *Internat. J. Numer. Meth. Engng* **54**, 1579–1603.

W. Huang, J. Ma and R. D. Russell (2008), 'A study of moving mesh PDE methods for numerical simulation of blowup in reaction diffusion equations', *J. Comput. Phys.* **227**, 6532–6552.

W. Huang, L. Kamenski and J. Lang (2009), Anisotropic mesh adaptation based upon *a posteriori* error estimates. Submitted.

J. M. Hyman and B. Larrouturou (1986), Dynamic rezone methods for partial differential equations in one space dimension. Technical Report LA-UR-86-1678, Los Alamos National laboratory, Los Alamos, NM.

J. M. Hyman and B. Larrouturou (1989), 'Dynamic rezone methods for partial

differential equations in one space dimension', *Appl. Numer. Math.* **5**, 435–450.
O.-P. Jacquotte (1988), 'A mechanical model for a new grid generation method in computational fluid dynamics', *Comput. Methods Appl. Mech. Engrg* **66**, 323–338.
O.-P. Jacquotte and G. Coussement(1992), 'Structured mesh adaption: Space accuracy and interpolation methods', *Comput. Methods Appl. Mech. Engrg* **101**, 397–432.
C. Johnson (1987), *Numerical Solution of Partial Differential Equations by the Finite Element Method*, Cambridge University Press.
T. Kaijser (1998), 'Computing the Kantorovich distance for images', *J. Math. Imaging Vision* **9**, 173–191.
J. Kautsky and N. K. Nichols (1980), 'Equidistributing meshes with constraints', *SIAM J. Sci. Statist. Comput.* **1**, 499–511.
J. Kautsky and N. K. Nichols (1982), 'Smooth regrading of discretized data', *SIAM J. Sci. Statist. Comput.* **3**, 145–159.
P. M. Knupp (1995), 'Mesh generation using vector fields', *J. Comput. Phys.* **119**, 142–148.
P. M. Knupp (1996), 'Jacobian-weighted elliptic grid generation', *SIAM J. Sci. Comput.* **17**, 1475–1490.
P. M. Knupp (2001), 'Algebraic mesh quality metrics', *SIAM J. Sci. Comput.* **23**, 193–218.
P. Knupp and N. Robidoux (2000), 'A framework for variational grid generation: Conditioning the Jacobian matrix with matrix norms', *SIAM J. Sci. Comput.* **21**, 2029–2047.
P. Knupp and S. Steinberg (1994), *Fundamentals of Grid Generation*, CRC Press, Boca Raton.
P. M. Knupp, L. Margolin and M. Shashkov (2002), 'Reference Jacobian optimization-based rezoning strategies for arbitrary Lagrangian Eulerian methods', *J. Comput. Phys.* **176**, 93–128.
N. Kopteva (2007), Convergence theory of moving grid methods, in *Adaptive Computations: Theory and Algorithms* (T. Tang and J. Xu, eds), Science Press, Beijing, pp. 159–210.
N. Kopteva and M. Stynes (2001), 'A robust adaptive method for a quasilinear one-dimensional convection–diffusion problem', *SIAM J. Numer. Anal.* **39**, 1446–1467.
R. Kozlov (2000), Symmetry applications to difference and differential-difference equations. PhD Thesis, Institut for matematiske fag, NTNU, Trondheim.
J. Lang, W. Cao, W. Huang and R. D. Russell (2003), 'A two-dimensional moving finite element method with local refinement based on *a posteriori* error estimates', *Appl. Numer. Math.* **46**, 75–94.
G. Lapenta and L. Chacón (2006), 'Cost-effectiveness of fully implicit moving mesh adaptation: A practical investigation in 1D', *J. Comput. Phys.* **219**, 86–103.
R. J. LeVeque (1990), *Numerical Methods for Conservation Laws*, Birkhäuser.
R. Li, T. Tang, and P.-W. Zhang (2002), 'A moving mesh finite element algorithm for singular problems in two and three space dimensions', *J. Comput. Phys.* **177**, 365–393.

S. T. Li and L. R. Petzold (1997), 'Moving mesh methods with upwinding schemes for time dependent PDEs', *J. Comput Phys.* **131**, 368–377.

S. T. Li, L. R. Petzold and Y. Ren (1998), 'Stability of moving mesh systems of partial differential equations', *SIAM J. Sci. Comput.* **20**, 719–738.

G. Liao and D. Anderson (1992), 'A new approach to grid generation', *Appl. Anal.* **44**, 285–297.

G. Liao and J. Xue (2006), 'Moving meshes by the deformation method', *J. Comput. Appl. Math.* **195**, 83–92.

V. D. Liseikin (1999), *Grid Generation Methods*, Springer, Berlin.

A. Liu and B. Joe (1994), 'Relationship between tetrahedron quality measures', *BIT* **34**, 268–287.

J. Mackenzie (1999), 'Uniform convergence analysis of an upwind finite-difference approximation of a convection–diffusion boundary value problem on an adaptive grid', *IMA J. Numer. Anal.* **19**, 233–249.

J. A. Mackenzie and W. R. Mekwi (2007a), On the use of moving mesh methods to solve PDEs, in *Adaptive Computations: Theory and Algorithms* (T. Tang and J. Xu, eds), Science Press, Beijing, pp. 242–278.

J. A. Mackenzie and W. R. Mekwi (2007b), 'An analysis of stability and convergence of a finite-difference discretization of a model parabolic PDE in 1D using a moving mesh', *IMA J. Numer. Anal.* **27**, 507–528.

J. A. Mackenzie and M. L. Robertson (2002), 'A moving mesh method for the solution of the one-dimensional phase-field equations', *J. Comput. Phys.* **181**, 526–544.

R. I. McLachlan (1994), 'Symplectic integration of Hamiltonian wave equations', *Numer. Math.* **66**, 465–492.

A. Marquina (1994), 'Local piecewise hyperbolic resolution of numerical fluxes for nonlinear scalar conservation laws', *SIAM J. Sci. Comput.* **15**, 894–904.

C. T. Miller, S. N. Gleyzer and P. T. Imhoff (1998), Numerical modeling of NAPL dissolution fingering in porous media, in *Physical Nonequilibrium in Soils: Modeling and Application* (H. M. Selim and L. Ma, eds), Ann Arbor Press.

K. Miller (1981), 'Moving finite elements II', *SIAM J. Numer. Anal.* **18**, 1033–1057.

K. Miller and R. N. Miller (1981), 'Moving finite elements I', *SIAM J. Numer. Anal.* **18**, 1019–1032.

P. K. Moore and J. E. Flaherty (1992), 'Adaptive local overlapping grid methods for parabolic system in two space dimensions', *J. Comput. Phys.* **98**, 54–63.

J. Moser (1965), 'On the volume elements of a manifold', *Trans. Amer. Math. Soc.* **120**, 286–294.

L. S. Mulholland, W. Huang and D. M. Sloan (1998), 'Pseudospectral solution of near-singular problems using numerical coordinate transformations based on adaptivity', *SIAM J. Sci. Comput.* **19**, 1261–1298.

N. Nakamura (1994), 'Nonlinear equilibriation of two-dimensional Eady waves', Simulations with viscous geostrophic momentum equations', *J. Atmos. Sci.* **51**, 1023–1035.

V. I. Oliker and L. D. Prussner (1988), 'On the numerical solution of the equation $(\partial^2 z/\partial x^2)(\partial^2 z/\partial y^2) - ((\partial^2 z/\partial x \partial y))^2 = f$ and its discretizations I', *Numer. Math.* **54**, 271–293.

P. J. Olver (1986), *Applications of Lie Groups to Differential Equations*, Springer, New York.

L. R. Petzold (1982), A description of DASSL: A differential/algebraic system solver. Technical report SAND82-8637, Sandia National Labs, Livermore, CA.

L. R. Petzold (1987), 'Observations on an adaptive moving grid method for one-dimensional systems for partial differential equations', *Appl. Numer. Math.* **3**, 347–360.

J. Pryce (1989), 'On the convergence of iterated remeshing', *IMA J. Numer. Anal.* **9**, 315–335.

Y. Qiu and D. M. Sloan (1998), 'Numerical solution of Fisher's equation using a moving mesh method', *J. Comput. Phys.* **146**, 726–746.

Y. Qiu and D. M. Sloan (1999), 'Analysis of difference approximations to a singularly perturbed two-point boundary value problem on an adaptively generated grid', *J. Comput. Appl. Math.* **101**, 1–25.

Y. Qiu, D. M. Sloan and T. Tang (2000), 'Numerical solution of a singularly perturbed two-point boundary value problem using equidistribution: Analysis of convergence', *J. Comput. Appl. Math.* **116**, 121–143.

S. T. Rachev and L. Rüschendorf (1998), *Mass Transportation Problems I: Theory*, Probability and its Applications, Springer, New York.

W. Ren and X. Wang (2000), 'An iterative grid redistribution method for singular problems in multiple dimensions', *J. Comput. Phys.* **159**, 246–273.

Y. G. Reshetnyak (1989), *Space Mappings with Bounded Distortion*, Vol. 73 of Translations of Mathematical Monographs, AMS, Providence, RI.

R. D. Russell, J. F. Williams, and X. Xu (2007), 'MOVCOL4: A moving mesh code for fourth-order time-dependent partial differential equations', *SIAM J. Sci. Comput.* **29**, 197–220

A. A. Samarskii, V. A. Galaktionov, S. P. Kurdyumov and A. P. Mikhailov (1995), *Blow-up in Quasilinear Parabolic Equations*, Vol. 19 of De Gruyter Expositions in Mathematics, Walter de Gruyter.

G. Sapiro (2003), Introduction to partial differential equations and variational formulations in image processing, in *Foundations of Computational Mathematics* (F. Cucker, ed.), Vol. 1, pp. 383–461.

P. Saucez, A. Vande Vouwer and P. A. Zegeling (2005), 'Adaptive method of lines solutions for the extended fifth order Korteveg–De Vries equation', *J. Comput. Math.* **183**, 343–357.

B. Semper and G. Liao (1995), 'A moving grid finite-element method using grid deformation', *Numer. Methods in PDEs* **11**, 603–615.

M. J. Sewell (1978), 'On Legendre transformations and umbilic catastrophes', *Math. Proc. Camb. Phil. Soc.* **83**, 273–288.

M. J. Sewell (2002), Some applications of transformation theory in mechanics, in *Large Scale Atmosphere–Ocean Dynamics*, Vol. II (J. Norbury and I. Roulstone, eds), Cambridge University Press, pp. 143–223.

R. Shewchuk (2002), Constrained Delaunay tetrahedralizations and provably good boundary recovery, in *IMR 2002*, Sandia National Laboratories, pp. 193–204.

G. E. Shilov and B. L. Gurevich (1978), *Integral, Measure and Derivative: A Unified Approach*, Dover.

J. H. Smith (1996), Analysis of moving mesh methods for dissipative partial differential equations. PhD Thesis, Department of Computer Science, Stanford University.

J. Stockie, J. A. Mackenzie, and R. D. Russell (2000), 'A moving mesh method for one-dimensional hyperbolic conservation laws', *SIAM J. Sci. Comput.* **22**, 1791–1813.

C. Sulem and P. L. Sulem (1999), *The Nonlinear Schrödinger Equation: Self-Focusing and Wave Collapse*, Springer.

M. H. M. Sulman (2008) Optimal mass transport for adaptivity and image registration. PhD Thesis, Simon Fraser University.

Z. Tan (2007), 'Adaptive moving mesh methods for two-dimensional resistive magneto-hydrodynamic PDE models', *Computers and Fluids* **36**, 758–771.

Z. Tan, K. M. Lim and B. C. Khoo (2007), 'An adaptive mesh redistribution method for the incompressible mixture flows using phase-field model', *J. Comput. Phys.* **225**, 1137–1158.

H. Z. Tang and T. Tang (2003), 'Adaptive mesh methods for one- and two-dimensional hyperbolic conservation laws', *SIAM J. Numer. Anal.* **41**, 487–515.

T. Tang (2005), Moving mesh methods for computational fluid dynamics, in *Recent Advances in Adaptive Computations*, Vol. 383 of *Contemporary Mathematics*, AMS, pp. 141–173.

T. Tang and J. Xu, eds (2007), *Adaptive Computations: Theory and Algorithms*, Science Press, Beijing.

T. W. Tee and L. N. Trefethen (2006), 'A rational spectral collocation method with adaptively transformed Chebyshev grid points', *SIAM J. Sci. Comput.* **28**, 1798–1811.

J. F. Thompson (1985), 'A survey of dynamically-adaptive grids in the numerical solution of partial differential equations', *Appl. Numer. Math.* **1**, 3–27.

J. F. Thompson and N. P. Weatherill (1992), 'Structured and unstructured grid generation', *Critical Reviews Biomed. Eng.* **20**, 73–120.

J. F. Thompson, Z. U. A. Warsi, and C. W. Mastin (1982), 'Boundary-fitted coordinate systems for numerical solution of partial differential equations: A review', *J. Comput. Phys.* **47**, 1–108.

J. F. Thompson, Z. U. A. Warsi, and C. W. Mastin (1985), *Numerical Grid Generation*, North-Holland.

Y. Touringy and F. Hülseman (1998), 'A new moving mesh algorithm for the finite element solution of variational problems', *SIAM J. Numer. Anal.* **35**, 1416–1438.

A. E. P. Veldman and K. Rinzema (1992), 'Playing with nonuniform grids', *J. Engrg Math.* **26**, 119–130.

J. G. Verwer, J. G. Blom, R. M. Furzeland and P. A. Zegeling (1989), A moving-grid method for one-dimensional PDEs based on the method of lines, in *Adaptive Methods for Partial Differential Equations* (J. E. Flaherty, P. J. Paslow, M. S. Shepard and J. D. Vasilakis, eds), SIAM, Philadelphia, pp. 160–175.

C. Villani (2003), *Topics in Optimal Transportation*, Vol. 58 of *Graduate Studies in Mathematics*, AMS.

E. Walsh, C. J. Budd and J. F. Williams (2009), The PMA method for grid generation applied to the Eady problem in meteorology. University of Bath report.

L.-L. Wang and J. Shen (2005), 'Error analysis for mapped Jacobi spectral methods', *J. Sci. Comput.* **24**, 183–218.

A. J. Wathen and M. J. Baines (1985), 'On the structure of the moving finite-element equations', *IMA J. Numer. Anal.* **5**, 161–182.

A. M. Winslow (1967), 'Numerical solution of the quasilinear Poisson equation in a nonuniform triangle mesh', *J. Comput. Phys.* **2**, 149–172.

A. M. Winslow (1981), Adaptive mesh rezoning by the equipotential method. Technical report UCID-19062, Lawrence Livermore Lab.

X. Xu, W.-H. Huang, R. D. Russell and J. F. Williams (2009), Convergence of de Boor's algorithm for generation of equidistributing meshes. Submitted.

N. N. Yanenko, E. A. Kroshko, V. V. Liseikin, V. M. Fomin, V. P. Shapeev and Y. A. Shitov (1976), *Methods for the Construction of Moving Grids for Problems of Fluid Dynamics with Big Deformations*, Vol. 59 of *Lecture Notes in Physics*, Springer.

P. A. Zegeling (1993), Moving-grid methods for time-dependent partial differential equations. CWI Tract 94.

P. A. Zegeling (2005), 'On resistive MHD models with adaptive moving meshes', *J. Sci. Comput.* **24**, 263–284.

P. A. Zegeling (2007), Theory and application of adaptive moving grid methods, in *Adaptive Computations: Theory and Algorithms*, Science Press, Beijing, pp. 279–332.

P. A. Zegeling and H. P. Kok (2004), 'Adaptive moving mesh computations for reaction–diffusion systems', *J. Comput. Appl. Math.* **168**, 519–528.

Z.-R Zhang and T. Tang (2002), 'An adaptive mesh redistribution algorithm for convection-dominated problems', *Comm. Pure Appl. Anal.* **1**, 341–357

B. Zitova and J. Flusser (2003), 'Image registration methods: A survey', *Image and Vision Comput.* **21**, 977–1000.

M. Zlamal (1968), 'On the finite element method', *Numer. Math.* **12**, 394–409.

Fast direct solvers for integral equations in complex three-dimensional domains

Leslie Greengard*
*Courant Instiute of Mathematical Sciences, New York University,
New York, NY 10012, USA
E-mail: greengard@courant.nyu.edu*

Denis Gueyffier[†]
*Courant Institute of Mathematical Sciences, New York University,
New York, NY 10012, USA
E-mail: dgueyffier@giss.nasa.gov*

Per-Gunnar Martinsson[‡]
*Department of Applied Mathematics, University of Colorado at Boulder,
526 UCB, Boulder, CO 80309-0526, USA
E-mail: per-gunnar.martinsson@colorado.edu*

Vladimir Rokhlin[§]
*Department of Mathematics and Department of Computer Science,
Yale University, 10 Hillhouse Avenue, New Haven CT 06511, USA
E-mail: rokhlin@cs.yale.edu*

---

* This work was supported in part by the Applied Mathematical Sciences Program of the US Department of Energy under Contract DEFG0200ER25053.
[†] This work was supported by DARPA through the Protein Design Processes Program (DSO contract HR0011-05-1-0044). Present address: NASA GISS and Columbia University, 2880 Broadway, New York, NY 10025.
[‡] This work was supported in part by National Science Foundation grants DMS-0748488 and DMS-0610097.
[§] This work was supported in part by AFOSR grant FA9550-07-1-0541.

Methods for the solution of boundary integral equations have changed significantly during the last two decades. This is due, in part, to improvements in computer hardware, but more importantly, to the development of fast algorithms which scale linearly or nearly linearly with the number of degrees of freedom required. These methods are typically iterative, based on coupling fast matrix-vector multiplication routines with conjugate-gradient-type schemes. Here, we discuss methods that are currently under development for the fast, direct solution of boundary integral equations in three dimensions. After reviewing the mathematical foundations of such schemes, we illustrate their performance with some numerical examples, and discuss the potential impact of the overall approach in a variety of settings.

CONTENTS

1	Introduction	244
2	The rapid application of operators	246
3	Fast direct solvers for structured matrices	252
4	Potential theory for the Poisson equation	254
5	A single-level fast solver	258
6	Numerical results I	262
7	Local geometric perturbations	263
8	Numerical results II	265
9	Conclusions	267
	Appendix: Rapid inversion of operators	268
	References	272

1. Introduction

Modern numerical methods for the solution of boundary integral equations with large numbers of degrees of freedom are currently based on the availability of fast algorithms for matrix-vector multiplication (application of the discretized integral operator). These include fast multipole methods, panel clustering methods, the method of local corrections, multigrid, precorrected-FFT and wavelet-based methods. Briefly stated, discretization of a boundary integral equation yields a dense $N \times N$ matrix, where N denotes the number of degrees of freedom used in describing the surface density defined on the given geometry. As a result, classical solution techniques, based on Gaussian elimination, require $O(N^3)$ work to factor the system matrix. The fast algorithms listed above provide the ability to apply the discretized integral operator to a vector in $O(N)$ or $O(N \log N)$ operations rather than $O(N^2)$. The combination of such a scheme with modern iterative methods – such as GMRES (Saad and Schultz 1986) – and well-conditioned integral equation formulations has reduced the net work required to $O(N \log N)$, bringing large-scale simulations within practical reach. The literature on this subject is now vast, and we refer the reader to only a few relevant publications: Carrier, Greengard and Rokhlin (1988), Chew, Jin, Michielssen and Song (2001), Darve and Have (2004), Greengard and Helsing (1998), Greengard and Rokhlin (1987, 1997), Hackbusch and Nowak (1989), Kapur and Long (1997), Nabors and White (1991), and Nishimura (2002).

Nevertheless, there are a number of areas where significant improvement can still be made. First, iterative methods deal poorly with ill-conditioned problems or with multiple right-hand sides. This is in marked contrast to classical direct methods which, following the $O(N^3)$ factorization step, require only $O(N^2)$ work to solve the linear system for each subsequent right-hand side. A second motivation for direct methods is that they are extremely effective at handling low-rank perturbations of the system matrix. In particular, the solution to the perturbed system can also be obtained in $O(N^2)$ work. An obvious question, then, is whether direct methods can be accelerated in the same way that iterative solvers have been.

In the last few years, a number of groups have been developing fast direct solvers for boundary integral equations in two and three dimensions that first compute a new type of 'compressed' factorization using $O(N^\alpha \log^\beta N)$ operations, with $1 \leq \alpha \leq 2$ and $0 \leq \beta \leq 2$. Application of the factored inverse then typically requires only $O(N)$ or $O(N \log N)$ operations for each right-hand side and/or low-rank perturbation of the system matrix (Canning and Rogovin 1998, Chandrasekaran *et al.* 2006, Gope, Chowdhury and Jandhyala 2005, Hackbusch 1999, Hackbusch and Khoromskij 2000, Martinsson and Rokhlin 2005, Pals 2004, Zhu and White 2005). Earlier work on direct solvers had focused primarily on volume scattering (Chen 2002, Chew 1989) or the one-dimensional or quasi-one-dimensional case (Eidelman and Gohberg 1999, Greengard and Rokhlin 1991, Lee and Greengard 1997, Martinsson and Rokhlin 2007, Michielssen, Boag and Chew 1996, Starr and Rokhlin 1994).

The main purpose of this paper is to state the obvious; fast, direct methods which result in compressed or 'data sparse' representations of the inverse matrix should have a dramatic impact on simulation and design. Stress analysis, electromagnetic analysis (including radar cross-section calculations), biophysical simulation, and a host of other tasks can be formulated in terms of a given integral equation with multiple right-hand sides (Martinsson 2006). Further, when solving time-dependent partial differential equations in a fixed geometry, a computational bottleneck is often the solution of a potential problem such as the Laplace equation with a new right-hand side at each time step. A second motivation for fast direct solvers is that the solution to 'nearby' problems can be computed very efficiently. In particular, if the geometry undergoes a local perturbation, this introduces a low-rank perturbation of the system matrix and, as indicated above, rapid updating becomes feasible.

In the next two sections, we review the basic idea that allows for the construction of fast solvers. We then show how to apply these ideas to dielectric interface problems and the exterior Neumann problem in three dimensions, both governed by the Laplace equation. For this, we extend the approach introduced by Martinsson and Rokhlin (2005) for two-dimensional problems. Our formalism, however, is a bit different and leads to some simplification in both presentation and implementation. We illustrate the power of the method for both molecular electrostatics and incompressible potential flow and conclude with some remarks about future directions of research.

2. The rapid application of operators

For uniformly discretized convolution operators, the problem of excessive cost of applying (or inverting) dense matrices was solved with the development of the Fast Fourier Transform (FFT) and related algorithms (Cooley and Tukey 1965). These methods were discovered during the 1960s, and attain their computational speed by exploiting algebraic properties of the operator. Such schemes are exact in exact arithmetic, and are fragile in the sense that they depend on regularity of the input and output data for their applicability. While the FFT revolutionized digital signal processing and a number of other fields, the ability to handle irregular data is often essential. Moreover, many of the integral operators encountered in mathematical physics are not translation-invariant.

Twenty years later, it was observed that many operators originating from physics admit a different type of 'fast' method. Specifically, the kernels of many such operators are smooth in the far field. When operators of this type are discretized, the resulting matrices contain large submatrices whose rank is low (to high precision). Combining this fact with simple structures from computer science (adaptive quadtrees, oct-trees, *etc.*), it is possible to construct algorithms for the application of such matrices to arbitrary vectors for a cost proportional to N in some situations, and $N \log(N)$ in others. Curiously, the first several algorithms of this type did not use the rank argument explicitly; they were restricted to cases where the operator had some special analytical structure, and used the corresponding special functions, for instance, multipole expansions for the Laplace equation in Greengard and Rokhlin (1987), Hermite polynomials in Greengard and Strain (1991), Laguerre polynomials in Strain (1992), *etc.* Naturally, each of these schemes was limited to a narrow class of operators, though some of them were quite efficient.

Another early group of 'fast' techniques replaced the kernel-dependent special functions with some appropriately chosen bases: Chebyshev poly-

Figure 2.1. A continuous charge distribution acting on a separated interval.

nomials in Alpert and Rokhlin (1991) and Rokhlin (1988), wavelets in Beylkin, Coifman and Rokhlin (1991), wavelet-like objects in Alpert, Beylkin, Coifman and Rokhlin (1993), *etc.* This approach is more general and easier to use, since a single scheme is applicable to a wider class of operators. Fast multipole and FFT-based variants have also been developed in a kernel-independent manner. They rely on sampling the governing Green's function in a systematic fashion and thereby building efficient representations of distant interactions (Phillips and White 1997, Ying, Biros, Zorin and Langston 2003, Gimbutas 1999).

Finally, it was observed that the preceding approaches, which build representations that are universally applicable and do not make use of the detailed distributions of sources, were suboptimal. That is, the numbers of basis elements used significantly exceed the ranks of the specific block matrices being approximated. Optimized low-rank approximations can be computed explicitly (via the SVD, QR, or the more recently introduced interpolative decompositions), and the resulting schemes can be noticeably faster than the original FMMs. Early work in this direction includes Hackbusch (1999) and Kapur and Long (1997).

It is instructive to examine this situation in some simple one-dimensional settings.

Example 1. Consider the integral operator defined by the formula

$$f(x) = P(\varphi)(x) = \int_{-1}^{1} \log(x-t)\, \varphi(t)\, dt, \qquad (2.1)$$

with $x \in [3,5]$ (Figure 2.1). Using the language of classical potential theory, one could say that P is the operator mapping the charge distribution on the interval $[-1,1]$ to the induced potential created on the interval $[3,5]$. Decomposing the function $\log(x-t)$ into a Taylor series with respect to t around the point $t=0$, we find that

$$f(x) = \left(\int_{-1}^{1} \varphi(t)\, dt\right) \log(x) - \sum_{k=1}^{\infty} \frac{1}{k} \left(\int_{-1}^{1} t^k \varphi(t)\, dt\right) \frac{1}{x^k}. \qquad (2.2)$$

Truncating the series on the right-hand side of (2.2) after an appropriately chosen number of terms, we obtain an approximation to f with any desired precision. It is easily seen that the error of a k-term approximation is slightly better than $1/3^k$, so that the number of terms required to obtain a specified

Table 2.1. Numbers of terms (Taylor, Legendre, and SVD) required to obtain several accuracies (in the L^2-norm) in Example 1.

ε	n_{Taylor}	n_{Legendre}	n_{SVD}
10^{-30}	61	39	18
10^{-24}	48	32	14
10^{-20}	40	26	12
10^{-16}	31	21	10
10^{-12}	23	16	7
10^{-10}	19	13	6
10^{-6}	10	8	4
10^{-4}	7	6	3

accuracy ε is of the order $-\log_3(\varepsilon)$. In the second column of Table 2.1 we list the numbers of terms actually required to obtain selected accuracies. We also note that (2.2) is a simplified version of the approximation used by most early versions of the FMM.

Obviously, there is nothing magical about Taylor series: many other approximations could be used. For example, decomposing the function φ into a Legendre series on the interval $[-1, 1]$ and performing elementary manipulations, we have

$$f(x) = \int_{-1}^{1} \log(x-t) \left(\sum_{k=0}^{\infty} P_k(t) \int_{-1}^{1} P_k(\tau) \varphi(\tau) \, d\tau \right)$$

$$= \sum_{k=0}^{\infty} \left(\int_{-1}^{1} P_k(\tau) \varphi(\tau) \, d\tau \right) \left(\int_{-1}^{1} P_k(t) \log(x-t) \, dt \right) \quad (2.3)$$

$$= 2 \sum_{k=0}^{\infty} \frac{1}{2k+1} \left(\int_{-1}^{1} P_k(\tau) \varphi(\tau) \, d\tau \right) (Q_{k+1}(x) - Q_{k-1}(x)),$$

where Q_k is the 'second' Legendre function (see, for example, Gradshteyn and Ryzhik (2000)). While the exact expression for the convergence rate of the series (2.3) is somewhat cumbersome (it involves standard estimates on the behaviour of Q_k), it is much higher than the convergence rate of (2.2). The numbers of terms required to obtain selected accuracies using this approach are listed in the third column of Table 2.1.

Finally, one can construct the Singular Value Decomposition (SVD) of the operator P (defined in (2.1) above), representing it in the form

$$f(x) = P(\varphi)(x) = \sum_{k=1}^{\infty} \lambda_k \, (v_k, \varphi) \, u_k(x), \quad (2.4)$$

Figure 2.2. Discrete charge distribution acting on an adjacent interval.

where $u_k : [3, 5] \to R$, $v_k : [-1, 1] \to R$, are the kth left and right singular vectors, respectively, and λ_k is the kth singular value. Truncating the expansion (2.4) after k terms, we obtain a k-term approximation to P that is optimal in the obvious sense. For this case, the numbers of terms required to obtain selected accuracies are listed in the third column of Table 2.1.

Table 2.1 is quite revealing. It turns out that the approximation used by the original FMM is surprisingly inefficient, and the orthogonal expansion is only marginally better. For example, at 16 digits, the optimal number of terms is 10 (which is quite good); the Taylor and Legendre series require 31 and 21 terms respectively.

Example 2. Here, we consider the $n \times n$ matrix $\mathbf{A}_n$ defined by the formula

$$\mathbf{A}_n(i, j) = \log(x_i - t_j), \tag{2.5}$$

with the points $t_1, t_2, \ldots, t_n, x_1, x_2, \ldots, x_n \in R$ defined by the formulae

$$\begin{aligned} t_i &= -1 + \frac{(i-1)}{n}, \\ x_i &= \frac{(i-1)}{n}. \end{aligned} \tag{2.6}$$

In other words, the points $\{t_i\}$ are equispaced on the interval $[-1, 0]$, and the points $\{x_i\}$ are equispaced on $[0, 1]$. None of the classical expansions are applicable in this case (since the 'targets' in this case are not separated from the 'charges'). On the other hand, the following lemma shows that the rank of $\mathbf{A}_n$ defined in (2.5) is small compared to n. The fact that compression does not require the separation of sources and targets will be essential in constructing efficient direct solvers.

Lemma 2.1. To any fixed precision ε, the rank of the matrix $\mathbf{A}_n$ defined in (2.5) is of the order $-\log(n) \cdot \log_3(\varepsilon)$.

Outline of proof. We start by subdividing the interval $[0, 1]$ into two subintervals of equal size: $[0, 1/2]$ and $[1/2, 1]$ (see Figure 2.3). Obviously, the interval $[1/2, 1]$ is separated from the interval $[-1, 0]$ by its own size, and the rank of their interactions is of the order $\log_3(\varepsilon)$ (see Example 1 above). We proceed by subdividing the interval $[0, 1/2]$ into subintervals $[0, 1/4]$ and $[1/4, 1/2]$; again, the interval $[1/4, 1/2]$ is separated from the interval $[-1, 0]$ by its own size, and the rank of their interactions is bounded by $\log_3(\varepsilon)$. On the next step, we subdivide the interval $[0, 1/4]$ into subintervals $[0, 1/8]$,

Figure 2.3. Recursive subdivision of an interval used in Lemma 1.1.

Table 2.2. Number of SVD terms for various accuracies in L^2 in Example 2.

ε	50	100	200	400	800	1600	3200	6400	12800	25600
10^{-30}	27	32	36	41	45	50	54	58	62	67
10^{-24}	23	26	30	34	37	40	44	47	51	54
10^{-20}	20	23	26	29	31	34	37	40	42	45
10^{-16}	17	19	21	23	26	28	30	32	34	36
10^{-12}	13	15	17	18	20	21	23	24	25	27
10^{-10}	11	13	14	15	17	18	19	20	21	22
10^{-6}	8	8	9	10	10	11	11	12	12	12
10^{-4}	5	6	6	6	7	7	7	7	7	7

$[1/8, 1/4]$, and observe that the rank of interactions between the intervals $[-1, 0]$, $[1/8, 1/4]$ is of the order $\log_3(\varepsilon)$. This process will terminate after roughly $\log_2(n)$ steps, and the total rank of interactions between the intervals $[-1, 0]$, $[0, 1]$ is bounded by $-\log(n) \cdot \log_3(\varepsilon)$. □

In Table 2.2, we list the ranks of the matrix $\mathbf{A}_n$ in (2.5) to accuracies varying from 10^{-4} to 10^{-30}, with n varying from 50 to 25600. The ranks were obtained via 'brute force' numerical calculation of the SVD, representing the matrix in the form

$$\mathbf{A}_n = \mathbf{U}_{n,k} \, \mathbf{D}_{k,k} \, (\mathbf{V}_{n,k})^* + O(\epsilon), \qquad (2.7)$$

with the columns of each of the matrices $\mathbf{U}_{n,k}$, $\mathbf{V}_{n,k}$ orthonormal, $\mathbf{D}_{k,k}$ a diagonal matrix with positive elements, and k tabulated in Table 2.2 as a function of the L^2-norm of the error. Obviously, once the expansion (2.7) has been constructed, it can be used as a primitive 'fast' algorithm for the application of the matrix $\mathbf{A}_n$ to arbitrary vectors, reducing the cost from n^2 to $(2n + k)\,k$.

Example 3. To illustrate the situation in two dimensions, consider the integral operator $P : L^2(\Omega_1) \to L^2(\Omega_2)$ defined by the formula

$$f(x, y) = P(\sigma)(x, y) = \int_{\Omega_1} \log \|(x, y) - (u, v)\| \, \sigma(u, v) \, du \, dv, \qquad (2.8)$$

FAST DIRECT SOLVERS FOR INTEGRAL EQUATIONS IN 3D DOMAINS 251

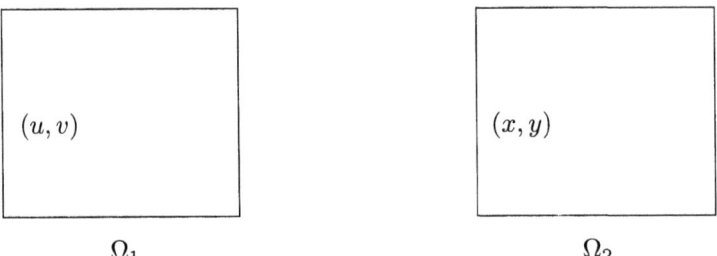

Figure 2.4. Two squares in the complex plane.

Table 2.3. Numbers of terms (Multipole and SVD) required to obtain various accuracies in Example 3.

ε	$n_{\text{multipole}}$	n_{SVD}
10^{-30}	91	31
10^{-24}	73	25
10^{-20}	61	21
10^{-16}	48	17
10^{-12}	36	13
10^{-10}	30	11
10^{-6}	18	7
10^{-4}	12	5

with Ω_1 the square with vertices $(1,-1)$, $(1,1)$, $(-1,1)$, $(-1,-1)$, and Ω_2 the square with vertices $(5,-1)$, $(5,1)$, $(3,1)$, $(3,-1)$, shown in Figure 2.4.

This situation is standard in the design of FMMs in two dimensions, where P is compressed via the Taylor expansion

$$f(x,y) = \text{Re}\left[\left(\int_{\Omega_1} \sigma(u,v)\right) \log z + \sum_{k=1}^{\infty} \frac{1}{k}\left(\int_{\Omega_1}(u+iv)^k \sigma(u,v)\right)\frac{1}{z^k}\right], \quad (2.9)$$

$z = x + iy$ and $\text{Re}[\cdot]$ denotes the real part of the complex-valued quantity inside the brackets. An obvious alternative is to construct the SVD of P (viewed as an operator mapping $L^2(\Omega_1) \to L^2(\Omega_2)$). The resulting dimensionalities are tabulated in Table 2.3 as a function of the required accuracy ε, together with the numbers of terms in the multipole expansion (2.9) achieving similar accuracy.

Observation 2.1. An examination of Tables 2.1–2.3 indicates that when a 'fast' approximate algorithm is to be constructed, it might be much more efficient to utilize decompositions obtained numerically (via SVD, QR, or a related procedure) than to use one of the classical expansions. This is quite understandable, since the SVD constructs an *optimal* expansion for the operator in question, including detailed knowledge about the locations of all sources, while the classical approximations are *general* tools. The obvious cost of the purely numerical approach is the need to construct the representations, *i.e.*, the left and right singular vectors, *etc*. In many cases, such as when solving a fixed linear system with multiple right-hand sides, this cost is trivial: the matrices in question are factored once and used repeatedly (see, for example, Martinsson (2006)).

3. Fast direct solvers for structured matrices

The systematic use of techniques of the type described above will allow us to efficiently approximate the (dense) system matrices that arise from integral equation discretizations of surfaces in three dimensions. Before entering into those details, however, let us consider, from a purely linear-algebraic point of view, the solution of a dense linear system with low-rank off-diagonal blocks. For this, we let the linear system $\mathbf{Ax} = \mathbf{b}$ be defined in block matrix form by the equations

$$\begin{aligned} \mathbf{A}_{11}\mathbf{x}_1 + \mathbf{A}_{12}\mathbf{x}_2 + \cdots + \mathbf{A}_{1N}\mathbf{x}_N &= \mathbf{b}_1, \\ \mathbf{A}_{21}\mathbf{x}_1 + \mathbf{A}_{22}\mathbf{x}_2 + \cdots + \mathbf{A}_{2N}\mathbf{x}_N &= \mathbf{b}_2, \\ &\vdots \\ \mathbf{A}_{N1}\mathbf{x}_1 + \mathbf{A}_{N2}\mathbf{x}_2 + \cdots + \mathbf{A}_{NN}\mathbf{x}_N &= \mathbf{b}_N, \end{aligned} \quad (3.1)$$

where $\mathbf{x}_i, \mathbf{b}_i \in \mathbb{R}^{n_i}$ and $\mathbf{A}_{ij} \in \mathbb{R}^{n_i \times n_j}$. Solution of the full linear system of dimension $N_{\text{tot}} = \sum_{i=1}^{N} n_i$ requires $O(N_{\text{tot}}^3)$ work.

Definition 3.1. When needed, the entry in the lth row and mth column of $\mathbf{A}_{ij}$ will be denoted by $\mathbf{A}_{ij}(l,m)$. Likewise, the entry in the lth row and mth column of $\mathbf{A}$ itself will be referred to as $\mathbf{A}(l,m)$.

Suppose now that the diagonal blocks $\mathbf{A}_{ii}$ are full-rank but that each of the $\mathbf{A}_{ij}$ can be decomposed as the product of three low-rank matrices:

$$\mathbf{A}_{ij} = \mathbf{L}_i \mathbf{S}_{ij} \mathbf{R}_j, \quad (3.2)$$

where $\mathbf{L}_i \in \mathbb{R}^{n_i \times k_i}$, $\mathbf{S}_{ij} \in \mathbb{R}^{k_i \times k_j}$, and $\mathbf{R}_j \in \mathbb{R}^{k_j \times n_j}$, with $k_i \ll n_i$ and $k_j \ll n_j$. It is worth noting that the factorization (3.2) is somewhat special in that $\mathbf{L}_i$ depends only on the index i and $\mathbf{R}_j$ depends only on the index j. We will see how such a factorization arises in the next section, but for now we assume it is given.

Under these conditions, we can introduce the auxiliary variables

$$\mathbf{y}_j = \mathbf{R}_j \mathbf{x}_j \quad (3.3)$$

and rewrite the original linear system (3.1) in the form

$$\mathbf{A}_{11}\mathbf{x}_1 + \mathbf{L}_1\mathbf{S}_{12}\mathbf{y}_2 + \cdots + \mathbf{L}_1\mathbf{S}_{1N}\mathbf{y}_N = \mathbf{b}_1,$$
$$\mathbf{L}_2\mathbf{S}_{21}\mathbf{y}_1 + \mathbf{A}_{22}\mathbf{x}_2 + \cdots + \mathbf{L}_2\mathbf{S}_{2N}\mathbf{y}_N = \mathbf{b}_2,$$
$$\vdots \qquad (3.4)$$
$$\mathbf{L}_N\mathbf{S}_{N1}\mathbf{y}_1 + \mathbf{L}_N\mathbf{S}_{N2}\mathbf{y}_2 + \cdots + \mathbf{A}_{NN}\mathbf{x}_N = \mathbf{b}_N.$$

The coupled linear system (3.3), (3.4), can be written in block matrix form:

$$\begin{pmatrix} \mathbf{A}_{11} & 0 & \cdots & 0 & 0 & \mathbf{L}_1\mathbf{S}_{12} & \cdots & \mathbf{L}_1\mathbf{S}_{1N} \\ 0 & \mathbf{A}_{22} & \cdots & 0 & \mathbf{L}_2\mathbf{S}_{21} & 0 & \cdots & \mathbf{L}_2\mathbf{S}_{2N} \\ 0 & 0 & \cdots & 0 & \cdots & \cdots & \cdots & \cdots \\ 0 & 0 & \cdots & \mathbf{A}_{NN} & \mathbf{L}_N\mathbf{S}_{N1} & \mathbf{L}_N\mathbf{S}_{N2} & \cdots & 0 \\ \hline \mathbf{R}_1 & 0 & \cdots & 0 & -\mathbf{I}_1 & 0 & \cdots & 0 \\ 0 & \mathbf{R}_2 & \cdots & 0 & 0 & -\mathbf{I}_2 & \cdots & 0 \\ 0 & 0 & \cdots & 0 & 0 & 0 & \cdots & 0 \\ 0 & 0 & \cdots & \mathbf{R}_N & 0 & 0 & \cdots & -\mathbf{I}_N \end{pmatrix} \begin{pmatrix} \mathbf{x}_1 \\ \mathbf{x}_2 \\ \cdots \\ \mathbf{x}_N \\ \hline \mathbf{y}_1 \\ \mathbf{y}_2 \\ \cdots \\ \mathbf{y}_N \end{pmatrix} = \begin{pmatrix} \mathbf{b}_1 \\ \mathbf{b}_2 \\ \cdots \\ \mathbf{b}_N \\ \hline 0 \\ 0 \\ \cdots \\ 0 \end{pmatrix},$$
$$(3.5)$$

where $\mathbf{I}_j$ is the identity matrix of dimension n_j.

In (3.5), we have replaced the dense linear system of (3.1) with N_{tot} unknowns by a sparse system of dimension $N_{\text{tot}} + K_{\text{tot}}$, where $K_{\text{tot}} = \sum_{i=1}^{N} k_i$. The sparsity, of course, depends strongly on how few auxiliary variables (K_{tot}) are needed. Note that the zero structure of the new system allows us to form the Schur complement in the $\{\mathbf{y}_j\}$ variables quite easily. Elementary row elimination leads from (3.5) to

$$\begin{pmatrix} \mathbf{E}_{11} & \mathbf{S}_{12} & \cdots & \mathbf{S}_{1N} \\ \mathbf{S}_{21} & \mathbf{E}_{22} & \cdots & \mathbf{S}_{2N} \\ \cdots & \cdots & \cdots & \cdots \\ \mathbf{S}_{N1} & \mathbf{S}_{N2} & \cdots & \mathbf{E}_{NN} \end{pmatrix} \begin{pmatrix} \mathbf{y}_1 \\ \mathbf{y}_2 \\ \cdots \\ \mathbf{y}_N \end{pmatrix} = \begin{pmatrix} \mathbf{C}_1\mathbf{b}_1 \\ \mathbf{C}_2\mathbf{b}_2 \\ \cdots \\ \mathbf{C}_N\mathbf{b}_N \end{pmatrix}, \qquad (3.6)$$

where

$$\mathbf{E}_{ii} = (\mathbf{R}_i\mathbf{A}_{ii}^{-1}\mathbf{L}_i)^{-1}, \ \mathbf{C}_i = \mathbf{E}_i\mathbf{R}_i\mathbf{A}_{ii}^{-1}. \qquad (3.7)$$

Combining (3.4), (3.6) and a little algebra, we can write

$$\mathbf{A}^{-1} = \mathbf{D} + \mathbf{B}(\mathbf{E} + \mathbf{S})^{-1}\mathbf{C}, \qquad (3.8)$$

where $\mathbf{S}$ is the dense matrix with off-diagonal blocks $\mathbf{S}_{ij}$ given by (3.2), $\mathbf{S}_{ii} \equiv 0$, $\mathbf{E}, \mathbf{C}$ are the block-diagonal matrices with diagonal entries $\mathbf{E}_{ii}, \mathbf{C}_i$, respectively, and $\mathbf{D}, \mathbf{B}$ are block-diagonal matrices with

$$\mathbf{D}_{ii} = \mathbf{A}_{ii}^{-1}(\mathbf{I}_i - \mathbf{L}_i\mathbf{C}_i),$$
$$\mathbf{B}_{ii} = \mathbf{A}_{ii}^{-1}\mathbf{L}_i\mathbf{E}_i. \qquad (3.9)$$

Note that $\mathbf{E} + \mathbf{S}$ is the Schur complement system in (3.6).

Finally, it is worth computing the cost of this procedure. For concreteness, suppose that $n_1 = n_2 = \cdots = n_N = m$, so that $N_{\text{tot}} = Nm$ and naive inversion would require $O(N^3 m^3)$ work. Suppose also that $k_1 = k_2 = \cdots = k_N = k$. In the new procedure, the initial computation of all the A_{ii}^{-1} then requires $O(Nm^3)$ work. Since the Schur complement system is of dimension Nk, inversion requires $O(N^3 k^3)$ work. The remaining work is dominated by the computation of the $\{\mathbf{x}_j\}$ once the $\{\mathbf{y}_j\}$ are known using (3.4) or (3.8). We leave it to the reader to verify that this requires $O(N^2 k^2 + Nmk^2 + Nm^2)$ work.

In the original paper (Martinsson and Rokhlin 2005), the solver is based directly on (3.8) and (3.9). The explicit introduction of the Schur complement makes the derivation a bit simpler. More importantly, however, the embedding of the system into a larger, sparse system allows for the immediate application of standard sparse matrix technology.

Remark 3.1. The algorithm described above is a 'single-level' scheme. That is, unknowns are assumed to have been grouped in some fashion so that off-diagonal blocks are of minimal rank. This idea can be used recursively. In multilevel versions of the method, the Schur complement $(\mathbf{E}+\mathbf{S})$ is treated like the original matrix $\mathbf{A}$. More precisely, the $\mathbf{y}_j$ unknowns are themselves grouped in such a way that off-diagonal blocks of the $\mathbf{E} + \mathbf{S}$ system are of minimal rank, *etc.*

It is the multilevel variants of the preceding analysis that form the basis for many of the fast, direct solvers currently under development. In the end, the method of choice will depend to a large extent on the constants implicit in the $O(N^\alpha \log^\beta N)$ notation, on robustness, and on ease of use. A multilevel scheme in the one-dimensional setting is presented in the appendix to illustrate the recursive nature of the computation.

4. Potential theory for the Poisson equation

We shift our attention now to the classical integral equation approach for solving the Poisson equation in the context of molecular electrostatics. We begin with the formulation of the problem as a partial differential equation:

$$-\nabla \cdot (\epsilon(\mathbf{x}) \nabla \Phi(\mathbf{x})) = \sum_{j=1}^{K} q_j \delta(\mathbf{x} - \mathbf{x}_j) \quad \text{in } \mathbb{R}^3, \tag{4.1}$$

where $\epsilon(\mathbf{x})$ equals ϵ_{in} within the molecular surface, and $\epsilon(\mathbf{x})$ equals ϵ_{out} outside. We will denote the interior region by Ω_{in}, the exterior region by Ω_{out}, and the surface by S (Figure 4.1). We assume the sources all lie in the interior of the molecule Ω_{in}. In order to recast the problem as an integral

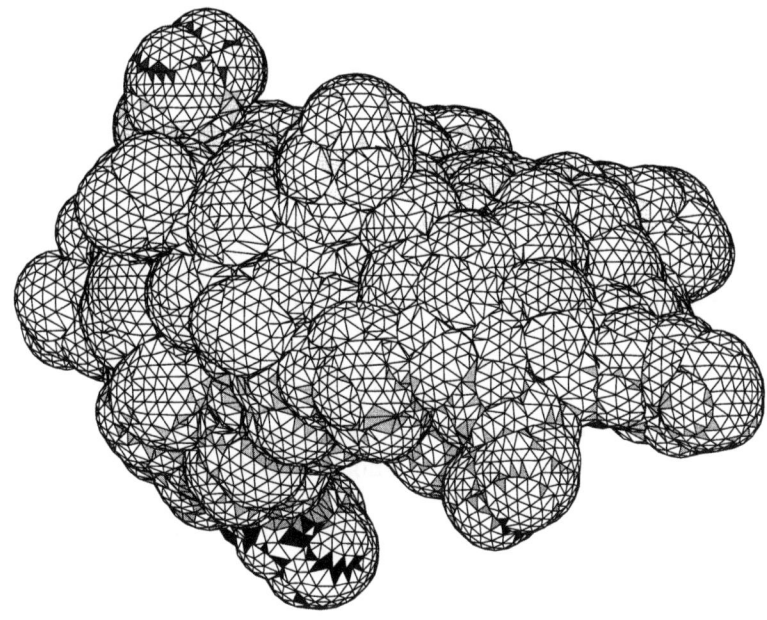

Figure 4.1. A typical molecular surface.

equation, we define the solution in terms of a source function Φ^{source} that accounts for the singular fields due to the point sources and a piecewise harmonic function Φ^{pol}. That is, we write

$$\Phi = \Phi^{\text{source}} + \Phi^{\text{pol}},$$

where

$$\Phi^{\text{source}} = \sum_{j=1}^{K} \frac{q_j}{4\pi\epsilon_{\text{in}}} \frac{1}{\|\mathbf{x} - \mathbf{x}_j\|}.$$

Remark 4.1. A more important model for biophysical applications is actually the linearized Poisson–Boltzmann equation (Davis and McCammon 1990, Rocchia *et al.* 2002, Sharp and Honig 1990), but it is easier to follow the main ideas in the present, simpler context.

Note that Φ^{source} satisfies the Poisson equation and is continuous across the surface S. However, the required interface conditions are that both the total potential Φ and its flux $\epsilon \frac{\partial \Phi}{\partial \nu}$ must be continuous across the dielectric boundary. For this, it is straightforward to see that Φ^{pol} must satisfy

$$\nabla^2 \Phi^{\text{pol}} = 0 \quad \text{in } \Omega_{\text{in}}, \Omega_{\text{out}}, \tag{4.2}$$

$$\left[\Phi^{\text{pol}}\right] = 0 \quad \text{on } S, \tag{4.3}$$

$$\left[\epsilon \frac{\partial \Phi^{\text{pol}}}{\partial \nu}\right] = -\left[\epsilon \frac{\partial \Phi^{\text{source}}}{\partial \nu}\right] \quad \text{on } S. \tag{4.4}$$

Here, the expression $[f]$ denotes the jump in the quantity f across S.

Following standard practice, we write Φ^{pol} as a single-layer potential due to a charge distribution σ defined on the surface:

$$\Phi^{\text{pol}}(\mathbf{x}) = \frac{1}{4\pi} \int_S \frac{\sigma(\mathbf{y})}{\|\mathbf{x} - \mathbf{y}\|} \, d\mathbf{y}. \tag{4.5}$$

The single-layer potential (4.5) is continuous across S and satisfies the following jump relations (Guenther and Lee 1988):

$$\frac{\partial \Phi^{\text{pol}}}{\partial \nu_-}(\mathbf{y}_0) \equiv \lim_{\substack{\mathbf{x} \to \mathbf{y}_0 \\ \mathbf{x} \in \Omega_{\text{in}}}} \frac{\partial \Phi^{\text{pol}}}{\partial \nu_0}(\mathbf{x}) = \tfrac{1}{2}\sigma(\mathbf{y}_0) + \int_S \frac{\partial G}{\partial \nu_0}(\mathbf{y}_0, \mathbf{y})\sigma(\mathbf{y}) \, d\mathbf{y}, \tag{4.6}$$

$$\frac{\partial \Phi^{\text{pol}}}{\partial \nu_+}(\mathbf{y}_0) \equiv \lim_{\substack{\mathbf{x} \to \mathbf{y}_0 \\ \mathbf{x} \in \Omega_{\text{out}}}} \frac{\partial \Phi^{\text{pol}}}{\partial \nu_0}(\mathbf{x}) = -\tfrac{1}{2}\sigma(\mathbf{y}) + \int_S \frac{\partial G}{\partial \nu_0}(\mathbf{y}_0, \mathbf{y})\sigma(\mathbf{y}) \, d\mathbf{y}, \tag{4.7}$$

where $\mathbf{y}_0$ denotes a point on S, $\frac{\partial \Phi^{\text{pol}}}{\partial \nu_0}(\mathbf{x})$ denotes the derivative in the direction of the outward normal at $\mathbf{y}_0$, and $G(\mathbf{x}, \mathbf{y})$ denotes the free-space Green's function $\frac{1}{4\pi|\mathbf{x}-\mathbf{y}|}$.

We will make use of the operator notation

$$\mathbf{K}\sigma(\mathbf{y}_0) = \int_S \frac{\partial G}{\partial \nu_0}(\mathbf{y}_0, \mathbf{y})\sigma(\mathbf{y}) \, d\mathbf{y} \tag{4.8}$$

to denote the normal derivative of the single-layer potential restricted to the interface.

It is clear from the representation (4.5) that equation (4.2) is automatically satisfied. Since the single-layer potential is continuous across the interface, (4.3) is also satisfied. Imposing the remaining condition (4.4) and using the jump relations, we obtain a Fredholm integral equation of the second kind for σ:

$$\frac{1}{2}\sigma(\mathbf{y}_0) + \lambda \int_S \frac{\partial G}{\partial \nu_0}(\mathbf{y}_0, \mathbf{y})\sigma(\mathbf{y}) \, d\mathbf{y} = -\lambda \frac{\partial \Phi^{\text{source}}}{\partial \nu}(\mathbf{y}_0), \tag{4.9}$$

where $\lambda = (\epsilon_{\text{in}} - \epsilon_{\text{out}})/(\epsilon_{\text{in}} + \epsilon_{\text{out}})$. In operator notation,

$$\frac{1}{2}\sigma + \lambda \mathbf{K}\sigma = -\lambda \frac{\partial \Phi^{\text{source}}}{\partial \nu}.$$

It remains only to discretize and solve (4.9). For this, we assume that S has been specified as a collection of N_T triangles. We approximate the potential due to the polarization charge as

$$\Phi^{\text{pol}}(\mathbf{x}) = \frac{1}{4\pi} \sum_{j=1}^{N_T} \int_{T_j} \frac{\sigma_j}{\|\mathbf{x}-\mathbf{y}\|} \, d\mathbf{y}, \tag{4.10}$$

where T_j denotes the jth triangle and σ_j is constant over T_j. Imposing the continuity of flux at each triangle centroid C_j results in the finite-dimensional linear system

$$\frac{1}{2}\sigma_j + \frac{\lambda}{4\pi} \sum_{k=1}^{N_T} \int_{T_k} \frac{\partial G}{\partial \nu_0}(C_j, \mathbf{y}) \sigma_k \, d\mathbf{y} = -\lambda \frac{\partial \Phi^{\text{source}}}{\partial \nu}(C_j). \tag{4.11}$$

This approach is generally referred to as a first-order accurate collocation scheme. We do not recommend such a quadrature rule in general, but higher-order accurate discretizations are rather involved and will distract us from our goals. For recent work on higher-order schemes in the context of molecular modelling, see Bardhan et al. (2005).

When $j = k$, the integral contribution in (4.11) is well known to vanish. That is, on a flat triangle,

$$\int_{T_j} \frac{\partial G}{\partial \nu_0}(C_j, \mathbf{y}) \sigma_j \, d\mathbf{y} = 0,$$

and the diagonal matrix entries $\mathbf{A}(i,i)$ of the linear system are simply equal to $\frac{1}{2}$. When j and k are distinct, the entry $\mathbf{A}(j,k)$ is given by

$$\mathbf{A}(j,k) = \frac{\lambda}{4\pi} \int_{T_k} \frac{\partial G}{\partial \nu_0}(C_j, \mathbf{y}) \, d\mathbf{y}, \tag{4.12}$$

namely, λ times the flux through the centroid of T_j due to a unit charge distribution on T_k. It is well defined, although nearly singular for nearby interactions. $\mathbf{A}(j,k)$ can easily be computed by a hybrid numerical/analytic scheme: the triangles T_k and T_j are rotated in space so that T_k lies in the xy-plane with one vertex at the origin and one segment lying along the x-axis. Integration over T_k in the x-direction is performed analytically and integration in the y-direction is performed numerically using Gaussian quadrature. Twenty Gaussian nodes yield more than six digits of accuracy.

Of particular interest is the evaluation of the total electrostatic energy

$$\frac{1}{2} \sum_{i=1}^{K} q_i \tilde{\Phi}(\mathbf{x}_i), \tag{4.13}$$

where

$$\tilde{\Phi}(\mathbf{x}_i) = \Phi^{\text{pol}}(\mathbf{x}_i) + \sum_{\substack{j=1 \\ j \neq i}}^{K} \frac{q_j}{4\pi \epsilon_{\text{in}}} \frac{1}{\|\mathbf{x}_i - \mathbf{x}_j\|}.$$

Remark 4.2. We do not intend to survey the literature on numerical methods for the Poisson or Poisson–Boltzmann equation here, except to note that the standard approach in molecular electrostatics has been based on the finite difference or finite element solution of the governing equations (Davis and McCammon 1990, Holst, Baker and Wang 2000, Rocchia *et al.* 2002, Sharp and Honig 1990). There has also been a great deal of work on integral equation methods, particularly since the advent of fast algorithms for computing the matrix vector product in the solution of linear systems like (4.11). In the Poisson–Boltzmann case, these include Altman, Bardhan, Tidor and White (2006), Boschitsch, Fenley and Olson (1999), Huang and Greengard (2002), Kuo *et al.* (2002), Lu, Cheng, Huang and McCammon (2006) and Liang and Subramaniam (1997).

5. A single-level fast solver

We are now in a position to bring the linear algebra of Section 3 together with the potential theory of Section 4. The essential step is the grouping of unknowns in the integral equation so that off-diagonal blocks can be well-approximated by low-rank matrices. In the one-dimensional case, the ordering of unknowns on the interval is sufficient, but in higher dimensions it is not. We begin by sorting all the triangles $\{T_j\}$ that define the surface S in (4.11) into cubic boxes. For this, we first enclose the surface S in a cube B, and subdivide B into $M \times M \times M$ boxes. Each of the N_T triangles is identified with the box containing its centroid. After this sorting step, we let N denote the number of boxes that are non-empty (denoted by $B_1, B_2, \ldots, B_N$). We then reorder the original triangles so that the first n_1 belong to B_1, the next n_2 belong to B_2, etc. The triangles in box B_i will be referred to as *surface patch* (or *patch*) P_i.

Definition 5.1. Two disjoint boxes B_j and B_k are called *neighbours* if they share a boundary point. Otherwise, they are called *well-separated*. Two surface patches P_j and P_k are called neighbours or well-separated in accordance with the relation between the boxes B_j and B_k.

Suppose now that box B_i and box B_j are well-separated and that $\mathbf{A}_{ij}$ is the block matrix describing the flux through 'target' triangles in B_i due to charge distributions on 'source' triangles in B_j. From standard multipole estimates (Greengard and Rokhlin 1997), it is clear that, for any fixed precision, the rank of $\mathbf{A}_{ij}$ is bounded independent of the number of triangles n_i and n_j. To be precise, for $\mathbf{x} \in B_i$,

$$\Phi(\mathbf{x}) = \sum_{l=0}^{p} \sum_{m=-l}^{l} M_l^m \frac{Y_l^m(\theta, \phi)}{r^{l+1}} + O\left(\left(\frac{1}{\sqrt{3}}\right)^p\right),$$

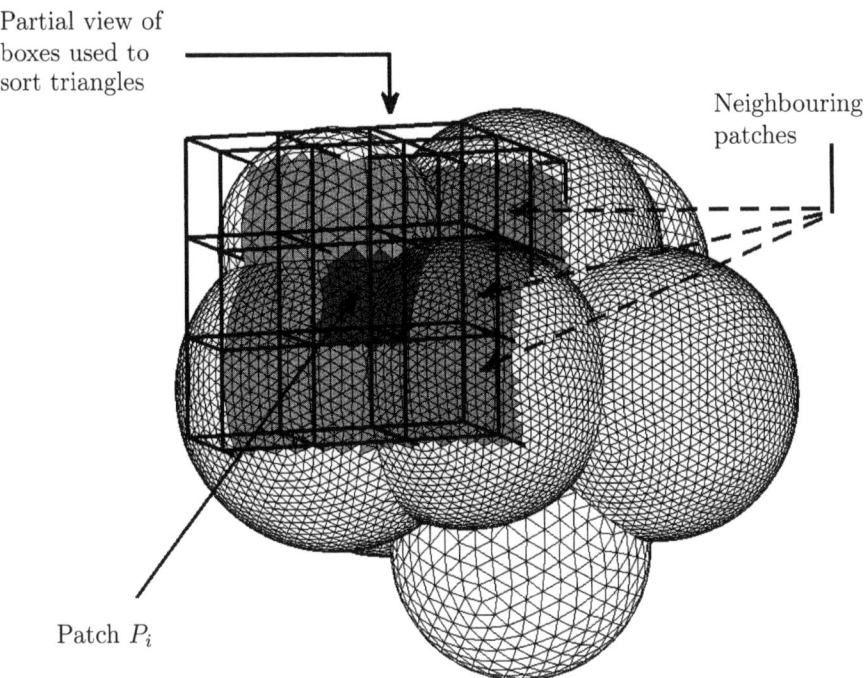

Figure 5.1. A grid of $M \times M \times M$ boxes encloses the surface S. The triangles whose centroids are located in subcube B_i are identified as surface patch P_i (dark grey). Its near-neighbour patches are indicated in light grey. The remaining patches are well-separated.

where (r, θ, ϕ) are the spherical coordinates of $\mathbf{x}$ with respect to the centre of box B_j, $Y_l^m(\theta, \phi)$ denotes the spherical harmonic of order l and degree m, and the $\{M_l^m\}$ are the net multipole moments of the charge distributions on the 'source' triangles in B_j. For a precision ϵ, then, one requires no more than $k = O((\log_{\sqrt{3}} \epsilon)^2)$ degrees of freedom to represent $\Phi(\mathbf{x})$. In short, the field induced in B_i has guaranteed smoothness properties due entirely to the separation between sources and targets and, to precision ϵ, the rank of $\mathbf{A}_{ij}$ is bounded by $(\log_{\sqrt{3}} \epsilon)^2$.

Rather than using multipole expansions, however, one could ask whether there is a purely linear-algebraic construct that expresses this fact. In particular, one could ask whether a subset of k_j source triangles could be used to represent the field in B_i with charge strengths computed from the n_j values σ_j. Similarly, one could ask whether a subset of k_i triangles in B_i could be used to sufficiently sample the field to precision ϵ, in the sense that the field at all n_i triangles in B_i can be computed by some kind of interpolation procedure. This is not a classical factorization like the SVD used in Section 2, but will be more suitable for our purposes. There are some remarkable results in this direction, for instance, Cheng, Gimbutas, Martinsson and Rokhlin (2005), Goreinov, Yrtyshnikov and Zamarashkin (1997) and Gu and Eisenstat (1996).

Theorem 5.1. Suppose that the matrix $\mathbf{A}_{ij}$ of dimension $n_i \times n_j$ has rank k to precision ϵ. Then there exists a factorization of $\mathbf{A}_{ij}$ of the form

$$\mathbf{A}_{ij} = \mathbf{L}_i \mathbf{S}_{ij} \mathbf{R}_j + O(\epsilon),$$

where $\mathbf{S}_{ij}$ is a $k \times k$ *submatrix* of $\mathbf{A}_{ij}$, $\mathbf{R}_j$ is a $k \times n_j$ matrix and $\mathbf{L}_i$ is an $n_i \times k$ matrix. Further, this decomposition is well-conditioned; the norms of $\mathbf{L}_i$ and $\mathbf{R}_j$ are bounded by $\sqrt{k(n_1 + n_2)}$.

Proof. This is simply a restatement of Theorem 3 of Cheng et al. (2005). □

The preceding theorem is not constructive. It does not say which k row indices and which k column indices to select in the extraction of the submatrix $\mathbf{S}_{ij}$ from $\mathbf{A}_{ij}$. Fortunately, Cheng et al. (2005) describe a QR-like algorithm which computes such a factorization using $O(n_1 n_2 k)$ work. (It should be noted that this factorization is not unique, but that is of no concern to us here.) We will refer to the k triangles in B_i so obtained as the *incoming skeleton* and to the k triangles in B_j as the *outgoing skeleton* (Cheng et al. 2005, Goreinov et al. 1997). Note that this is precisely what we need. $\mathbf{R}_j$ serves as the mapping from the original vector of n_j charges to the k charges on the outgoing skeleton and $\mathbf{L}_i$ serves as the mapping from the field sampled on the k incoming triangles to the full set of n_i triangles in subcube B_i. We view the skeletonization procedures as *black-box* routines and refer the reader to the original papers for further explanation.

Unfortunately, we are not yet done, since the well-separatedness criterion does not apply to block matrices $\mathbf{A}_{ij}$ when B_i and B_j are neighbours. In this context, simple *a priori* estimates do not apply. Nevertheless, Theorem 5.1 and the construction of the skeletons still apply – it is just that the rank k needs to be determined on the fly, given the desired precision ϵ. (This was a principal motivation of the earlier paper, Gu and Eisenstat (1996).) Recall that we considered precisely such a case in one dimension in Example 2 of Section 2.

To see why the rank of interactions between neighbouring boxes might be low, despite the absence of strict separation, readers familiar with potential theory will recall that on a flat surface, the operator $\mathbf{K}$ in (4.8) is identically zero. It is perhaps reasonable, then, to expect that the block matrix of interactions of two surface patches is in general not full-rank, even at high precision.

5.1. Global skeletonization

The reader may have noticed that a strong form of 'skeletonization' is required in the set-up of the system (3.4). In particular, for each block of unknowns corresponding to patch P_i, $i = 1, \ldots, N$, the same number of triangles k_i is used for both the incoming and outgoing skeletons. One way to accomplish this for patch P_i is to insist that the incoming and outgoing skeletons actually be the same. For the computation, we concatenate all of the $\mathbf{A}_{ij}$ submatrices followed by all of the $\mathbf{A}_{ji}^T$ submatrices, excluding the diagonal blocks:

$$\mathbf{A}_i \equiv \begin{bmatrix} \mathbf{A}_{i1} \mathbf{A}_{i2} \cdots \mathbf{A}_{i,i-1} \mathbf{A}_{i,i+1} \cdots \mathbf{A}_{iN} \ \mathbf{A}_{1i}^T \mathbf{A}_{2i}^T \cdots \mathbf{A}_{i-1,i}^T \mathbf{A}_{i+1,i}^T \cdots \mathbf{A}_{Ni}^T \end{bmatrix}. \quad (5.1)$$

The $n_i \times 2(N_{\text{tot}} - n_i)$ matrix $\mathbf{A}_i$ so-formed can then be passed to the skeletonizing machinery of Cheng *et al.* (2005). The number of row indices k_i obtained through skeletonization is not known *a priori*, but returned as part of the calculation.

Lemma 5.2. (Martinsson and Rokhlin 2005) Let $\mathbf{A}_i$ denote the $n_i \times 2(N_{\text{tot}} - n_i)$ matrix defined in (5.1) and let the interpolatory decomposition for $\mathbf{A}_i$ be given by

$$\mathbf{A}_i = \mathbf{L}_i \mathbf{S}_i \mathbf{V}_i + O(\epsilon), \quad (5.2)$$

for a specified precision ϵ, with $\mathbf{L}_i \in \mathbb{R}^{n_i \times k_i}$, $\mathbf{S}_i \in \mathbb{R}^{k_i \times k_i}$ a *submatrix* of $\mathbf{A}_{ij}$, and $\mathbf{V}_i \in \mathbb{R}^{k_i \times 2(N_{\text{tot}} - n_i)}$. Then, the triangles corresponding to the k_i row indices can serve as both the incoming and outgoing skeletons. Moreover, in the notation of Theorem 5.1,

$$\mathbf{A}_{ij} = \mathbf{L}_i \mathbf{S}_{ij} \mathbf{L}_j^T + O(\epsilon).$$

Sketch of proof. Note that skeletonization of the first $N - 1$ blocks of A_i guarantees that the k_i row indices which have been returned serve as a satisfactory incoming skeleton. In other words, the incoming field is sufficiently sampled to the specified precision ϵ at those k_i triangle centroids, and L_i is the mapping from those values to the values on all n_i triangles in patch P_i. The remainder of the theorem follows from considering the transpose of (5.2). That is, simultaneous skeletonization of the second $N - 1$ block matrices guarantees that the k_i indices which have been returned also serve as a satisfactory outgoing skeleton. Further, $\mathbf{L}_i^T$ must serve as the mapping from the given charges on all n_i triangles in patch P_i to effective charges on the skeleton triangles that correctly represent the field on all other patches to precision ϵ. □

As noted in Martinsson and Rokhlin (2005), for the problems of potential theory, one can reduce the cost of this step dramatically. One need not explore all the pairwise interactions between the triangles in patch P_i and those which are well-separated, since any potential field induced inside B_i

by well-separated triangles could equally well be induced by a charge distribution located on the interface separating B_i from its well-separated boxes. That interface is simply the outer boundary of the $3 \times 3 \times 3$ 'supercell' of boxes centred on B_i (the set of boxes shown in Figure 5.1). In other words, an artificial set of triangles on the surface of the supercell can replace all of the A_{ij} blocks in the construction (5.1) corresponding to well-separated patches P_j. The number of triangles needed is determined only by the desired precision ϵ. Likewise, if a subset of triangles inside B_i matches the field on the surface of the supercell to precision ϵ, it will also do so in any well-separated box. Thus, the same artificial set of triangles on the surface of the supercell can replace all of blocks A_{ij}^T corresponding to well-separated patches P_j in the construction of (5.1) as well. Computationally, this means that the dimension of the matrix which must be skeletonized for patch P_i is of the order $n_i \times 2(N_i + M)$, where N_i denotes the number of triangles in the neighbours of B_i only and M denotes the number of triangles on the supercell surface.

6. Numerical results I

We have implemented the one-level direct solver described in the preceding sections, which takes as input an arbitrary triangulated surface and uses the simple piecewise constant approximation of the surface charge density to solve the dielectric interface problem

As an example, we selected atomic coordinates of the protein BPTI from the Protein Data Bank (PDB) and used the STLib package to create a surface triangulation (depicted in Figure 4.1). 20600 triangles were created in the original discretization, and compression with a tolerance of 10^{-2} yielded 7049 'skeleton' triangles (the dimension of the Schur complement in (3.6)). Factorization (in this naive one-level approach) required about 20 minutes, but the subsequent application of A^{-1} required about 0.3 seconds on a single-processor 1.9 GHz workstation. A useful point of comparison is that an FMM-based iterative solution would have required several seconds: much shorter than the factorization time but more than an order of magnitude slower than the application of the compressed inverse. Compression with a tolerance of 10^{-1} yielded 1596 'skeleton' triangles, factorization required about 10 minutes, and the application of A^{-1} required about 0.1 seconds.

The power of the direct solver, of course, is made much more apparent when solving the integral equation repeatedly. In molecular electrostatics, for example, one is often interested in the electrostatic energy as the interior charge distribution is modified. Each new set of charge locations induces a different right-hand side in (4.9). The total cost (after the initial factorization) for M interior charge distributions (solving M PDEs) is then $M \times 0.1$ seconds.

7. Local geometric perturbations

As implied in the Introduction, one of the striking areas of application of fast direct solvers lies in their ability to accelerate the solution of 'nearby' problems. By nearby, we mean that the new system matrix is a low-rank modification of the one for which we already have an efficient factorization. In linear algebra, this idea is generally attributed to Sherman, Morrison and Woodbury (1949, 1950). There is a host of applications of this idea. In the context of integral equations, it has been used successfully, for example, in both electromagnetics (Kastner 1989) and elasticity (James and Pai 1999).

One of our goals is to allow for electrostatic modelling in protein design using the one- or two-body decomposable method of Marshall, Vizcarra and Mayo (2005). This method requires the solution of many electrostatic problems with a fixed protein backbone, with local modifications of only one or two amino acid side chains. Each such modification requires the deletion and insertion of a modest number of triangles in the definition of the molecular surface. Since we will rely heavily on the Sherman–Morrison–Woodbury approach, we briefly summarize it here in the context of boundary integral equations.

7.1. Addition of triangles

Suppose that the matrix $\mathbf{A}$ of system (4.11) has been constructed for a surface with N triangles, but that we now wish to solve the same interface problem with an extra collection of p triangles *added*. We will denote the added triangles by $T_{N+1}, \ldots, T_{N+p}$. (For the moment, we can assume that we are adding a second surface component.) Then, the new linear system that describes the perturbed problem takes the form

$$\begin{pmatrix} \mathbf{A} & \mathbf{B}_+ \\ \mathbf{C}_+ & \mathbf{D}_+ \end{pmatrix} \begin{pmatrix} \sigma \\ \sigma_p \end{pmatrix} = \begin{pmatrix} \mathbf{f} \\ \mathbf{f}_p \end{pmatrix}. \tag{7.1}$$

Here, $\mathbf{B}_+ \in \mathbb{R}^{N \times p}$, $\mathbf{C}_+ \in \mathbb{R}^{p \times N}$, and $\mathbf{D}_+ \in \mathbb{R}^{p \times p}$ with

$$\mathbf{B}_+(i,j) = A(T_i, T_{N+j}),$$

$$\mathbf{C}_+(i,j) = A(T_{N+j}, T_i),$$

$$\mathbf{D}_+(i,j) = \begin{cases} \frac{1}{2} & \text{if } i = j, \\ A(T_{N+i}, T_{N+j}) & \text{if } i \neq j, \end{cases}$$

$$\mathbf{f}_p(j) = -\lambda \frac{\partial \Phi^{\text{source}}}{\partial \nu}(C_{N+j}),$$

where C_{N+j} is the centroid of the jth added triangle, and $A(T_i, T_j)$ is defined in (4.12).

Block Gaussian elimination yields the Schur complement formula:
$$(\mathbf{I} - \mathbf{D}_+^{-1}\mathbf{C}_+\mathbf{A}^{-1}\mathbf{B}_+)\sigma_p = \mathbf{D}_+^{-1}\mathbf{f}_p - \mathbf{D}_+^{-1}\mathbf{C}_+\mathbf{A}^{-1}\mathbf{f}. \tag{7.2}$$
It is easy to check that setting up this $p \times p$ linear system requires p applications of $\mathbf{A}^{-1}$ to compute $(\mathbf{A}^{-1}\mathbf{B}_+)$, p^2 inner products of N-vectors to compute $\mathbf{C}_+(\mathbf{A}^{-1}\mathbf{B}_+)$, and p^3 operations to compute $\mathbf{D}_+^{-1}(\mathbf{C}_+(\mathbf{A}^{-1}\mathbf{B}_+))$. The right-hand side requires one application of $\mathbf{A}^{-1}$, p inner products of N vectors and one application of $\mathbf{D}_+^{-1}$. Given σ_p, the first N entries of the solution, namely σ, can be obtained from
$$\sigma = \mathbf{A}^{-1}\mathbf{f} - \mathbf{A}^{-1}\mathbf{B}_+\sigma_p.$$

7.2. Deletion of triangles

Suppose now that one wants to subtract q triangles $T_{j_1}, T_{j_2}, \ldots, T_{j_q}$ from the original surface instead of adding them. This can be done with a simple trick; one adds new triangles $T_{N+1}, T_{N+2}, \ldots, T_{N+q}$ to the discretization in exactly the same spatial location as $T_{j_1}, T_{j_2}, \ldots, T_{j_q}$, imposing the condition that the charge density on the new triangles $\{\sigma_{N+1}, \sigma_{N+2}, \ldots, \sigma_{N+q}\}$ negates the charge density on the original ones $\{\sigma_{j_1}, \sigma_{j_2}, \ldots, \sigma_{j_q}\}$.

We leave it to the reader to verify the following lemma.

Lemma 7.1. Let the original linear system of (4.11) be denote by $\mathbf{A}\sigma = \mathbf{f}$ and let $\{j_1, j_2, \ldots, j_q\}$ denote the indices of q triangles which are to be deleted. Consider the augmented linear system
$$\begin{pmatrix} \mathbf{A} & \mathbf{B}_- \\ \mathbf{C}_- & \mathbf{D}_- \end{pmatrix} \begin{pmatrix} \sigma \\ \sigma_- \end{pmatrix} = \begin{pmatrix} \mathbf{f} \\ 0 \end{pmatrix}, \tag{7.3}$$
where $\mathbf{D}_- \in \mathbb{R}^{q \times q}$, $\mathbf{B}_- \in \mathbb{R}^{N \times q}$, and $\mathbf{C}_- \in \mathbb{R}^{q \times N}$ with
$$\mathbf{D}_-(i, j) = \delta_{ij},$$
$$\mathbf{B}_-(i, l) = \begin{cases} 0 & \text{if } i = j_l, \\ \mathbf{A}(i, j_l) & \text{if } i \neq j_l, \end{cases}$$
$$\mathbf{C}_-(l, i) = \begin{cases} 1 & \text{if } i = j_l, \\ 0 & \text{if } i \neq j_l. \end{cases}$$
Then the solution components in σ are equal to those that would be obtained from the linear system (4.11) with the triangles $\{T_{j_1}, T_{j_2}, \ldots, T_{j_q}\}$ deleted.

7.3. Simultaneous addition and deletion of triangles

The low-rank modification of the linear system (4.11) corresponding to adding p triangles $\{T_{N+1}, \ldots, T_{N+p}\}$ and subtracting q triangles with indices $\{j_1, \ldots, j_q\}$ can be carried out simultaneously. It is straightforward to see that the final system has the following structure:

$$\begin{pmatrix} \mathbf{A} & \mathbf{B}_+ & \mathbf{B}_- \\ \mathbf{C}_+ & \mathbf{D}_+ & \mathbf{B}_* \\ \mathbf{C}_- & 0 & \mathbf{D}_- \end{pmatrix} \begin{pmatrix} \sigma \\ \sigma_+ \\ \sigma_- \end{pmatrix} = \begin{pmatrix} \mathbf{f} \\ \mathbf{f}_p \\ 0 \end{pmatrix}, \qquad (7.4)$$

where $\mathbf{B}_+, \mathbf{B}_-, \mathbf{C}_+, \mathbf{C}_-, \mathbf{D}_+, \mathbf{D}_-, \mathbf{f}_p$ are defined above and

$$\mathbf{B}_*(l, m) = A(T_{N+i}, T_{j_m}).$$

8. Numerical results II

Let us now consider a triangulated aircraft with 28252 elements (Figure 8.1). To compute potential flow around the aircraft, we assume the velocity field is given by

$$\nabla \Phi = \nabla \Phi^{\text{in}} + \nabla \Phi^{\text{scat}},$$

where

$$\nabla \Phi^{\text{in}} = (1, 0, 0).$$

That is, we assume the incoming velocity is uniform and oriented along the

Figure 8.1. A triangulated aircraft.

Figure 8.2. The induced potential on the aircraft surface.

x-axis. Φ^{scat} is the 'scattered' field induced by the aircraft in response to the zero normal flow (Neumann) boundary condition:

$$\frac{\partial \Phi}{\partial \nu} = 0$$

or

$$\frac{\partial \Phi^{\text{scat}}}{\partial \nu} = -\frac{\partial \Phi^{\text{in}}}{\partial \nu}.$$

For this the single-layer potential representation yields the integral equation (4.9) with $\lambda = -1$.

After compression to two digits of accuracy, the skeletonized Schur complement structure contains 4759 triangles. Our single-level scheme requires about 15 minutes for this step, while the cost for each subsequent application of the inverse is 0.2 seconds.

Suppose now that we introduce an auxiliary 'flap' (Figure 8.3) with fourteen new triangles. This is not a well-resolved structure: it is simply being used to illustrate the capability of the algorithm discussed in the preceding section. Using the updating method of the preceding section, the inverse matrix had to be applied to fourteen right-hand sides at a total cost of 14×0.2 seconds.

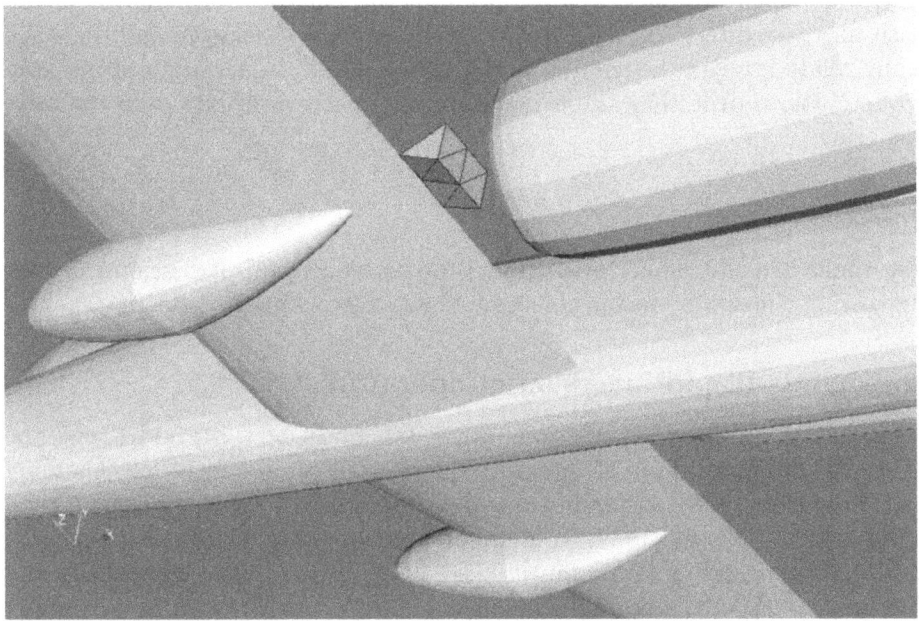

Figure 8.3. A small flap is added to the geometry.

When the number of triangles being added or subtracted is large, the Sherman–Morrison–Woodbury approach becomes inefficient. Rather than doing the update exactly, one can solve the system (7.4) iteratively, with preconditioner

$$\begin{pmatrix} \mathbf{A}^{-1} & 0 & 0 \\ 0 & \mathbf{I} & 0 \\ 0 & 0 & \mathbf{I} \end{pmatrix}. \tag{8.1}$$

9. Conclusions

During the past twenty years, many previously intractable problems in computational physics and engineering were brought within reach by the combination of hardware improvements and fast algorithms. For computational tools to be well-integrated into design processes, however, we will need more. Fast, direct methods are natural candidates for this, since they lead to very fast schemes for problems involving multiple right-hand sides, and rapid updating for modest changes to the system matrix. It is also worth noting that this technology will have an impact in many time-dependent simulations where implicit methods play a role. Such methods typically require the solution of some partial differential equation at each time step, often in a fixed geometry.

In this paper, we have tried to present some of the ideas that form the basis for these direct solvers. This is an active area of research and there are many related schemes currently under development in a variety of research groups. We expect that these methods will mature rapidly over the next few years.

Acknowledgements

We would like to thank David Bindel for several useful discussions and for suggesting the sparse linear-algebraic viewpoint outlined in Section 3.

Appendix: Rapid inversion of operators.

As indicated in the Introduction, there is a body of work addressing the efficient, direct solution of integral equations: see, for instance, Canning and Rogovin (1998), Chandrasekaran et al. (2006), Chen (2002), Cheng et al. (2005), Chew (1989), Eidelman and Gohberg (1999), Gope et al. (2005), Greengard and Rokhlin (1991), Hackbusch (1999), Hackbusch and Khoromskij (2000), Lee and Greengard (1997), Martinsson and Rokhlin (2005, 2007), Michielssen et al. (1996), Pals (2004), Starr and Rokhlin (1994), and Zhu and White (2005).

Here, we illustrate the basic idea of these multilevel schemes. For this, suppose that we would like to solve the system of linear-algebraic equations

$$AX = Y, \tag{A.1}$$

with the $n \times n$ matrix A defined by the formula

$$A_{ij} = \log(|y_i - x_j|), \tag{A.2}$$

whenever $i \neq j$, and

$$A_{ii} = 1, \tag{A.3}$$

with the points $x_1, x_2, \ldots, x_n \in R$ defined by the formulae

$$x_i = -1 + \frac{2(i-1)}{(n-1)}. \tag{A.4}$$

In other words, the points $\{x_i\}$ are equispaced on the interval $[-1, 1]$ (note the similarity to the Example 2 above). Clearly, the matrix A is dense, and solving the system (2.9) directly will require order n^3 operations. Below, we outline a simple multilevel direct procedure that will solve (2.9) in order $n(\log(n))^2$ operations.

For simplicity, we will assume that $n = m\,2^k$, where m is a small integer number ($m \sim 20$ is a reasonable choice), and k is a positive integer. We start by subdividing the interval $[-1, 1]$ into two subintervals $[a, j], [j, q]$, with $a = -1, j = 0$, and $q = 1$. Each of the intervals $[a, j], [j, q]$ is subdivided into

Figure A.1. The interval $[-1, 1]$ is recursively subdivided four times.

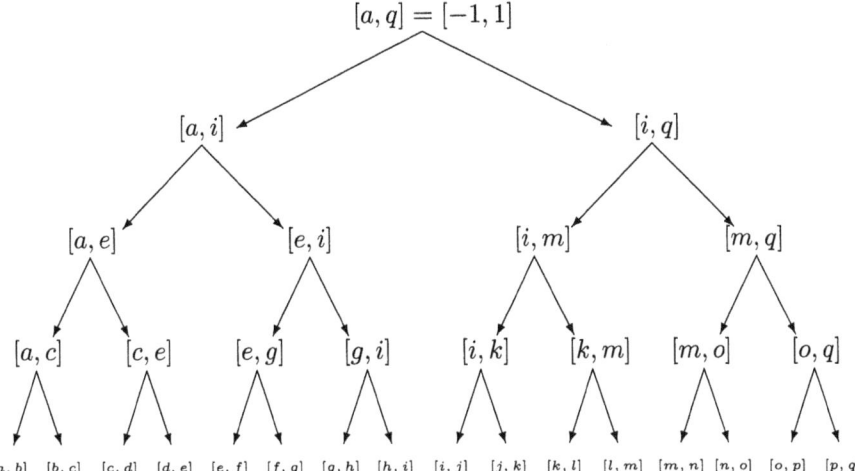

Figure A.2. A depiction of the tree structure created by recursive subdivision.

two subintervals of equal length: $[a, j] = [a, f] \cup [f, j]$, $[j, q] = [j, n] \cup [n, q]$. After continuing this process of recursive splitting for two more steps, we obtain the situation depicted in Figure A.1.

We impose a tree structure on the obtained subdivision, as depicted in Figure A.2; now the intervals $[a, j]$, $[j, q]$, are children of the interval $[a, q]$; the intervals $[a, f]$, $[f, j]$ are children of the interval $[a, j]$, etc.

Figure A.4(a) depicts the matrix A, subdivided into a collection of submatrices; with the exception of the diagonal blocks (marked with X), each of the submatrices is responsible for the interaction of a pair of subintervals depicted in Figure A.1 having the same parent (as depicted in Figure A.2).

Observation A.1. Due to Lemma 2.1 above, the numerical rank of each of the submatrices marked with L in Figure A.4 is roughly $\log(n) \cdot \log(1/\varepsilon)$, and we will assume that it has been represented in the form

$$L = \sum_{i=1}^{p} \alpha_i \beta_i^*, \tag{A.5}$$

where $p \sim \log(n) \cdot \log(1/\varepsilon)$, and $\alpha_1, \alpha_2, \ldots \alpha_p$, $\beta_1, \beta_2, \ldots \beta_p$, are vectors of appropriate dimensionality.

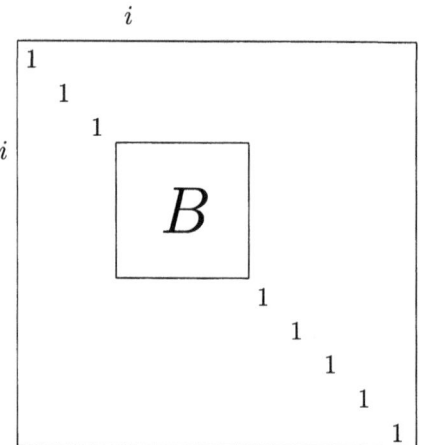

Figure A.3. The matrix $B^{n,i}$ is derived from the $n \times n$ identity matrix by replacing the diagonal block of dimension $m \times m$ beginning at the (i, i) entry with B.

The above observation leads to a 'fast' procedure for the solution of (2.9) in a fairly straightforward manner; below is a description of a rudimentary scheme of this type. We start with a definition.

Definition A.1. Suppose that m, n, i are three positive integers, such that $m \leq n$, $i + m \leq n + 1$, and that B is an $m \times m$ matrix. We will denote by $B^{n,i}$ the $n \times n$ matrix obtained from the $n \times n$ identity matrix by replacing with B the $m \times m$ diagonal block starting at the ith row, and ending at the $(i+m-1)$st row (see Figure A.3). In other words, when the operator $B^{n,i} : R^n \to R^n$ acts on the vector $x \in R^n$, it applies the operator $B : R^m$ to R^m to a subset of the vector x consisting of the elements $x_i, x_{i+1}, x_{i+2}, \ldots, x_{i+m-1}$, and leaves the rest of the elements of x unchanged.

Step 1. We start by inverting each of the diagonal blocks $X_1, X_2, \ldots X_k$ in A (see Figure A.4(a)), obtaining their inverses $Y_1, Y_2, \ldots Y_k$, and multiplying A by the product

$$Y = Y_1^{n,1} \circ Y_2^{n,m+1} \circ Y_3^{n,2m+1} \circ \cdots \circ Y_{2^k}^{n,(2^k-1)m+1}. \qquad (A.6)$$

We observe that the matrices $Y_1^{n,1}, Y_2^{n,m+1}, \ldots, Y_{2^k}^{n,(2^k-1)m+1}$ commute with each other, and that the product $B_1 = YA$ has the form depicted in Figure A.4(a) (the 'I' in the diagonal blocks denote identity matrices). We also observe that the cost of this step is of the order $2^k \cdot m^3 + \log_2(n) \cdot m^2$.

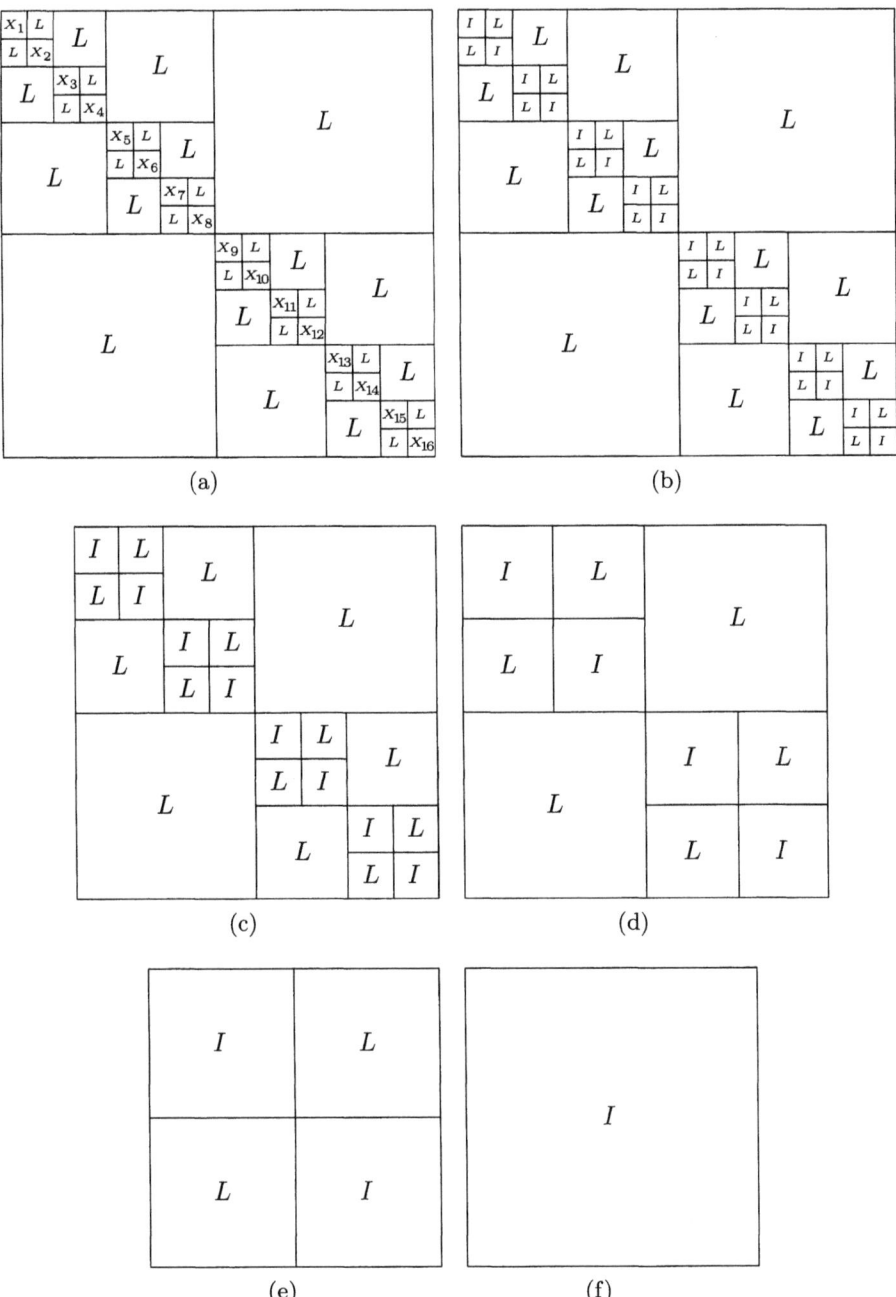

Figure A.4. The system matrix for Example 4 is shown in (a). Each off-diagonal block matrix corresponds to the interactions of a pair of subintervals having the same parent. A graphical depiction of the inversion process is shown in (b)–(f).

Step 2. We observe that each of the diagonal blocks of dimensionality $2m \times 2m$ in B_1 (depicted in Figure A.1) is a sum of the identity and a matrix of rank $2p$ (see (A.5) above) and that the total number of such blocks in Figure A.4(a) is 2^{k-1}. Denoting the inverse of the jth of these blocks by Z_j, we multiply the matrix B_1 from the left by the product

$$Z = Z_1^{n,1} \circ Z_2^{n,m+1} \circ Z_3^{n,2m+1} \circ \cdots \circ Z_{2^k}^{n,(2^k-1)m+1}, \tag{A.7}$$

obtaining the matrix B_2 depicted in Figure A.4(b). We also apply Z to the right-hand side, so that the resulting system of linear equations is equivalent to the initial one. The cost of this step is of the order $2^{(k-1)} \cdot m^2 + \log_2(n) \cdot m^2$.

Step 3. This step is identical to Step 2, except the dimensionality of the diagonal blocks is doubled, and the number of these blocks is halved; the result is depicted in Figure A.4(c), and the cost remains of the order $2^{(k-1)} \cdot m^2 + \log_2(n) \cdot m^2$.

Step k. This final step is illustrated in Figures A.4(e) and A.4(f), and its result is the identity matrix in dimension $m \cdot 2^k = n$; its cost is the same as that of Steps 1 and 2.

Summing up the costs of all Steps 1 to k, and observing that $k \sim \log_2(n)$, we obtain the total cost $O(n \log^4(n))$.

Remark A.1. While the cost of the simple scheme described above is of the order $n \log^4(n)$, fairly simple modifications produce algorithms whose cost is proportional to n (see, for example, Martinsson and Rokhlin (2005)). In higher dimensions, logarithmic factors can not be eliminated entirely (at least, not with the currently available techniques), and one ends up with algorithms costing $O(n \log^\alpha(n))$ operations, with $\alpha \in [1, 3]$. This is an active area of research, and both the algorithms and the resulting estimates are undergoing rapid evolution.

REFERENCES

B. Alpert and V. Rokhlin (1991), 'A fast algorithm for the evaluation of Legendre expansions', *SIAM J. Sci. Statist. Comput.* **12**, 158–179.

B. Alpert, G. Beylkin, R. Coifman and V. Rokhlin (1993), 'Wavelet-like bases for the fast solution of second-kind integral equations', *SIAM J. Sci. Comput.* **14**, 159–184.

M. D. Altman, J. P. Bardhan, B. Tidor and J. K. White (2006), 'FFTSVD: A fast multiscale boundary-element method solver suitable for bio-MEMS and biomolecule simulation', *IEEE Trans. Computer-Aided Design of Integrated Circuits and Systems* **25**, 274–284.

J. P. Bardhan, M. D. Altman, S. M. Lippow, B. Tidor and J. K. White (2005), 'A curved panel integration technique for molecular surfaces', in *Proc. NSTI-Nanotech 2005 Conf.*, Vol. 1, pp. 512–515.

G. Beylkin, R. Coifman and V. Rokhlin (1991), 'Fast wavelet transforms and numerical algorithms I', *Comm. Pure Appl. Math.* **14**, 141–183.

A. H. Boschitsch, M. O. Fenley and W. K. Olson (1999), 'A fast adaptive multipole algorithm for calculating screened Coulomb (Yukawa) interactions', *J. Comput. Phys.* **151**, 212.

F. X. Canning and K. Rogovin (1998), 'Fast direct solution of standard moment-method matrices', *IEEE Antennas Propag.* **40**, 15–26.

J. Carrier, L. Greengard and V. Rokhlin (1988), 'A fast adaptive multipole algorithm for particle simulations', *SIAM J. Sci. Statist. Comput.* **9**, 669–686.

S. Chandrasekaran, P. Dewilde, M. Gu, W. Lyons and T. Pals (2006), 'A fast solver for HSS representations via sparse matrices', *SIAM J. Matrix Anal. Appl.* **29**, 67–81.

Y. Chen (2002). 'Fast direct solver for the Lippman–Schwinger equation', *Adv. Comput. Math.* **16**, 175–190.

H. Cheng, Z. Gimbutas, P. G. Martinsson and V. Rokhlin (2005), 'On the compression of low rank matrices', *SIAM J. Sci. Comput.* **26**, 1389–1404.

W. C. Chew (1989). 'An N^2 algorithm for the multiple scattering solution of N scatterers', *Micro. Opt. Tech. Lett.* **2**, 380–383.

W. C. Chew, J.-M. Jin, E. Michielssen and J. Song, editors (2001). *Fast and Efficient Algorithms in Computational Electromagnetics*, Artech House, Boston.

J. W. Cooley and J. W. Tukey (1965), 'An algorithm for the machine calculation of complex Fourier series', *Math. Comput.* **19**, 297–301.

E. Darve and P. Have (2004), 'Fast multipole method for Maxwell equations stable at all frequencies', *Royal Soc. London, Trans. Ser. A* **362**, 603–628.

M. E. Davis and J. A. McCammon (1990), 'Electrostatics in biomolecular structure and dynamics' *Chem. Rev.* **90**, 509–521.

Y. Eidelman and I. Gohberg (1999). 'Linear complexity inversion algorithms for a class of structured matrices', *Integral Equations Operator Theory* **35**, 28–52.

Z. Gimbutas (1999), A generalized fast multipole method for non-oscillatory kernels. PhD Dissertation, Yale University.

D. Gope, I. Chowdhury and V. Jandhyala (2005), 'DiMES: Multilevel fast direct solver based on multipole expansions for parasitic extraction of massively coupled 3D microelectronic structures', in *Proc. 42nd Annual Conference on Design Automation*, pp. 159–162.

S. A. Goreinov, E. E. Yrtyshnikov and N. L. Zamarashkin (1997), 'A theory of pseudoskeleton approximations', *Linear Algebra Appl.* **261**, 1–21.

I. S. Gradshteyn and I. M. Ryzhik (2000), *Table of Integrals, Series, and Products*, Academic Press, New York.

L. Greengard and J. Helsing (1998), 'On the numerical evaluation of elastostatic fields in locally isotropic two-dimensional composites', *J. Mech. Phys. Solids* **46**, 1441–1462.

L. Greengard and V. Rokhlin (1987), 'A fast algorithm for particle simulations', *J. Comput. Phys.* **73**, 325–348.

L. Greengard and V. Rokhlin (1991), 'On the numerical solution of two-point boundary value problems', *Comm. Pure Appl. Math.* **44**, 419–452.

L. Greengard and V. Rokhlin (1997), A new version of the fast multipole method for the Laplace equation in three dimensions. In *Acta Numerica*, Vol. 6, Cambridge University Press, pp. 229–269.

L. Greengard and J. Strain (1991). 'The Fast Gauss Transform', *SIAM J. Sci. Statist. Comput.* **12**, 79–94.

M. Gu and S. C. Eisenstat (1996), 'Efficient algorithms for computing a strong rank-revealing QR factorization', *SIAM J. Sci. Comput.* **17**, 848–869.

R. B. Guenther and J. W. Lee (1988), *Partial Differential Equations of Mathematical Physics and Integral Equations*, Prentice-Hall, Englewood Cliffs, NJ.

W. Hackbusch (1999), 'A sparse matrix arithmetic based on H-matrices I: Introduction to H-matrices', *Computing* **62**, 89–108.

W. Hackbusch and B. N. Khoromskij (2000), 'A sparse H-matrix arithmetic II: Application to multi-dimensional problems', *Computing* **64**, 21–47.

W. Hackbusch and Z. P. Nowak (1989), 'On the fast matrix multiplication in the boundary element method by panel clustering', *Numer. Math.* **54**, 463–491.

M. J. Holst, N. A. Baker and F. Wang (2000), 'Adaptive multilevel finite element solution of the Poisson–Boltzmann equation I: Algorithms and examples', *J. Comput. Chem.* **21**, 1319–1342.

J. Huang and L. Greengard (2002), 'A new version of the fast multipole method for screened Coulomb interactions interactions in three dimensions', *J. Comput. Phys.* **180**, 642–658.

D. L. James and D. K. Pai (1999), 'ArtDefo: Accurate real time deformable objects', in *Proc. SIGGRAPH 99*, pp. 65–72.

S. Kapur and D. Long (1997), 'IES^3: A fast integral equation solver for efficient 3-D extraction', in *Proc. IEEE International Conference on Computer Aided Design*, pp. 448–455.

R. Kastner (1989), 'An 'add-on method' for the analysis of scattering from large planar structures', *IEEE Trans. Antennas Propag.* **37**, 353–361.

S. S. Kuo, M. D. Altman, J. P. Bardhan, B. Tidor and J. K. White (2002), 'Fast methods for simulation of biomolecular electrostatics', in *Proc. IEEE–ACM Int. Conf. Comput. Aided Design*, pp. 466–473.

J.-Y. Lee and L. Greengard (1997), 'A fast adaptive numerical method for stiff two-point boundary value problems', *SIAM J. Sci. Comput.* **18**, 403–429.

J. Liang and S. Subramaniam (1997), 'Computation of molecular electrostatics with boundary element methods', *Biophys. J.* **73**, 1830.

B. Lu, X. Cheng, J. Huang and J. A. McCammon (2006), 'Order N algorithm for computation of electrostatic interactions in biomolecular systems', *Proc. Nat. Acad. Sci.* **103**, 19314–19319.

S. A. Marshall, C. L. Vizcarra and S. L. Mayo (2005), 'One- and two-body decomposable Poisson–Boltzmann methods for protein design calculations', *Protein Science* **14**, 1293–1304.

P.-G. Martinsson (2006), 'Fast evaluation of electrostatic interactions in multiphase dielectric media', *J. Comput. Phys.* **211**, 289–299.

P.-G. Martinsson and V. Rokhlin (2005), 'A fast direct solver for boundary integral equations in two dimensions', *J. Comput. Phys.* **205**, 1–23.

P.-G. Martinsson and V. Rokhlin (2007), 'A fast direct solver for scattering problems involving elongated structures', *J. Comput. Phys.* **221**, 288–302.

E. Michielssen, A. Boag and W. C. Chew (1996), 'Scattering from elongated objects: Direct solution in $O(N \log^2 N)$ operations', *IEEE Proc. H* **143**, 277–283.

K. Nabors and J. White (1991), 'FASTCAP: A multipole accelerated 3-D capacitance extraction program', *IEEE Trans. Computer-Aided Design of Integrated Circuits and Systems* **10**, 1447–1459.

N. Nishimura (2002), 'Fast multipole accelerated boundary integral equation methods', *Appl. Mech. Rev.* **55**, 299–324.

T. Pals (2004), Multipole for scattering computations: Spectral discretization, stabilization, fast solvers. PhD Dissertation, Department of Electrical and Computer Engineering, University of California, Santa Barbara.

J. R. Phillips and J. White (1997), 'A precorrected-FFT method for electrostatic analysis of complicated 3-D structures', *IEEE Trans. Computer-Aided Design* **16**, 1059–1072.

W. Rocchia, S. Sridharan, A. Nicholls, E. Alexov, A. Chiabrera and B. Honig (2002), 'Rapid grid-based construction of the molecular surface for both molecules and geometric objects: Applications to the finite difference Poisson–Boltzmann method', *J. Comput. Chem.* **23**, 128–137.

V. Rokhlin (1985), 'Rapid solution of integral equations of classical potential theory', *J. Comput. Phys.* **60**, 187–207.

V. Rokhlin (1988), 'A fast algorithm for the discrete Laplace transformation', *J. Complexity* **4**, 12–32.

Y. Saad and M. H. Schultz (1986), 'GMRES: A generalized minimum residual algorithm for solving nonsymmetric linear systems', *SIAM J. Sci. Statist. Comput.* **7**, 856–869.

K. A. Sharp and B. Honig (1990), 'Electrostatic interactions in macromolecules: Theory and applications', *Ann. Rev. Biophys. Biophys. Chem.* **19**, 301–332.

J. Sherman and W. J. Morrison (1949), 'Adjustment of an inverse matrix corresponding to changes in the elements of a given column or a given row of the original matrix', *Ann. Math. Statist.* **20**, 621.

P. Starr and V. Rokhlin (1994). 'On the numerical solution of two-point boundary value problems II', *Comm. Pure Appl. Math.* **47**, 1117–1159.

J. Strain (1992), 'The fast Laplace transform based on Laguerre functions', *Math. Comp.* **58**, 275–284.

M. A. Woodbury (1950), Inverting modified matrices. Memorandum Report 42, Statistical Research Group, Princeton University.

L. Ying, G. Biros, D. Zorin and M. H. Langston (2003), 'A new parallel kernel-independent fast multipole method', in *Proc. ACM/IEEE Conf. on Supercomp.*, p. 14.

Z. Zhu and J. K. White (2005), 'FastSies: A fast stochastic integral equation solver for modeling the rough surface effect', in *Proc. 2005 IEEE/ACM Int. Conference on Computer-Aided Design*, pp. 675–682.

Online references

Protein Data Bank: www.rcsb.org
STLib package: www.cacr.caltech.edu/~sean/projects/stlib/html/mst

Blow-up or no blow-up? A unified computational and analytic approach to 3D incompressible Euler and Navier–Stokes equations

Thomas Y. Hou
Applied and Computational Mathematics, 217-50,
Caltech, Pasadena, CA 91125, USA
E-mail: hou@acm.caltech.edu

Whether the 3D incompressible Euler and Navier–Stokes equations can develop a finite-time singularity from smooth initial data with finite energy has been one of the most long-standing open questions. We review some recent theoretical and computational studies which show that there is a subtle dynamic depletion of nonlinear vortex stretching due to local geometric regularity of vortex filaments. We also investigate the dynamic stability of the 3D Navier–Stokes equations and the stabilizing effect of convection. A unique feature of our approach is the interplay between computation and analysis. Guided by our local non-blow-up theory, we have performed large-scale computations of the 3D Euler equations using a novel pseudo-spectral method on some of the most promising blow-up candidates. Our results show that there is tremendous dynamic depletion of vortex stretching. Moreover, we observe that the support of maximum vorticity becomes severely flattened as the maximum vorticity increases and the direction of the vortex filaments near the support of maximum vorticity is very regular. Our numerical observations in turn provide valuable insight, which leads to further theoretical breakthrough. Finally, we present a new class of solutions for the 3D Euler and Navier–Stokes equations, which exhibit very interesting dynamic growth properties. By exploiting the special nonlinear structure of the equations, we prove nonlinear stability and the global regularity of this class of solutions.

CONTENTS

1. Introduction   278
2. Dynamic depletion of vortex stretching in 3D Euler equations   282
3. Dynamic stability of 3D Navier–Stokes equations   305
4. Stabilizing effect of convection for 3D Navier–Stokes   320
5. Concluding remarks   341
References   343

1. Introduction

The question of whether the 3D incompressible Navier–Stokes equations can develop a finite-time singularity from smooth initial data is one of the most long-standing open problems in fluid dynamics and mathematics. This is also one of the seven Millennium Open Problems posted by the Clay Mathematical Institute (see www.claymath.org). The understanding of this problem could improve our understanding on the onset of turbulence and the intermittency properties of turbulent flows.

The 3D incompressible Navier–Stokes equations are given by

$$\mathbf{u}_t + (\mathbf{u} \cdot \nabla)\mathbf{u} = -\nabla p + \nu \Delta \mathbf{u}, \qquad (1.1)$$

$$\nabla \cdot \mathbf{u} = 0, \qquad (1.2)$$

with initial condition $\mathbf{u}(\mathbf{x}, 0) = \mathbf{u}_0(\mathbf{x})$. Here $\mathbf{u}$ is velocity, p is pressure, and ν is viscosity. We consider only the initial value problem and assume that the solution decays rapidly at infinity. Defining vorticity by $\boldsymbol{\omega} = \nabla \times \mathbf{u}$, then $\boldsymbol{\omega}$ is governed by

$$\boldsymbol{\omega}_t + (\mathbf{u} \cdot \nabla)\boldsymbol{\omega} = \nabla \mathbf{u} \cdot \boldsymbol{\omega} + \nu \Delta \boldsymbol{\omega}. \qquad (1.3)$$

The first term on the right-hand side of (1.3) is called the vortex stretching term, which is absent in the two-dimensional problem. Note that $\nabla \mathbf{u}$ is formally of the same order as $\boldsymbol{\omega}$. Thus the vortex stretching term has a formal quadratic scaling with respect to vorticity. This formal quadratic nonlinearity in the vortex stretching term is the main difficulty in studying the dynamic stability and global regularity of the 3D Navier–Stokes equations. Under suitable smallness assumptions on the initial condition, global existence and regularity results have been obtained for some time (Ladyzhenskaya 1970, Constantin and Foias 1988, Temam 2001, Majda and Bertozzi 2002). But these methods based on energy estimates do not generalize to the 3D Navier–Stokes with large data. Energy estimates seem to be too crude to give a definite answer to whether diffusion is strong enough to control the nonlinear growth due to vortex stretching. A more refined analysis seems to be needed, which takes into account the special nature of the nonlinearities and their local interactions.

We believe that the global regularity of the 3D Navier–Stokes equations is closely related to that of the 3D Euler equations. Since the nonlinearity of the 3D Navier–Stokes equations is supercritical, the balance among different nonlinear terms in the Euler equations may play an even more important role than the diffusion term. Thus, it makes sense to investigate the mechanism which may lead to finite-time blow-up or dynamic depletion of the nonlinear vortex stretching in the 3D Euler equations.

There has been some interesting development in the theoretical understanding of the 3D incompressible Euler equations. In particular, Constantin, Fefferman and Majda have shown that the local geometric regularity of vortex lines can play an important role in depleting nonlinear vortex stretching (Constantin 1994, Constantin, Fefferman and Majda 1996). Inspired by their work, Deng, Hou and Yu (2005, 2006a) recently showed that geometric regularity of vortex lines, even in an extremely localized region containing the maximum vorticity, can lead to depletion of nonlinear vortex stretching, thus avoiding finite-time singularity formation of the 3D Euler equations. To obtain these results, Deng, Hou and Yu used a Lagrangian approach and explored the connection between the stretching of local vortex lines and the growth of vorticity. In particular, they showed that if the vortex lines near the region of maximum vorticity satisfy some local geometric regularity conditions and the maximum velocity field is integrable in time, then no finite-time blow-up is possible. These localized non-blow-up criteria provide stronger constraints on the local geometry of a potential finite-time singularity.

There have been many computational attempts to find finite-time singularities of the 3D Euler and Navier–Stokes equations: see, *e.g.*, Chorin (1982), Pumir and Siggia (1990), Kerr and Hussain (1989), Grauer and Sideris (1991), Shelley, Meiron and Orszag (1993), Kerr (1993), Caflisch (1993), Boratav and Pelz (1994), Fernandez, Zabusky and Gryanik (1995), Pelz (1997), Grauer, Marliani and Germaschewski (1998), Kerr (2005). One example that has been studied extensively is the interaction of two perturbed antiparallel vortex tubes. This example is interesting because of the vortex reconnection observed for the corresponding Navier–Stokes equations. It is natural to ask whether the 3D Euler equations would develop a finite-time singularity in the limit of vanishing viscosity. Kerr (1993, 2005) presented numerical evidence which suggested a finite-time singularity of the 3D Euler equations for two perturbed antiparallel vortex tubes. Kerr's blow-up scenario is consistent with the non-blow-up criterion of Beale, Kato and Majda (1984) and that of Constantin, Fefferman and Majda (1996). But it falls into the critical case of Deng, Hou and Yu's local non-blow-up criteria (Deng, Hou and Yu 2005, 2006a).

Guided by this local geometric non-blow-up analysis, Hou and Li (2006) performed *extremely large-scale computations* with resolution up to 1536 × 1024 × 3072 to re-examine Kerr's blow-up scenario (Kerr 1993). They used a novel pseudo-spectral method with a 36th-order Fourier smoothing function which keeps a significant portion of the Fourier modes beyond the 2/3 cut-off point in the Fourier spectrum for the 2/3 de-aliasing rule. Their extensive numerical results demonstrated that the pseudo-spectral method with the high-order Fourier smoothing gives a much better performance than the pseudo-spectral method with the 2/3 de-aliasing rule. In particular, they showed that the Fourier smoothing method captures about 12 ∼ 15% more effective Fourier modes than the 2/3 de-aliasing method in each dimension. For 3D Euler equations, the total number of effective modes in the Fourier smoothing method is about 20% more than that in the 2/3 de-aliasing method. This is a very significant increase in the resolution for a large-scale computation.

There were several interesting findings in the large-scale computations of Hou and Li (2006) for the 3D Euler equations using the initial data for the antiparallel vortex tubes. First, they discovered a surprising dynamic cancellation in the vortex stretching term due to the local geometric regularity of the vortex filaments. Vortex stretching was found to deplete dynamically from a formally quadratic nonlinearity to a much weaker $O(\omega \log(\omega))$ type of nonlinearity, which leads to only double exponential growth in the maximum vorticity. Secondly, they showed that the velocity field is bounded up to $T = 19$, beyond the alleged singularity time $T = 18.7$ of Kerr (2005). With a bounded velocity field, the non-blow-up criterion of Deng, Hou and Yu (2005) applies, which provides theoretical support for their computational results. Thirdly, they found that the vorticity vector near the point of maximum vorticity aligns almost perfectly with the second eigenvector of the rate of strain tensor. The second eigenvalue of the rate of strain tensor is the smallest eigenvalue and does not seem to grow dynamically, while the first and third eigenvalues grow very rapidly in time. This is further strong evidence for the dynamic depletion of vortex stretching.

Inspired by the numerical findings of their paper of 2006, Hou and Li (2008a) investigated the dynamic stability of the 3D Navier–Stokes equations by introducing an exact 1D model of the axisymmetric Navier–Stokes equations along the symmetry axis. This 1D model is exact in the sense that one can construct a family of exact solutions for the 3D Navier–Stokes equations from this 1D model. Thus the 1D model preserves some essential features of the 3D Navier–Stokes equations. What is surprising is that they obtained a Lyapunov function which satisfies a new maximum principle. This provides a pointwise estimate on the dynamic stability of the Navier–Stokes equations. The traditional energy estimates are incapable of capturing such subtle cancellation effects. Based on the global regularity of the 1D model, they constructed a new class of solutions for the 3D Euler and Navier–Stokes equations, which exhibit very interesting dynamic growth properties, but remain smooth for all times.

Motivated by the work of Hou and Li (2008a), Hou and Lei (2009b) further proposed a new 3D model to study the stabilizing effect of convection. This model was derived by neglecting the convection term from a reformulated axisymmetric Navier–Stokes equations. It shares almost all the properties of the 3D Navier–Stokes equations. In particular, the strong solution of the model satisfies an energy identity similar to that of the full 3D Navier–Stokes equations. They proved a non-blow-up criterion of Beale–Kato–Majda type as well as a non-blow-up criterion of Prodi–Serrin type for the model. Moreover, they proved that, for any suitable weak solution of the 3D model in an open set in space-time, the one-dimensional Hausdorff measure of the associated singular set is zero (Hou and Lei 2009a). This partial regularity result is an analogue of the Caffarelli–Kohn–Nirenberg theory (Caffarelli, Kohn and Nirenberg 1982) for the 3D Navier–Stokes equations.

Despite the striking similarity at the theoretical level between the 3D model and the Navier–Stokes equations, the former has a completely different behaviour from the full Navier–Stokes equations. Hou and Lei's study showed that the 3D model seems to form a finite-time singularity, while the mechanism of generating such a finite-time singularity is removed when convection is added back to the 3D model. Convection seems to play a very important role in stabilizing the potential blow-up of the Navier–Stokes equations. This result may have an important impact on future global regularity analysis of 3D Navier–Stokes equations. Up to now, most analysis uses energy estimates in which convection plays no role at all. Such global methods of analysis are too crude. Their studies suggest that one needs to develop a new localized analysis which can in essence exploit the stabilizing effect of convection.

There has been some interesting development in the study of the 3D incompressible Navier–Stokes equations and related models. By exploiting the special structure of the governing equations, Cao and Titi (2007) proved the global well-posedness of the 3D viscous primitive equation for large-scale ocean and atmospheric dynamics. For the axisymmetric Navier–Stokes equations, Chen, Strain, Tsai and Yau (2008, 2009) and Koch, Nadirashvili, Seregin and Sverak (2009) recently proved that if $|\mathbf{u}(x,t)| \leq C_*|t|^{-1/2}$, where C_* is allowed to be large, then the velocity field $\mathbf{u}$ is regular at time zero. The 2D Boussinesq equations are closely related to the 3D axisymmetric Navier–Stokes equations with swirl (away from the symmetry axis). Recently, Hou and Li (2005) and Chae (2006) proved independently the global existence of the 2D Boussinesq equations with partial viscosity. By taking advantage of the limiting property of some rapidly oscillating operators and using nonlinear averaging, Babin, Mahalov and Nicolaenko (2001) proved global regularity of the 3D Navier–Stokes equations for some initial data characterized by uniformly large vorticity.

The rest of the paper is organized as follows. In Section 2, we study the dynamic depletion of vortex stretching for the 3D Euler equations. We also discuss at length how to design an effective high-resolution pseudo-spectral method to compute potentially singular solutions of the 3D Euler equations. Section 3 is devoted to studying the dynamic stability of the 3D Navier–Stokes equations. In Section 4, we investigate the stabilizing effect of convection for the 3D Navier–Stokes equations. Some concluding remarks are made in Section 5.

2. Dynamic depletion of vortex stretching in 3D Euler equations

Due to the supercritical nature of the nonlinearity of the 3D Navier–Stokes equations, the 3D Navier–Stokes equations with large initial data are convection-dominated, instead of diffusion-dominated. For this reason, we believe that the understanding of whether the corresponding 3D Euler equations would develop a finite-time blow-up could shed useful light on the global regularity of the Navier–Stokes equations.

Let us consider the 3D Euler equations in the vorticity form. One important observation is that when we consider the convection term together with the vortex stretching term, the two nonlinear terms can be actually represented as a commutator or a Lie derivative:

$$\boldsymbol{\omega}_t + (\mathbf{u} \cdot \nabla)\boldsymbol{\omega} - (\boldsymbol{\omega} \cdot \nabla)\mathbf{u} = 0. \tag{2.1}$$

It is reasonable to believe that the commutator would lead to some cancellation among the two nonlinear terms, thus weakening the nonlinearity dynamically. This points to the potential important role of convection in the 3D Euler equations. Another way to realize the importance of convection is to use the Lagrangian formulation of the vorticity equation. When we consider the two terms together, we preserve the Lagrangian structure of the solution (Chorin and Marsden 1993),

$$\boldsymbol{\omega}(X(\alpha,t),t) = X_\alpha(\alpha,t)\boldsymbol{\omega}_0(\alpha), \tag{2.2}$$

where $X_\alpha = \frac{\partial X}{\partial \alpha}$ and $X(\alpha, t)$ is the flow map,

$$\frac{\mathrm{d}X}{\mathrm{d}t}(\alpha,t) = \mathbf{u}(X(\alpha,t),t), \quad X(\alpha,0) = \alpha. \tag{2.3}$$

Therefore, vorticity increases in time only through the dynamic deformation of the Lagrangian flow map. On the other hand, due to the divergence-free property of the velocity field, the flow map is volume-preserving, that is,

$\det(X_\alpha(\alpha, t)) \equiv 1$. Thus, as vorticity increases dynamically, the parallelepiped spanned by the three vectors, $(X_{\alpha_1}, X_{\alpha_2}, X_{\alpha_3})$, will experience severe deformation and become flattened dynamically. A formal asymptotic analysis shows that the support of maximum vorticity also experiences a similar deformation and becomes severely flattened as vorticity increases. This is confirmed by our numerical experiments: see Section 2.5. Such deformation tends to weaken the nonlinearity of vortex stretching dynamically.

We remark that convection plays an essential role in deforming the support of maximum vorticity and induces an anisotropic scaling in the collapse of the support of maximum vorticity. By exploiting the anisotropic scaling of the support of maximum vorticity, Hou, Lei and Li (2008) recently proved the global regularity of the axisymmetric Navier–Stokes equations with a family of very large anisotropic initial data: see Section 2.8 for more discussions. On the other hand, if we ignore the convection term in the Euler equations, the vortex stretching term may indeed achieve the $O(|\omega|^2)$ scaling dynamically and develop an isotropic singularity in finite time: see Section 4 for more discussions.

2.1. A brief review

We begin with a brief review of the subject. Due to the formal quadratic nonlinearity in vortex stretching, only short time existence is known for the 3D Euler equations (Majda and Bertozzi 2002). One of the most well-known results on the 3D Euler equations is due to Beale, Kato and Majda (1984), who showed that the solution of the 3D Euler equations blows up at T if and only if $\int_0^T \|\omega\|_\infty(t)\, dt = \infty$, where ω is vorticity.

There have been some interesting recent theoretical developments. In particular, Constantin, Fefferman and Majda (1996) showed that local geometric regularity of the unit vorticity vector can lead to depletion of the vortex stretching. Let $\xi = \omega/|\omega|$ be the unit vorticity vector and let $\mathbf{u}$ be the velocity field. Roughly speaking, Constantin, Fefferman and Majda proved that if (1) $\|\mathbf{u}\|_\infty$ is bounded in a $O(1)$ region containing the maximum vorticity, (2) $\int_0^t \|\nabla \xi\|_\infty^2\, d\tau$ is uniformly bounded for $t < T$, then the solution of the 3D Euler equations remains regular up to $t = T$.

There has been some numerical evidence that suggests a finite-time blow-up of the 3D Euler equations. One of the most well-known examples is the finite-time collapse of two antiparallel vortex tubes by R. Kerr (1993, 2005). In his computations, Kerr used a pseudo-spectral discretization in the x- and y-directions, and a Chebyshev discretization in the z-direction with resolution of order $512 \times 256 \times 192$. His computations showed that the maximum vorticity blows up like $O((T-t)^{-1})$ with $T = 18.9$. In his subsequent paper, Kerr (2005) applied a high wavenumber filter to the data obtained in his original computations to 'remove the noise that masked the structures in earlier graphics' presented in the 1993 paper. With this filtered solution, he presented some scaling analysis of the numerical solutions up to $t = 17.5$. Two new properties were presented in the 2005 paper. First, the velocity field was shown to blow up like $O(T-t)^{-1/2}$ with T being revised to $T = 18.7$. Secondly, he showed that the blow-up is characterized by two anisotropic length scales, $\rho \approx (T-t)$ and $R \approx (T-t)^{1/2}$. It is worth noting that there is still a considerable gap between the predicted singularity time $T = 18.7$ and the final time $t = 17$ of Kerr's original computations, which he used as the primary evidence for the finite-time singularity.

Kerr's blow-up scenario is consistent with the non-blow-up criterion of Beale, Kato and Majda (1984) and that of Constantin, Fefferman and Majda (1996). But it falls into the critical case of Deng, Hou and Yu's local non-blow-up criteria (Deng, Hou and Yu 2005, 2006a). Below we describe the local non-blow-up criteria of Deng, Hou and Yu.

2.2. The local non-blow-up criteria of Deng, Hou and Yu (2005, 2006a)

Motivated by the result of Constantin, Fefferman and Majda (1996), Deng, Hou and Yu (2005) have obtained a sharper non-blow-up condition which uses only very localized information of the vortex lines. Assume that at each time t there exists some vortex line segment L_t on which the local maximum vorticity is comparable to the global maximum vorticity. Further, we denote $L(t)$ as the arclength of L_t, **n** the unit normal vector of L_t, and κ the curvature of L_t.

Theorem 2.1. (Deng, Hou and Yu 2005) Assume that (1) $\max_{L_t}(|\mathbf{u} \cdot \boldsymbol{\xi}| + |\mathbf{u} \cdot \mathbf{n}|) \leq C_U(T-t)^{-A}$ with $A < 1$, and (2) $C_L(T-t)^B \leq L(t) \leq C_0/\max_{L_t}(|\kappa|, |\nabla \cdot \boldsymbol{\xi}|)$ for $0 \leq t < T$. Then the solution of the 3D Euler equations remains regular up to $t = T$ provided that $A + B < 1$.

In Kerr's computations, the first condition of Theorem 2.1 is satisfied with $A = 1/2$ if we use $\|\mathbf{u}\|_\infty \leq C(T-t)^{-1/2}$ as alleged in Kerr (2005). Kerr's computations suggested that κ and $\nabla \cdot \boldsymbol{\xi}$ are bounded by $O((T-t)^{-1/2})$ in the inner region of size $(T-t)^{1/2} \times (T-t)^{1/2} \times (T-t)$ (Kerr 2005). Moreover, the length of the vortex tube in the inner region is of order $(T-t)^{1/2}$. If we choose a vortex line segment of length $(T-t)^{1/2}$ (i.e., $B = 1/2$), then the second condition is satisfied. However, we violate the condition $A + B < 1$. Thus Kerr's computations fall into the critical case of Theorem 2.1. In a subsequent paper, Deng, Hou and Yu (2006a) improved the non-blow-up condition to include the critical case, $A + B = 1$.

Theorem 2.2. (Deng, Hou and Yu 2006a) Under the same assumptions as Theorem 2.1, in the case of $A + B = 1$, the solution of the 3D Euler equations remains regular up to $t = T$ if the scaling constants C_U, C_L and C_0 satisfy an algebraic inequality, $f(C_U, C_L, C_0) > 0$.

We remark that this algebraic inequality can be checked numerically if we obtain a good estimate of these scaling constants. For example, if $C_0 = 0.1$, which seems reasonable since the vortex lines are relatively straight in the inner region, Theorem 2.2 would imply no blow-up up to T if $2C_U < 0.43C_L$. Unfortunately, there was no estimate available for these scaling constants in Kerr (1993). One of our original motivations for repeating Kerr's computations using higher resolutions was to obtain a good estimate for these scaling constants.

2.3. Computing potentially singular solutions using pseudo-spectral methods

Computing Euler singularities numerically is an extremely challenging task. First of all, it requires huge computational resources. Tremendous resolutions are required to capture the nearly singular behaviour of the Euler equations. Secondly, one has to perform a careful convergence study. It is dangerous to interpret the blow-up of an under-resolved computation as evidence of finite-time singularities for the 3D Euler equations. Thirdly, if we believe that the numerical solution we compute leads to a finite-time blow-up, we need to demonstrate the validity of the asymptotic blow-up rate, *i.e.*, is the blow-up rate $\|\omega\|_{L^\infty} \approx \frac{C}{(T-t)^\alpha}$ asymptotically valid as $t \to T$? If a numerical solution is well resolved only up to T_0 and there is still an order-one gap between T_0 and the predicted singularity time T, then one can not apply the Beale–Kato–Majda criterion (Beale, Kato and Majda 1984) to this fitted singularity, since the most significant contribution to $\int_0^T \|\omega(t)\|_{L^\infty}\,dt$ comes from the time interval $[T_0, T]$, but there is no accuracy in the extrapolated solution in this time interval if $(T - T_0) = O(1)$. Finally, one also needs to check if the blow-up rate of the numerical solution is consistent with other non-blow-up criteria (Constantin, Fefferman and Majda 1996, Deng, Hou and Yu 2005, Deng, Hou and Yu 2006a) which provide additional constraints on the blow-up rate of the velocity field and the local geometric regularity on the vortex filaments. The interplay between theory and numerics is clearly essential in our search for Euler singularities.

Hou and Li (2006, 2007) repeated Kerr's computations using two pseudo-spectral methods. The first pseudo-spectral method used the standard 2/3 de-aliasing rule to remove the aliasing error. For the second pseudo-spectral method, they used a novel 36th-order Fourier smoothing to remove the aliasing error. For the Fourier smoothing method, they used a Fourier smoother along the x_j-direction as follows: $\rho(2k_j/N_j) \equiv \exp(-36(2k_j/N_j)^{36})$, where k_j is the wavenumber ($|k_j| \leq N_j/2$). The time integration was performed by using the classical fourth-order Runge–Kutta scheme. Adaptive time-stepping was used to satisfy the CFL stability condition with CFL number equal to $\pi/4$. In order to perform a careful resolution study, they used a sequence of resolutions: 768 × 512 × 1536, 1024 × 768 × 2048 and 1536 × 1024 × 3072 in their computations. They computed the solution up to $t = 19$, beyond the alleged singularity time $T = 18.7$ by Kerr (2005). Their computations were carried out on the PC cluster LSSC-II in the Institute of Computational Mathematics and Scientific/Engineering Computing of Chinese Academy of Sciences and the Shenteng 6800 cluster in the Super Computing Center of the Chinese Academy of Sciences. The maximal memory consumption in their computations was about 120 Gbytes. The largest number of grid points is close to 5 billion.

2.4. Convergence study of spectral methods for the Burgers equation

As a first step, we demonstrate that the two pseudo-spectral methods can be used to compute a singular solution arbitrarily close to the singularity time. For this purpose, we perform a careful convergence study of the two pseudo-spectral methods in both physical and spectral spaces for the 1D inviscid Burgers equation. The advantage of using the inviscid 1D Burgers equation is that it shares some essential difficulties with the 3D Euler equations, yet we have a semi-analytic formulation for its solution. By using the Newton iterative method, we can obtain an approximate solution to the exact solution up to 13 digits of accuracy. Moreover, we know exactly when a shock singularity will form in time. This enables us to perform a careful convergence study in both physical space and spectral space very close to the singularity time. This provides a solid foundation to the convergence study of the two spectral methods.

We consider the inviscid 1D Burgers equation

$$u_t + \left(\frac{u^2}{2}\right)_x = 0, \quad -\pi \leq x \leq \pi, \tag{2.4}$$

with an initial condition given by

$$u|_{t=0} = u_0(x).$$

We impose a periodic boundary condition over $[-\pi, \pi]$. By the method of characteristics, it is easy to show that the solution of the 1D Burgers equation is given by

$$u(x,t) = u_0(x - tu(x,t)). \tag{2.5}$$

The above implicit formulation defines a unique solution for $u(x,t)$ up to the time when the first shock singularity develops. After the shock singularity develops, equation (2.5) gives a multi-valued solution. An entropy condition is required to select a unique physical solution beyond the shock singularity (LeVeque 1992).

We now use a standard pseudo-spectral method to approximate the solution. Let N be an integer, and let $h = \pi/N$. We denote by $x_j = jh$

($j = -N, \ldots, N$) the discrete grid points over the interval $[-\pi, \pi]$. To describe the pseudo-spectral methods, we recall that the discrete Fourier transform of a periodic function $u(x)$ with period 2π is defined by

$$\hat{u}_k = \frac{1}{2N} \sum_{j=-N+1}^{N} u(x_j) e^{-ikx_j}.$$

The inversion formula reads

$$u(x_j) = \sum_{k=-N+1}^{N} \hat{u}_k e^{ikx_j}.$$

We note that $\hat{u}_k$ is periodic in k with period $2N$. This is an artifact of the discrete Fourier transform, and the source of the aliasing error. To remove the aliasing error, one usually applies some kind of de-aliasing filtering when we compute the discrete derivative. Let $\rho(k/N)$ be a cut-off function in the spectrum space. A discrete derivative operator may be expressed in the Fourier transform as

$$\widehat{(D_h u)}_k = ik\rho(k/N)\hat{u}_k, \quad k = -N+1, \ldots, N. \tag{2.6}$$

Both the 2/3 de-aliasing rule and the Fourier smoothing method can be described by a specific choice of the high-frequency cut-off function, ρ (also known as Fourier filter). For the 2/3 de-aliasing rule, the cut-off function is chosen to be

$$\rho(k/N) = \begin{cases} 1, & \text{if } |k/N| \leq 2/3, \\ 0, & \text{if } |k/N| > 2/3. \end{cases} \tag{2.7}$$

In our computations, in order to obtain an alias-free computation on a grid of M points for a quadratic nonlinear equation, we apply the above filter to the high wavenumbers so as to retain only $(2/3)M$ unfiltered wavenumbers before making the coefficient-to-grid Fast Fourier Transform. This de-aliasing procedure is alternatively known as the 3/2 de-aliasing rule because to obtain M unfiltered wavenumbers one must compute nonlinear products in physical space on a grid of $(3/2)M$ points: see p. 229 of Boyd (2000) for more discussions.

For the Fourier smoothing method, we choose ρ as follows:

$$\rho(k/N) = e^{-\alpha(|k|/N)^m}, \tag{2.8}$$

with $\alpha = 36$ and $m = 36$. In our implementation, both filters are applied to the numerical solution at every time step. Thus, for the 2/3 de-aliasing rule, the Fourier modes with wavenumbers $|k| \geq 2/3N$ are always set to zero. Thus there is no aliasing error being introduced in our approximation of the nonlinear convection term. For the Fourier smoothing method, the nonlinear term will have some non-zero modes beyond the 2/3 point cut-off

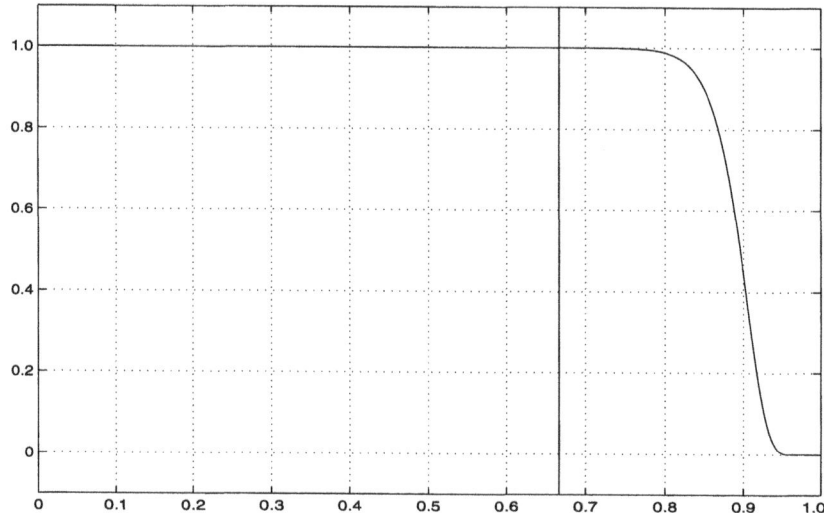

Figure 2.1. The profile of the Fourier smoothing, $\exp(-36(x)^{36})$, as a function of x. The vertical line corresponds to the cut-off point in the Fourier spectrum in the 2/3 de-aliasing rule. We can see that using this Fourier smoothing we keep about $12 \sim 15\%$ more modes than those using the 2/3 de-aliasing rule.

point in the Fourier space. However, these non-zero modes will accumulate in time to pollute the solution.

The Fourier smoothing method we choose is based on three considerations. The first one is that the aliasing instability is introduced by the highest-frequency Fourier modes. As demonstrated in Goodman, Hou and Tadmor (1994), as long as one can damp out a small portion of the highest-frequency Fourier modes, the mild instability caused by the aliasing error can be under control. The second observation is that the magnitude of the Fourier coefficient is decreasing with respect to the wavenumber $|k|$ for a function that has a certain degree of regularity. Typically, we have $|\hat{u}_k| \leq C/(1+|k|^m)$ if the mth derivative of a function u is bounded in L^1. Thus the high-frequency Fourier modes have a relatively smaller contribution to the overall solution than the low- to intermediate-frequency modes. The third observation is that one should not cut off high-frequency Fourier modes abruptly to avoid the Gibbs phenomenon and the loss of the L^2-energy associated with the solution. This is especially important when we compute a nearly singular solution whose high-frequency Fourier coefficient has a very slow decay.

Based on the above considerations, we choose a smooth cut-off function which decays exponentially fast with respect to the high wavenumber. In our cut-off function, we choose the parameters $\alpha = 36$ and $m = 36$. These two parameters are chosen to achieve two objectives. (i) When $|k|$ is close to N, the cut-off function reaches machine precision, *i.e.*, 10^{-16}. (ii) The cut-off function remains very close to 1 for $|k| < 4N/5$, and decays rapidly and smoothly to zero beyond $|k| = 4N/5$. In Figure 2.1, we plot the cut-off function $\rho(x)$ as a function of x. The cut-off function used by the 2/3 de-aliasing rule is plotted on top of the cut-off function used by the Fourier smoothing method. We can see that the Fourier smoothing method keeps about $12 \sim 15\%$ more modes than the 2/3 de-aliasing method. In this paper, we will demonstrate by our numerical experiments that the extra modes we keep by the Fourier smoothing method give an accurate approximation of the correct high-frequency Fourier modes.

We have performed a sequence of resolution studies with the largest resolution being $N = 16384$ (Hou and Li 2007). Our extensive numerical results demonstrate that the pseudo-spectral method with the high-order Fourier smoothing (the Fourier smoothing method for short) gives a much more accurate approximation than the pseudo-spectral method with the 2/3 de-aliasing rule (the 2/3 de-aliasing method for short). One of the interesting observations is that the unfiltered high-frequency coefficients in the Fourier smoothing method approximate accurately the corresponding exact Fourier coefficients. Moreover, we observe that the Fourier smoothing method captures about $12 \sim 15\%$ more effective Fourier modes than the 2/3 de-aliasing method in each dimension: see Figure 2.2. The gain is even higher for the 3D Euler equations since the number of effective modes in the Fourier smoothing method is higher in three dimensions. Further, we find that the error produced by the Fourier smoothing method is highly localized near the region where the solution is most singular. In fact, the pointwise error decays exponentially fast away from the location of the shock singularities. On the other hand, the error produced by the 2/3 de-aliasing method spreads out to the entire domain as we approach the singularity time: see Figure 2.3.

2.5. The high-resolution 3D Euler computations of Hou and Li (2006, 2007)

Hou and Li (2006) performed high-resolution computations of the 3D Euler equations using the initial data for the two antiparallel vortex tubes. They used the same initial condition whose analytic formula was given by Kerr (see Section III of Kerr (1993), and also Hou and Li (2006) for corrections of some typos in the description of the initial condition in Kerr (1993)). However, there was some minor difference between their discretization and Kerr's discretization. Hou and Li used a pseudo-spectral discretization in all three directions, while Kerr used a pseudo-spectral discretization only in the x- and y-directions and used a Chebyshev discretization in the z-direction. Based on the results of early tests, positive vorticity in the symmetry plane was imposed in the initial condition of Kerr (1993). How this was imposed

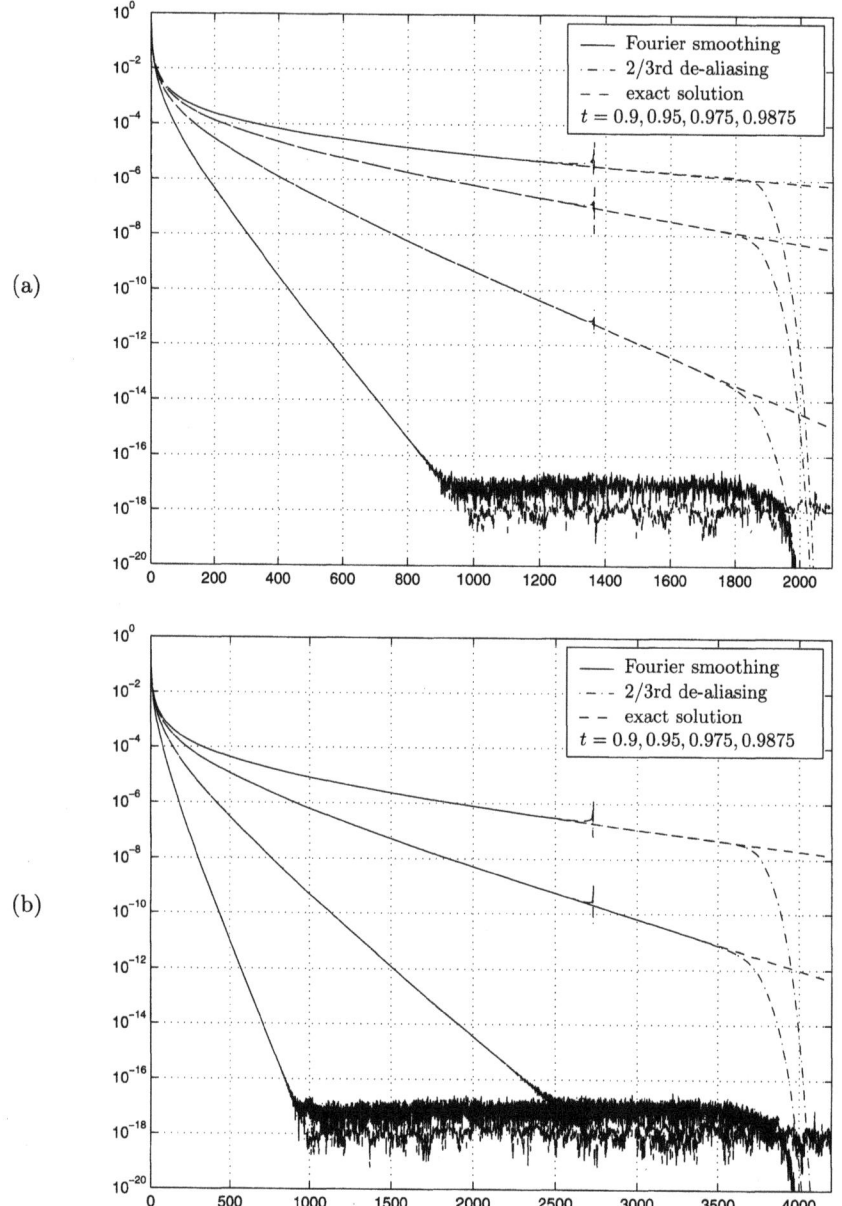

Figure 2.2. Comparison of Fourier spectra of the two methods on different resolutions at a sequence of times. (a) $N = 4096$, (b) $N = 8192$. Dashed lines, 'exact' spectra; solid lines, Fourier smoothing method; dash-dotted lines, 2/3 de-aliasing method. Times, $t = 0.9, 0.95, 0.975, 0.9875$ respectively (from bottom to top). Initial condition, $u_0(x) = \sin(x)$. Singularity time for this initial condition, $T = 1$.

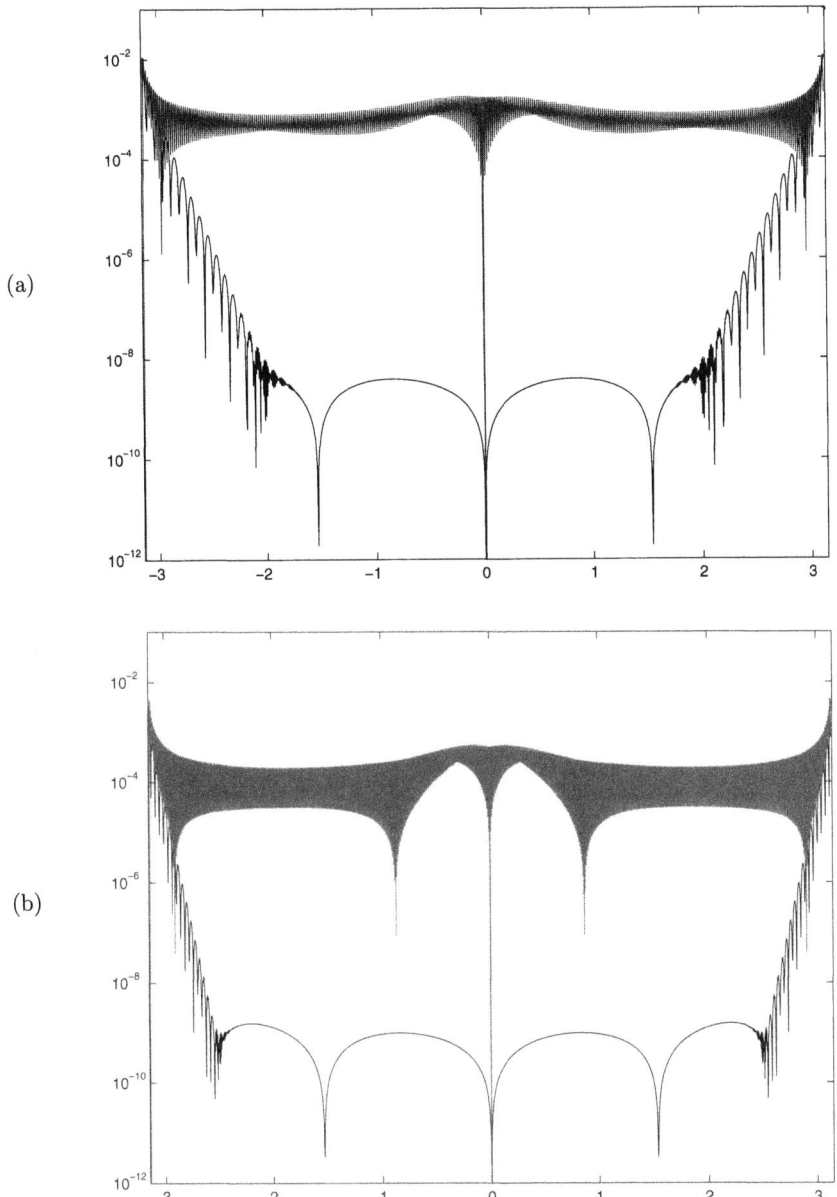

Figure 2.3. The pointwise errors of the two pseudo-spectral methods as a function of time using three different resolutions. The plot is in a log scale. (a) $N = 1024$, (b) $N = 2048$, both at $t = 0.9875$. Initial condition, $u_0(x) = \sin(x)$. The error of the 2/3 de-aliasing method (upper curve) is highly oscillatory and spreads out over the entire domain, while the error of the Fourier smoothing method (lower curve) is highly localized near the location of the shock singularity.

as the vorticity field was mapped onto the Chebyshev mesh was not documented by Kerr (1993). This has led to some ambiguity in reproducing that initial condition which is being resolved by Kerr's group (private communication).

We will summarize the main findings of Hou and Li (2006) in the rest of Section 2. We first illustrate the dynamic evolution of the vortex tubes. In Figure 2.4, we plot the isosurface of the 3D vortex tubes at $t = 0$ and $t = 6$ respectively. As we can see, the two initial vortex tubes are very smooth and relatively symmetric. Due to the mutual attraction of the two antiparallel vortex tubes, the two vortex tubes approach each other and become flattened dynamically. By time $t = 6$, there is already a significant flattening near the centre of the tubes. In Figure 2.5, we plot the local 3D vortex structure of the upper vortex tube at $t = 17$. By this time, the 3D vortex tube has essentially turned into a thin vortex sheet with rapidly decreasing thickness. The vortex lines become relatively straight. The vortex sheet rolls up near the left edge of the sheet.

In order to see better the dynamic development of the local vortex structure, we plot a sequence of vorticity contours on the symmetry plane at $t = 17.5, 18, 18.5$, and 19, respectively, in Figure 2.6. From these results, we can see that the vortex sheet is compressed in the z-direction. It is clear that a thin layer (or a vortex sheet) is formed dynamically. The head of the vortex sheet is a bit thicker than the tail at the beginning. The head of the vortex sheet begins to roll up around $t = 16$. By the time $t = 19$, the head of the vortex sheet has travelled backward for quite a distance, and the vortex sheet has been compressed quite strongly along the z-direction.

We would like to make a few important observations. First of all, the maximum vorticity at a later stage of the computation is actually located near the rolled-up region of the vortex sheet and moves away from the bottom of the vortex sheet. Thus the mechanism of strong compression between the two vortex tubes becomes weaker dynamically at the later time. Secondly, the location of maximum strain and that of maximum vorticity separate as time increases. Thirdly, the relatively 'strong' growth of the maximum velocity between $t = 15$ and $t = 17$ becomes saturated after $t = 17$ when the location of maximum vorticity moves to the rolled-up region: see Figure 2.14. All these factors contribute to the dynamic depletion of vortex stretching.

We now perform a convergence study for the two numerical methods using a sequence of resolutions. For the Fourier smoothing method, we use the resolutions $768 \times 512 \times 1536$, $1024 \times 768 \times 2048$, and $1536 \times 1024 \times 3072$ respectively. Except for the computation on the largest resolution, $1536 \times 1024 \times 3072$, all computations are carried out from $t = 0$ to $t = 19$. The computation on the final resolution, $1536 \times 1024 \times 3072$, is started from $t = 10$ with the initial condition given by the computation with the

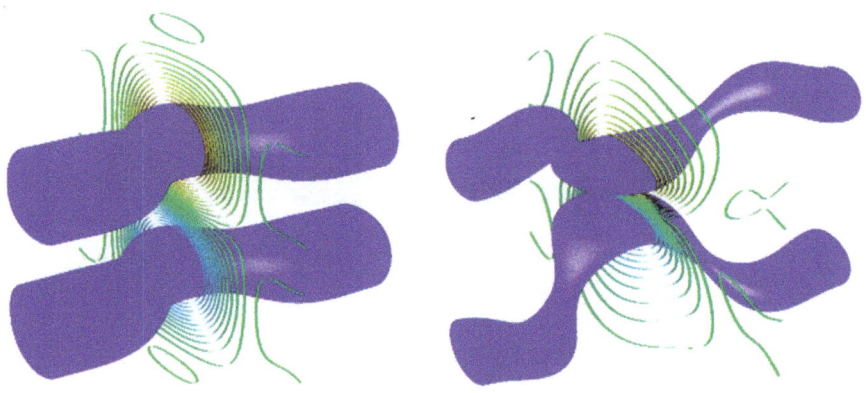

Figure 2.4. The 3D view of the vortex tube for $t = 0$ and $t = 6$. The tube is the isosurface at 60% of the maximum vorticity. The ribbons on the symmetry plane are the contours at other different values.

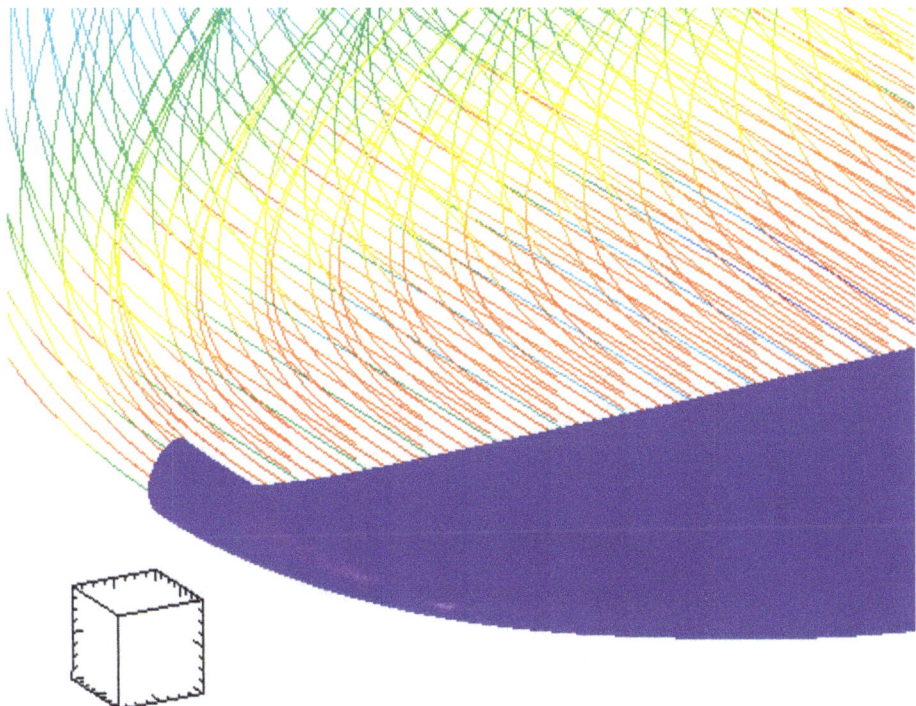

Figure 2.5. The local 3D vortex structures of the upper vortex tube and vortex lines around the maximum vorticity at $t = 17$.

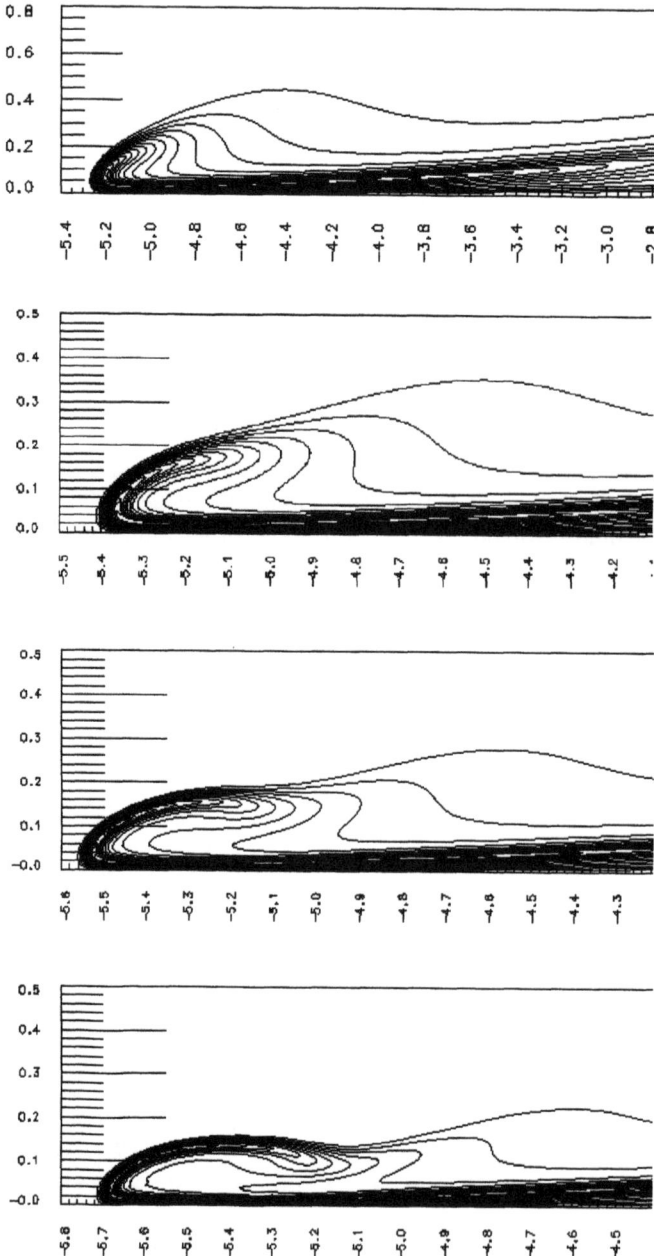

Figure 2.6. The contour of axial vorticity of the upper vortex tube around the maximum vorticity on the symmetry plane (the xz-plane) at $t = 17.5, 18, 18.5, 19$.

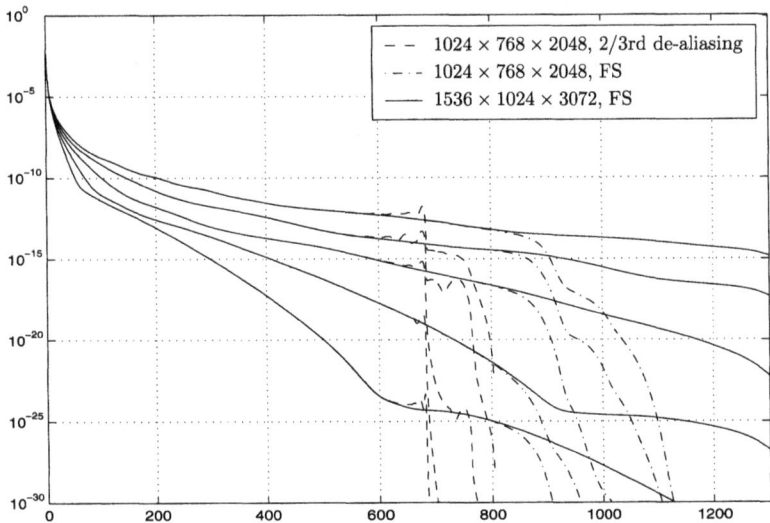

Figure 2.7. The energy spectra versus wavenumbers. The dashed lines and dash-dotted lines are the energy spectra, with resolution $1024 \times 768 \times 2048$, using the 2/3 de-aliasing rule and Fourier smoothing, respectively. The times for the spectra lines are $t = 15, 16, 17, 18, 19$ respectively.

resolution $1024 \times 768 \times 2048$. For the 2/3 de-aliasing method, we use the resolutions $512 \times 384 \times 1024$, $768 \times 512 \times 1536$ and $1024 \times 768 \times 2048$ respectively. The computations using these three resolutions are all carried out from $t = 0$ to $t = 19$. See Hou and Li (2006, 2007) for more details.

In Figure 2.7, we compare the Fourier spectra of the energy obtained by using the 2/3 de-aliasing method with those obtained by the Fourier smoothing method. For a fixed resolution, $1024 \times 768 \times 2048$, we can see that the Fourier spectra obtained by the Fourier smoothing method retain more effective Fourier modes than those obtained by the 2/3 de-aliasing method. This can be seen by comparing the results with the corresponding computations using a higher resolution, $1536 \times 1024 \times 3072$ (the solid lines). Moreover, the Fourier smoothing method does not give the spurious oscillations in the Fourier spectra. In comparison, the Fourier spectra obtained by the 2/3 de-aliasing method produce some spurious oscillations near the 2/3 cut-off point. We would like to emphasize that the Fourier smoothing method conserves the total energy extremely well. More studies including the convergence of the enstrophy spectra can be found in Hou and Li (2006, 2007).

It is worth emphasizing that a significant portion of those Fourier modes beyond the 2/3 cut-off position are still accurate for the Fourier smoothing method. This portion of the Fourier modes that go beyond the 2/3 cut-off point is about 12 ∼ 15% of total number of modes in each dimension. For 3D problems, the total number of effective modes in the Fourier smoothing method is about 20% more than that in the 2/3 de-aliasing method. For our largest resolution, we have about 4.8 billion unknowns. An increase of 20% in the effective Fourier modes represents a very significant increase in the resolution for a large-scale computation.

2.6. Comparison of the two spectral methods in physical space

Next, we compare the solutions obtained by the two methods in physical space for the velocity field and the vorticity. In Figure 2.8, we compare the maximum velocity as a function of time computed by the two methods using resolution $1024 \times 768 \times 2048$. The two solutions are almost indistinguishable. In Figure 2.9, we plot the maximum vorticity as a function of time. The two solutions also agree reasonably well. However, the comparison of the solutions obtained by the two methods at resolutions lower than $1024 \times 768 \times 2048$ shows more significant differences between the two methods: see Figure 2.10.

To understand better how the two methods differ in their performance, we examine the contour plots of the axial vorticity in Figures 2.11, 2.12 and 2.13. As we can see, the vorticity computed by the 2/3 de-aliasing method already develops small oscillations at $t = 17$. The oscillations grow bigger by $t = 18$ (see Figure 2.12), and bigger still at $t = 19$ (see Figure 2.13). We note that the oscillations in the axial vorticity contours concentrate near the region where the magnitude of vorticity is close to zero. Thus they have less of an effect on the maximum vorticity. On the other hand, the solution computed by the Fourier smoothing method is still relatively smooth.

2.7. Dynamic depletion of vortex stretching

In this section, we present some convincing numerical evidence which shows that there is a strong dynamic depletion of vortex stretching due to local geometric regularity of the vortex lines. We first present the result on the growth of the maximum velocity in time: see Figure 2.14. The growth rate of the maximum velocity plays a critical role in the non-blow-up criteria of Deng, Hou and Yu (2005, 2006a). As we can see from Figure 2.14, the maximum velocity remains bounded up to $t = 19$. This is in contrast to the claim in Kerr (2005) that the maximum velocity blows up like $O((T-t)^{-1/2})$ with $T = 18.7$. We note that the velocity field is smoother than the vorticity field. Thus it is easier to resolve the velocity field than the vorticity field. We observe an excellent agreement between the maximum velocity fields computed by the two largest resolutions. Since the velocity field is bounded, the first condition of Theorem 2.1 is satisfied by taking $A = 0$. Furthermore,

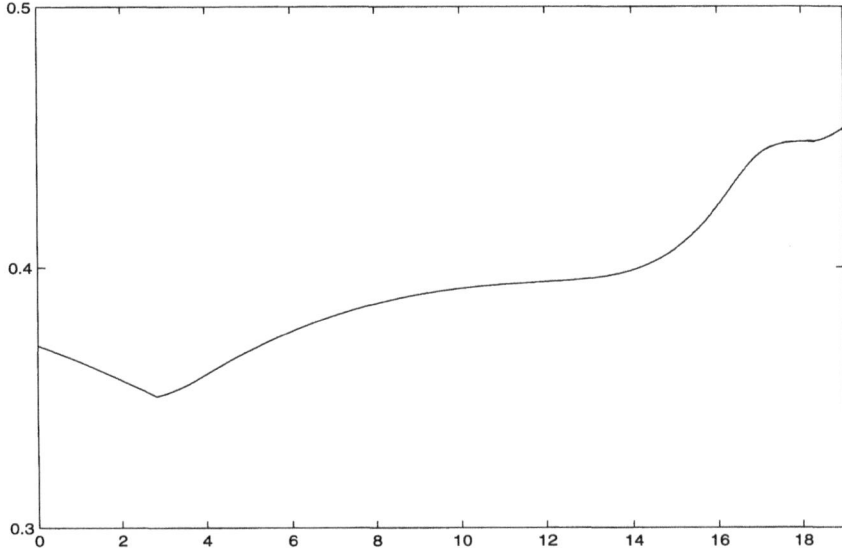

Figure 2.8. Comparison of maximum velocity as a function of time computed by two methods. Solid line, solution obtained by the Fourier smoothing method; dashed line, solution obtained by the 2/3 de-aliasing method. Resolution $1024 \times 768 \times 2048$ for both methods.

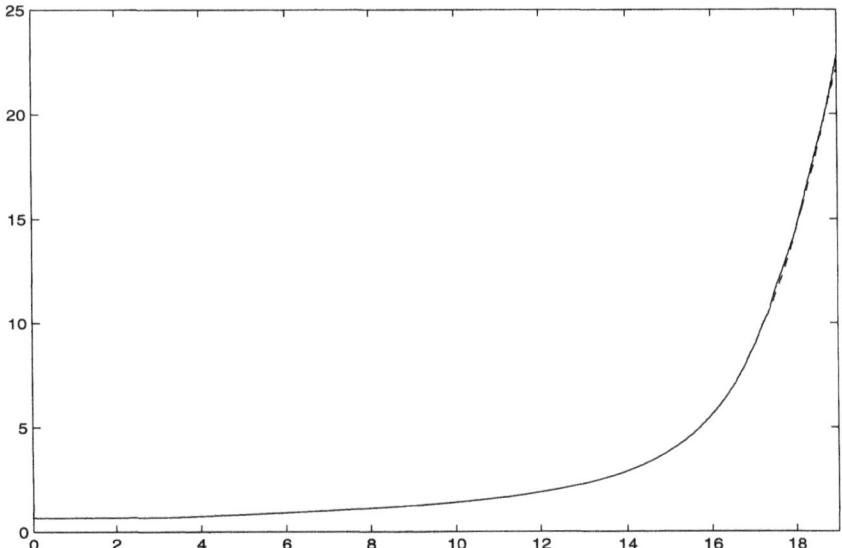

Figure 2.9. Comparison of maximum vorticity as a function of time computed by two methods. Solid line, solution obtained by the Fourier smoothing method; dashed line, solution obtained by the 2/3 de-aliasing method. Resolution $1024 \times 768 \times 2048$ for both methods.

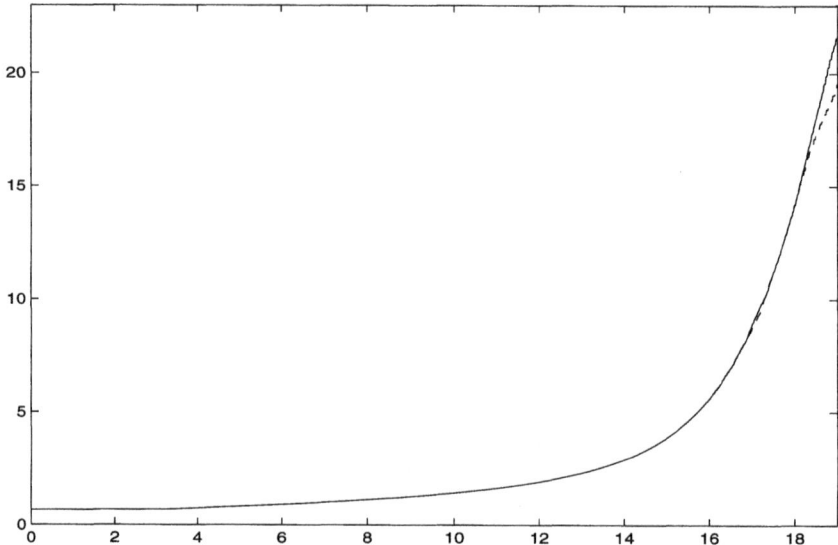

Figure 2.10. Comparison of maximum vorticity as a function of time computed by two methods. Solid line, solution obtained by the Fourier smoothing method; dashed line, solution obtained by the 2/3 de-aliasing method. Resolution 768 × 512 × 1024 for both methods.

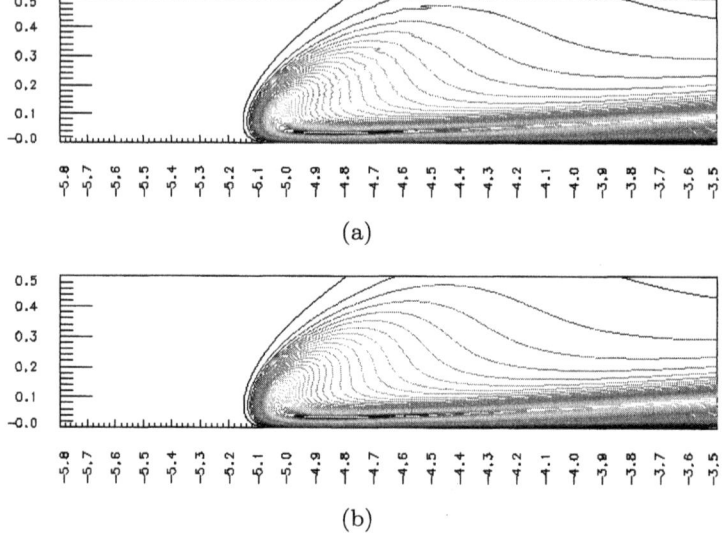

Figure 2.11. Comparison of axial vorticity contours at $t = 17$ computed by two methods. (a) Solution obtained by the 2/3 de-aliasing method, (b) solution obtained by the Fourier smoothing method. Resolution 1024 × 768 × 2048 for both methods.

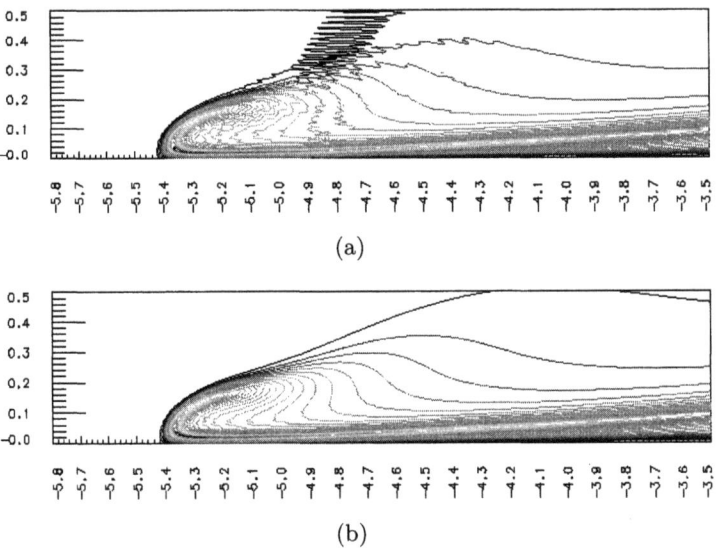

Figure 2.12. Comparison of axial vorticity contours at $t = 18$ computed by two methods. (a) Solution obtained by the 2/3 de-aliasing method, (b) solution obtained by the Fourier smoothing method. Resolution $1024 \times 768 \times 2048$ for both methods.

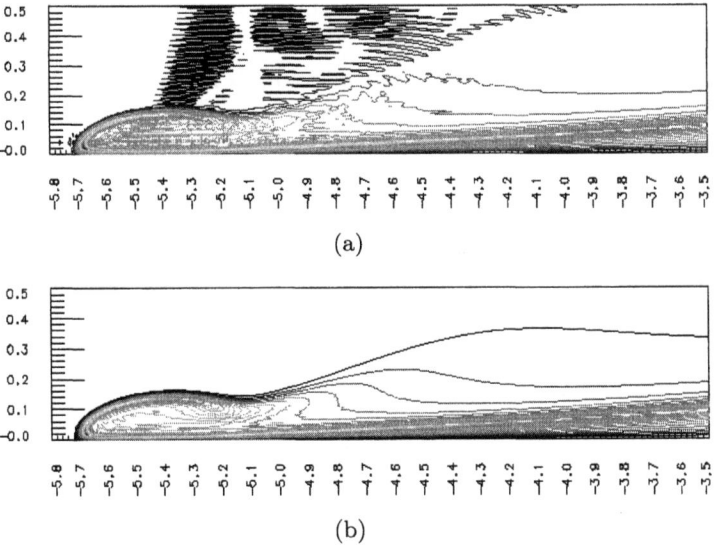

Figure 2.13. Comparison of axial vorticity contours at $t = 19$ computed by two methods. (a) Solution obtained by the 2/3 de-aliasing method, (b) solution obtained by the Fourier smoothing method. Resolution $1024 \times 768 \times 2048$ for both methods.

since both $\nabla \cdot \boldsymbol{\xi}$ and κ are bounded by $O((T-t)^{-1/2})$ in the inner region of size $(T-t)^{1/2} \times (T-t)^{1/2} \times (T-t)$ (Kerr 2005), the second condition of Theorem 2.1 is satisfied with $B = 1/2$ by taking a segment of the vortex line with length $(T-t)^{1/2}$ within this inner region. Thus Theorem 2.1 can be applied to our computation, which implies that the solution of the 3D Euler equations remains smooth at least up to $T = 19$.

We also study the maximum vorticity as a function of time. The maximum vorticity is found to increase rapidly from the initial value of 0.669 to 23.46 at the final time $t = 19$, a factor of 35 increase from its initial value. Our computations show no sign of finite-time blow-up of the 3D Euler equations up to $T = 19$, beyond the singularity time predicted by Kerr. The maximum vorticity computed by resolution $1024 \times 768 \times 2048$ agrees very well with that computed by resolution $1536 \times 1024 \times 3072$ up to $t = 17.5$. There is some mild disagreement towards the end of the computation. This indicates that a very high space resolution is needed to capture the rapid growth of maximum vorticity at the final stage of the computation.

In order to understand the nature of the dynamic growth in vorticity, we examine the degree of nonlinearity in the vortex stretching term. In Figure 2.15, we plot the quantity, $\|\boldsymbol{\xi} \cdot \nabla \mathbf{u} \cdot \boldsymbol{\omega}\|_\infty$, as a function of time. If the maximum vorticity indeed blew up like $O((T-t)^{-1})$, as alleged in Kerr (1993), this quantity should have been quadratic as a function of maximum vorticity. We find that there is tremendous cancellation in this vortex stretching term. It actually grows more slowly than $C\|\tilde{\omega}\|_\infty \log(\|\tilde{\omega}\|_\infty)$: see Figure 2.15. It is easy to show that $\|\boldsymbol{\xi} \cdot \nabla \mathbf{u} \cdot \boldsymbol{\omega}\|_\infty \leq C\|\tilde{\omega}\|_\infty \log(\|\tilde{\omega}\|_\infty)$ would imply at most doubly exponential growth in the maximum vorticity. Indeed, as demonstrated by Figure 2.16, the maximum vorticity does not grow more rapidly than doubly exponential in time. We have also generated a similar plot by extracting the data from Kerr (1993). We find that $\log(\log(\|\omega\|_\infty))$ basically scales linearly with respect to t from $14 \leq t \leq 17.5$ when Kerr's computations are still reasonably resolved. This implies that the maximum vorticity up to $t = 17.5$ in his computations does not grow more rapidly than doubly exponential in time. This is consistent with our conclusion.

We study the decay rate in the energy spectrum in Figure 2.17 at $t = 16, 17, 18, 19$. A finite-time blow-up of enstrophy would imply that the energy spectrum decays no more rapidly than $|k|^{-3}$. Our computations show that the energy spectrum approaches $|k|^{-3}$ for $|k| \leq 100$ as time increases to $t = 19$. This is in qualitative agreement with Kerr's results. Note that there are fewer than 100 modes available along the $|k_x|$- or $|k_y|$-direction in Kerr's computations: see Figure 18(a),(b) of Kerr (1993). On the other hand, our computations show that the high-frequency Fourier spectrum for $100 \leq |k| \leq 1300$ decays much more rapidly than $|k|^{-3}$, as one can see from Figure 2.17. This indicates that there is no blow-up in enstrophy.

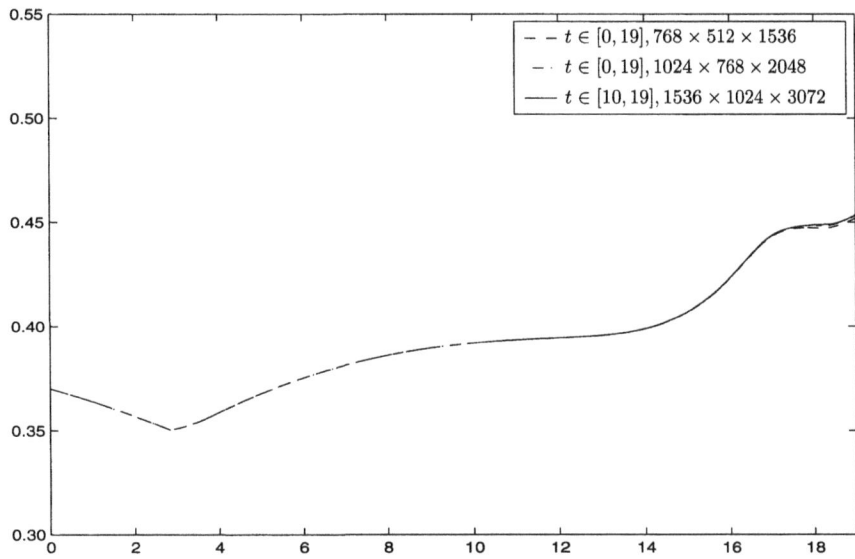

Figure 2.14. Maximum velocity $\|\mathbf{u}\|_\infty$ in time using three different resolutions.

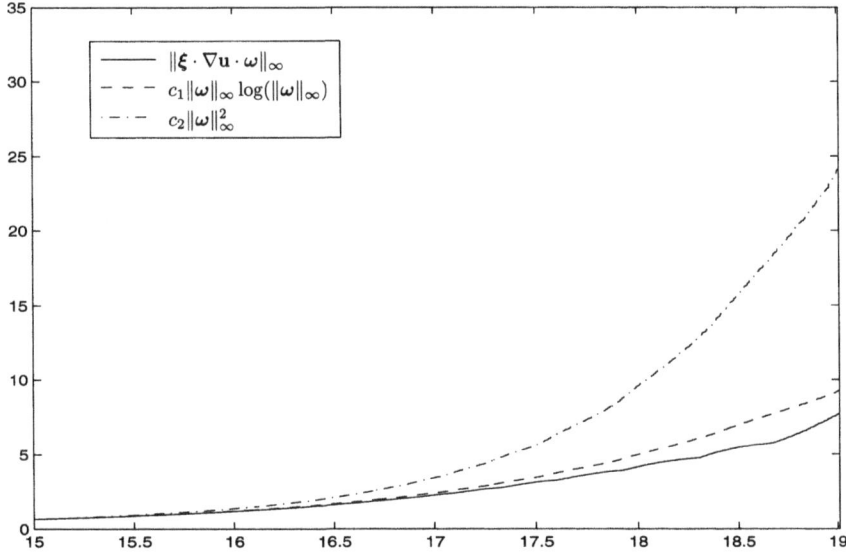

Figure 2.15. Study of the vortex stretching term in time, resolution $1536 \times 1024 \times 3072$. The inequality $|\boldsymbol{\xi} \cdot \nabla \mathbf{u} \cdot \boldsymbol{\omega}| \leq c_1 |\boldsymbol{\omega}| \log |\boldsymbol{\omega}|$ and $\frac{D}{Dt}|\boldsymbol{\omega}| = \boldsymbol{\xi} \cdot \nabla \mathbf{u} \cdot \boldsymbol{\omega}$ implies $|\boldsymbol{\omega}|$ bounded by a double exponential.

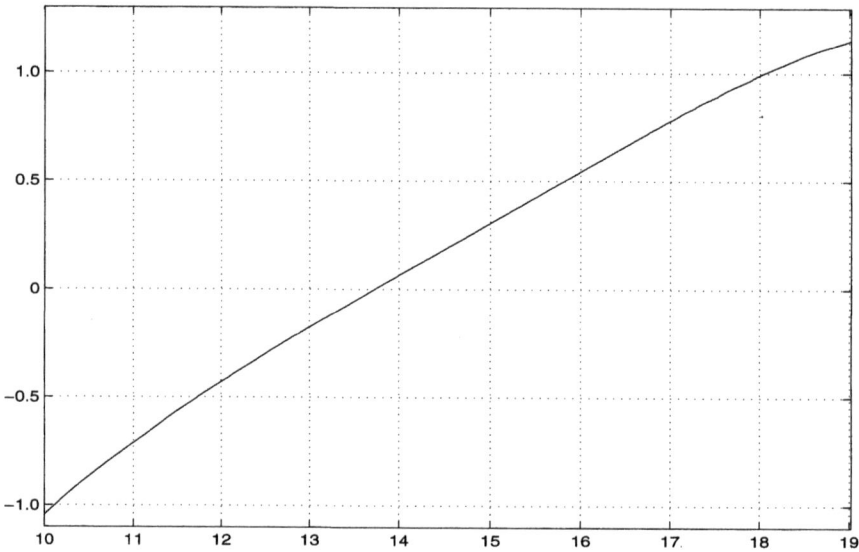

Figure 2.16. The plot of $\log\log \|\boldsymbol{\omega}\|_\infty$ versus time, resolution $1536 \times 1024 \times 3072$.

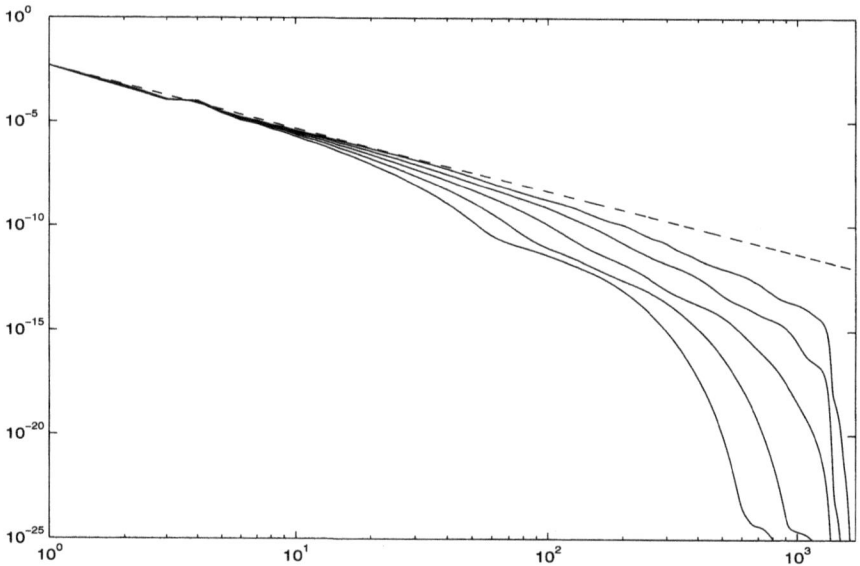

Figure 2.17. The energy spectra for velocity at $t = 15, 16, 17, 18, 19$ (from bottom to top) in log-log scale. Dashed line, k^{-3}.

Table 2.1. The alignment of the vorticity vector and the eigenvectors of S around the point of maximum vorticity with resolution $1536 \times 1024 \times 3072$. Here, θ_i is the angle between the ith eigenvector of S and the vorticity vector.

Time	$|\omega|$	λ_1	θ_1	λ_2	θ_2	λ_3	θ_3
16.012	5.628	−1.508	89.992	0.206	0.007	1.302	89.998
16.515	7.016	−1.864	89.995	0.232	0.010	1.631	89.990
17.013	8.910	−2.322	89.998	0.254	0.006	2.066	89.993
17.515	11.430	−2.630	89.969	0.224	0.085	2.415	89.920
18.011	14.890	−3.625	89.969	0.257	0.036	3.378	89.979
18.516	19.130	−4.501	89.966	0.246	0.036	4.274	89.984
19.014	23.590	−5.477	89.966	0.247	0.034	5.258	89.994

It is interesting to ask how the vorticity vector aligns with the eigenvectors of the deformation tensor. Recall that the vorticity equations can be written as

$$\frac{\partial}{\partial t}\omega + (\mathbf{u} \cdot \nabla)\omega = S \cdot \omega, \quad S = \frac{1}{2}(\nabla \mathbf{u} + \nabla^T \mathbf{u}) \quad (2.9)$$

(see Majda and Bertozzi (2002)). Let $\lambda_1 < \lambda_2 < \lambda_3$ be the three eigenvalues of S. The incompressibility condition implies that $\lambda_1 + \lambda_2 + \lambda_3 = 0$. If the vorticity vector aligns with the eigenvector corresponding to λ_3, which gives the maximum rate of stretching, then it is very likely that the 3D Euler equations would blow up in a finite time.

In Table 2.1, we document the alignment information of the vorticity vector around the point of maximum vorticity with resolution $1536 \times 1024 \times 3072$. In this table, θ_i is the angle between the ith eigenvector of S and the vorticity vector. One can see clearly that for $16 \leq t \leq 19$ the vorticity vector at the point of maximum vorticity is almost perfectly aligned with the second eigenvector of S. Note that the second eigenvalue, λ_2, is about 20 times smaller in magnitude than the largest eigenvalue λ_3, and does not grow much in time. The alignment of the vorticity vector with the second eigenvector of the deformation tensor is another indication that there is a strong dynamic depletion of vortex stretching.

2.8. Global regularity of large anisotropic initial data

The numerical studies of the 3D Euler equations by Hou and Li (2006) strongly suggest that the support of maximum vorticity becomes severely flattened and develops an anisotropic scaling as vorticity increases rapidly in time. This seems quite generic and is a consequence of the incompressibility and the Lagrangian structure of the vorticity equation. Convection plays an essential role in producing this anisotropic structure of the solution. Motivated by the desire to understand how the local anisotropic structure of the solution near the support of maximum vorticity may lead to the depletion of vortex stretching, Hou, Lei and Li (2008) recently studied the 3D axisymmetric Navier–Stokes equations with large anisotropic data. They proved the global regularity of the 3D Navier–Stokes equations for a family of large anisotropic initial data. Moreover, they obtained a global bound of the solution in terms of its initial data in some L^p-norm. Their results also revealed some interesting dynamic growth behaviour of the solution due to the interaction between the axial vorticity and the the derivative of vorticity.

Specifically, let u^θ and ω^θ be the angular velocity and vorticity components of the 3D axisymmetric Navier–Stokes equations. They considered initial data for u^θ and ω^θ that have the following scaling property:

$$u^\theta(r,z,0) = \frac{1}{\epsilon^{1-\delta}} U_0(\epsilon r, z), \quad \omega^\theta(r,z,0) = \frac{1}{\epsilon^{1-\delta}} W_0(\epsilon r, z), \qquad (2.10)$$

where $r = \sqrt{x^2 + y^2}$, δ and ϵ are some small positive parameters, and the rescaled profiles U_0/r and W_0/r are bounded in L^{2p} and L^{2q}, respectively, for some p and q with $p = 2q$; note that u^θ and ω^θ must satisfy a compatibility condition: $u^\theta|_{r=0} = 0 = \omega^\theta|_{r=0}$ (Liu and Wang 2006). We remark that these initial data are not small. In fact, we have

$$\|\mathbf{u}_0\|_{L^2(\mathbb{R}^2 \times [0,1])} \|\nabla \mathbf{u}_0\|_{L^2(\mathbb{R}^2 \times [0,1])} = \frac{C_0}{\epsilon^{4-2\delta}} \gg 1,$$

for ϵ small, where $\mathbf{u}_0$ is the initial velocity vector. Thus the classical regularity analysis for small initial data does not apply to these sets of anisotropic initial data.

Hou, Lei and Li (2008) proved the global regularity of the 3D axisymmetric Navier–Stokes equations for initial data (2.10) by exploring the anisotropic structure of the solution for ϵ small. They also obtained a global bound on $\|u^\theta/r\|_{L^{2p}}$ and $\|\omega^\theta/r\|_{L^{2q}}$ in terms of their initial data. Note that by using the scaling invariance property of the Navier–Stokes equations, their global regularity result also applies to the following rescaled initial data:

$$u^\theta(r,z,0) = \frac{1}{\epsilon^{2-\delta}} U_0\left(r, \frac{z}{\epsilon}\right), \quad \omega^\theta(r,z,0) = \frac{1}{\epsilon^{3-\delta}} W_0\left(r, \frac{z}{\epsilon}\right), \qquad (2.11)$$

and

$$u^\theta(r,z,0) = \frac{1}{\epsilon} U_0\left(\frac{r}{\epsilon^{1-\delta}}, \frac{z}{\epsilon}\right), \quad \omega^\theta(r,z,0) = \frac{1}{\epsilon^2} W_0\left(\frac{r}{\epsilon^{1-\delta}}, \frac{z}{\epsilon}\right). \qquad (2.12)$$

Note that the parameters ϵ in the initial data (2.10)–(2.11) and δ in (2.12) measure the degree of anisotropy of the initial data. If $\delta = 0$, then the initial data (2.12) become isotropic, *i.e.*,

$$\mathbf{u}_0(x,y,z) = \frac{1}{\epsilon}\mathbf{U}_0\left(\frac{x}{\epsilon}, \frac{y}{\epsilon}, \frac{z}{\epsilon}\right).$$

Their analysis would break down when there is no anisotropic scaling in the initial data, *i.e.*, $\delta = 0$. Clearly, if the analysis could be extended to the case of $\delta = 0$, one would prove the global regularity of the 3D axisymmetric Navier–Stokes equations for general initial data by using the scaling invariance property of the Navier–Stokes equations. It is interesting to note that by using an anisotropic scaling of the initial data, we turn the global regularity of the 3D Navier–Stokes equations into a critical case of $\delta = 0$.

We remark that the global regularity results of Hou, Lei and Li (2008) were obtained on a regular size domain, $\mathbb{R}^2 \times [0,1]$, for initial data (2.10). In this sense, their results are different from those global regularity results obtained for a thin domain, $\Omega_\epsilon = Q_1 \times [0, \epsilon]$ with Q_1 being a bounded domain in $\mathbb{R}^2$. The global regularity of the 3D Navier–Stokes equations in a thin domain of the form Ω_ϵ has been studied by Raugel and Sell in a series of papers (Raugel and Sell 1993a, 1994, 1993b). They proved the global regularity of the 3D Navier–Stokes equations under the assumption that $\|\nabla \mathbf{u}_0\|_{L^2(\Omega_\epsilon)}^2 \leq C_0 \ln \frac{1}{\epsilon}$. This is an improvement over the classical global regularity result for small data, which requires $\|\nabla \mathbf{u}_0\|_{L^2(\Omega_\epsilon)}^2 \leq C^* \epsilon$ (Raugel and Sell 1993a). One may interpret the global regularity result of Hou, Lei and Li with initial data (2.11) as a result on a generalized thin domain. Note that the initial data given by (2.11) satisfy the following bound: $\|\nabla \mathbf{u}_0\|_{L^2(\Omega_\epsilon)}^2 = C_0 \epsilon^{-5+2\delta}$ (here $\delta > 0$ can be made arbitrarily small), which is much larger than the corresponding bound $C_0 \ln \frac{1}{\epsilon}$ required by the global regularity analysis of Raugel and Sell (1993a, 1994, 1993b).

3. Dynamic stability of 3D Navier–Stokes equations

The axisymmetric 3D Navier–Stokes equation with swirl is perhaps the simplest form of the 3D Navier–Stokes equations, yet still retains the most essential difficulties of the 3D Navier–Stokes equations. It has attracted a lot of attention in recent years. Although some partial progress has been made in studying the global regularity of the axisymmetric Navier–Stokes equations with swirl using energy estimates (see, *e.g.*, Chae and Lee (2002) and references cited there), the question of global regularity for general initial data is still an open question.

Hou and Li (2008a) studied the dynamic stability of the axisymmetric Navier–Stokes equations with swirl via a new 1D model. This model is derived from the axisymmetric Navier–Stokes equations along the symmetry axis. Surprisingly, this model is an exact reduction of the 3D axisymmetric

Navier–Stokes equations along the symmetry axis. It captures the essential nonlinear features of the 3D Navier–Stokes equations. One of the important findings by Hou and Li (2008a) was that the convection term plays an essential role in cancelling some of the vortex stretching terms. Specifically, they found a positive Lyapunov function which satisfies a new conservation law and a maximum principle. This holds for both the viscous and inviscid cases. This *a priori* pointwise estimate plays a critical role in obtaining nonlinear stability and global regularity of the 1D model. Using this *a priori* estimate, they proved global regularity of the 3D Navier–Stokes equations for a family of large data, which can experience large transient dynamic growth but remain smooth for all times.

It is worth emphasizing that such subtle dynamic stability properties of the 3D Navier–Stokes equations would have been completely missed by using the traditional energy estimates. Traditional energy estimates are too crude to capture some of the most essential properties of the 3D incompressible Navier–Stokes equations. To illustrate its limitations, we briefly review how the energy estimates are used in proving global regularity of the 3D Navier–Stokes equations.

For incompressible Navier–Stokes equations, one of the most important *a priori* estimates is the energy identity. More precisely, for any strong solution $\mathbf{u}$, we have

$$\frac{1}{2}\frac{d}{dt}\int |\mathbf{u}|^2\,d\mathbf{x} + \nu \int |\nabla \mathbf{u}|^2\,d\mathbf{x} = 0, \tag{3.1}$$

by observing $\int \mathbf{u}\cdot(\mathbf{u}\cdot\nabla\mathbf{u})\,d\mathbf{x} = -\frac{1}{2}\int(\nabla\cdot\mathbf{u})|\mathbf{u}|^2\,d\mathbf{x} = 0$, since $\mathbf{u}$ is divergence-free. Unfortunately, this energy identity is not strong enough to rule out finite-time singularities. To prove global regularity, we need to obtain control in a stronger norm, either in $\|\mathbf{u}\|_{L^p}$ with $p \geq 3$ or in $\|\omega\|_{L^2}$. To illustrate the main difficulty of the traditional energy estimates, let us perform energy estimates for the vorticity equation:

$$\frac{1}{2}\frac{d}{dt}\int |\omega|^2\,d\mathbf{x} + \nu \int |\nabla\omega|^2\,d\mathbf{x} = \int \omega\cdot\nabla\mathbf{u}\cdot\omega\,d\mathbf{x}. \tag{3.2}$$

Again, the convection term does not contribute to the L^2-norm of vorticity (or any L^p-norm with $p > 1$). The main difficulty is to control the vortex stretching term. Using the Sobolev embedding theory, one can show that

$$\int \omega\cdot\nabla\mathbf{u}\cdot\omega\,d\mathbf{x} \leq C_\nu\left(\int |\omega|^2\,d\mathbf{x}\right)^3 + \frac{\nu}{2}\int |\nabla\omega|^2\,d\mathbf{x}, \tag{3.3}$$

which can not be improved. This implies that

$$\frac{1}{2}\frac{d}{dt}\int |\omega|^2\,d\mathbf{x} + \frac{\nu}{2}\int |\nabla\omega|^2\,d\mathbf{x} \leq C_\nu\left(\int |\omega|^2\,d\mathbf{x}\right)^3. \tag{3.4}$$

Unfortunately, the above estimate does not imply global regularity for large data even if we use the energy identity (3.1). However, the estimate (3.4) can be used to obtain global regularity for small initial data. To see this, we substitute the following interpolation inequality,

$$\left(\int |\omega|^2 \, d\mathbf{x}\right)^2 = \|\omega\|_{L^2}^4 \leq C_0 \|\mathbf{u}\|_{L^2}^2 \|\nabla \omega\|_{L^2}^2, \tag{3.5}$$

into (3.4) to obtain

$$\frac{1}{2}\frac{d}{dt}\|\omega\|_{L^2}^2 \leq \left(C_\nu C_0 \|\mathbf{u}\|_{L^2}^2 \|\omega\|_{L^2}^2 - \frac{\nu}{2}\right)\|\nabla \omega\|_{L^2}^2 \leq 0, \tag{3.6}$$

provided that

$$C_\nu C_0 \|\mathbf{u}_0\|_{L^2}^2 \|\omega_0\|_{L^2}^2 \leq \frac{\nu}{2}. \tag{3.7}$$

Since $\|\mathbf{u}(t)\|_{L^2}^2 \leq \|\mathbf{u}_0\|_{L^2}^2$ for all t, condition (3.7) and inequality (3.6) imply that $\|\omega(t)\|_{L^2}^2 \leq \|\omega_0\|_{L^2}^2$ for all times. Note that $\|\omega_0\|_{L^2}^2 = \|\nabla \mathbf{u}_0\|_{L^2}^2$. Thus we can also replace (3.7) by

$$C_\nu C_0 \|\mathbf{u}_0\|_{L^2}^2 \|\nabla \mathbf{u}_0\|_{L^2}^2 \leq \frac{\nu}{2}. \tag{3.8}$$

Due to the incompressibility condition, convection plays no role in the energy estimate. The same estimate can be also applied to the following nonlinear diffusion equation:

$$w_t = w^2 + \nu \Delta w. \tag{3.9}$$

An energy estimate gives

$$\frac{1}{2}\frac{d}{dt}\int |w|^2 \, d\mathbf{x} + \nu \int |\nabla w|^2 \, d\mathbf{x} = \int w^3 \, d\mathbf{x}. \tag{3.10}$$

Using an embedding inequality similar to (3.3), we get

$$\frac{1}{2}\frac{d}{dt}\int |w|^2 \, d\mathbf{x} + \frac{\nu}{2}\int |\nabla w|^2 \, d\mathbf{x} \leq C_\nu \left(\int |w|^2 \, d\mathbf{x}\right)^3, \tag{3.11}$$

which is identical to (3.4).

However, it is well known that (3.9) can develop a finite-time isotropic self-similar blow-up solution, which does not violate the energy identity (3.1), in the sense that $\int_0^T \|w(t)\|_{L^2}^2 \, dt < \infty$. The above analysis shows that energy estimates can not distinguish a nonlinear diffusion equation, which has a finite-time blow-up solution, from the 3D Navier–Stokes equations, which have completely different physical properties and may not necessarily blow up in finite time.

3.1. Reformulation of 3D axisymmetric Navier–Stokes equations

Consider the 3D axisymmetric incompressible Navier–Stokes equations

$$u^\theta_t + u^r u^\theta_r + u^z u^\theta_z = \nu\left(\nabla^2 - \frac{1}{r^2}\right)u^\theta - \frac{1}{r}u^r u^\theta, \tag{3.12}$$

$$\omega^\theta_t + u^r \omega^\theta_r + u^z \omega^\theta_z = \nu\left(\nabla^2 - \frac{1}{r^2}\right)\omega^\theta + \frac{1}{r}((u^\theta)^2)_z + \frac{1}{r}u^r\omega^\theta, \tag{3.13}$$

$$-\left(\nabla^2 - \frac{1}{r^2}\right)\psi^\theta = \omega^\theta, \tag{3.14}$$

where $r = \sqrt{x^2 + y^2}$, u^θ, ω^θ and ψ^θ are the angular components of the velocity, vorticity and stream function respectively, and

$$u^r = -(\psi^\theta)_z, \quad u^z = \frac{1}{r}(r\psi^\theta)_r.$$

Note that equations (3.12)–(3.14) completely determine the evolution of the 3D axisymmetric Navier–Stokes equations.

Hou and Li (2008a) introduced the following new variables,

$$u_1 = u^\theta/r, \quad \omega_1 = \omega^\theta/r, \quad \psi_1 = \psi^\theta/r, \tag{3.15}$$

and derived the following equivalent system that governs the dynamics of u_1, ω_1 and ψ_1:

$$\partial_t u_1 + u^r \partial_r u_1 + u^z \partial_z u_1 = \nu\left(\partial_r^2 + \frac{3}{r}\partial_r + \partial_z^2\right)u_1 + 2u_1\psi_{1z}, \tag{3.16a}$$

$$\partial_t \omega_1 + u^r \partial_r \omega_1 + u^z \partial_z \omega_1 = \nu\left(\partial_r^2 + \frac{3}{r}\partial_r + \partial_z^2\right)\omega_1 + (u_1^2)_z, \tag{3.16b}$$

$$-\left(\partial_r^2 + \frac{3}{r}\partial_r + \partial_z^2\right)\psi_1 = \omega_1, \tag{3.16c}$$

where $u^r = -r\psi_{1z}$, $u^z = 2\psi_1 + r\psi_{1r}$. Liu and Wang (2006) showed that if $\mathbf{u}$ is a smooth velocity field, then u^θ, ω^θ and ψ^θ must satisfy the compatibility condition $u^\theta|_{r=0} = \omega^\theta|_{r=0} = \psi^\theta|_{r=0} = 0$. Thus u_1, ψ_1 and ω_1 are well defined as long as the solution remains smooth.

3.2. An exact 1D model for 3D Navier–Stokes equations

Hou and Li (2008a) derived an exact 1D model along the symmetry axis by assuming the solution is more singular along the z-direction than along the r-direction (i.e., the solution has an locally anisotropic scaling). Along the symmetry axis $r = 0$, we have $u^r = 0$, $u^z = 2\psi_1$. Since the solution is more singular along the z-direction, one can drop the derivatives along the r-direction to the leading order in the reformulated Navier–Stokes equations (note that $\frac{3}{r}\partial_r$ is of the same order as ∂_r^2). This gives rise to the following

1D model:

$$(u_1)_t + 2\psi_1(u_1)_z = \nu(u_1)_{zz} + 2(\psi_1)_z u_1, \tag{3.17}$$
$$(w_1)_t + 2\psi_1(w_1)_z = \nu(w_1)_{zz} + (u_1^2)_z, \tag{3.18}$$
$$-(\psi_1)_{zz} = w_1. \tag{3.19}$$

Note that the system (3.17)–(3.19) is already a closed system. Let $\tilde{u} = u_1$, $\tilde{v} = -(\psi_1)_z$, and $\tilde{\psi} = \psi_1$. By integrating (3.18) with respect to z, one can further reduce the above system to

$$(\tilde{u})_t + 2\tilde{\psi}(\tilde{u})_z = \nu(\tilde{u})_{zz} - 2\tilde{v}\tilde{u}, \tag{3.20}$$
$$(\tilde{v})_t + 2\tilde{\psi}(\tilde{v})_z = \nu(\tilde{v})_{zz} + (\tilde{u})^2 - (\tilde{v})^2 + c(t), \tag{3.21}$$

where $\tilde{v} = -(\tilde{\psi})_z$, $\tilde{v}_z = \tilde{w}$, and $c(t)$ is an integration constant to enforce the mean of $\tilde{v}$ equal to zero. If we assume that the solution is periodic with respect to z with period 1, the integration constant $c(t)$ is equal to $3\int_0^1 (\tilde{v})^2\, dz - \int_0^1 (\tilde{u})^2\, dz$.

A surprising result is that the above 1D model is exact. This is stated in the following theorem.

Theorem 3.1. Let u_1, ψ_1 and w_1 be the solution of the 1D model (3.17)–(3.19) and define

$$u^\theta(r, z, t) = r u_1(z, t), \quad w^\theta(r, z, t) = r w_1(z, t), \quad \psi^\theta(r, z, t) = r \psi_1(z, t).$$

Then $(u^\theta(r, z, t), w^\theta(r, z, t), \psi^\theta(r, z, t))$ is an exact solution of the 3D Navier–Stokes equations.

Theorem 3.1 tells us that the 1D model (3.17)–(3.19) preserves some essential nonlinear structure of the 3D axisymmetric Navier–Stokes equations.

3.3. Properties of the model equation

In this section, we will study some properties of the 1D model. We first consider the properties of some further simplified models obtained from these equations. Both numerical and analytical studies are presented for these simplified models. Based on the understanding of the simplified models, we prove the global existence of the full 1D model.

The ODE model

To start with, we consider an ODE model by ignoring the convection and diffusion term:

$$(\tilde{u})_t = -2\tilde{v}\tilde{u}, \tag{3.22}$$
$$(\tilde{v})_t = (\tilde{u})^2 - (\tilde{v})^2, \tag{3.23}$$

with initial condition $\tilde{u}(0) = \tilde{u}_0$ and $\tilde{v}(0) = \tilde{v}_0$.

Clearly, if $\tilde{u}_0 = 0$, then $\tilde{u}(t) = 0$ for all $t > 0$. In this case, the equation for $\tilde{v}$ is decoupled from $\tilde{u}$ completely, and will blow up in finite time if $\tilde{v}_0 < 0$. In fact, if $\tilde{v}_0 < 0$ and $\tilde{u}_0$ is very small, then the solution can experience very large growth dynamically. The growth can be made arbitrarily large if we choose $\tilde{u}_0$ to be arbitrarily small. However, the special nonlinear structure of the ODE system has an interesting cancellation property which has a stabilizing effect on the solution for large times. This is described by the following theorem.

Theorem 3.2. Assume that $\tilde{u}_0 \neq 0$. Then the solution $(\tilde{u}(t), \tilde{v}(t))$ of the ODE system (3.22)–(3.23) exists for all times. Moreover, we have

$$\lim_{t \to +\infty} \tilde{u}(t) = 0, \quad \lim_{t \to +\infty} \tilde{v}(t) = 0. \tag{3.24}$$

Proof. Inspired by the work of Constantin, Lax and Majda (1985), we make the following change of variables: $w = \tilde{u} + i\tilde{v}$. Then the ODE system (3.22)–(3.23) is reduced to the following complex nonlinear ODE:

$$\frac{dw}{dt} = iw^2, \quad w(0) = w_0, \tag{3.25}$$

which can be solved analytically. The solution has the form

$$w(t) = \frac{w_0}{1 - iw_0 t}. \tag{3.26}$$

In terms of the original variables, we have

$$\tilde{u}(t) = \frac{\tilde{u}_0(1 + \tilde{v}_0 t) - \tilde{u}_0 \tilde{v}_0 t}{(1 + \tilde{v}_0 t)^2 + (\tilde{u}_0 t)^2}, \tag{3.27}$$

$$\tilde{v}(t) = \frac{\tilde{v}_0(1 + \tilde{v}_0 t) + \tilde{u}_0^2 t}{(1 + \tilde{v}_0 t)^2 + (\tilde{u}_0 t)^2}. \tag{3.28}$$

It is clear from (3.27)–(3.28) that the solution of the ODE system (3.22)–(3.23) exists for all times and decays to zero as $t \to +\infty$ as long as $\tilde{u}_0 \neq 0$. This completes the proof of Theorem 3.2. □

As we can see from (3.27)–(3.28), the solution can grow very fast in a very short time if $\tilde{u}_0$ is small, but $\tilde{v}_0$ is large and negative. For example, if we let $\tilde{v}_0 = -1/\epsilon$ and $\tilde{u}_0 = \epsilon$ for $\epsilon > 0$ small, we obtain at $t = \epsilon$

$$\tilde{u}(\epsilon) = 1/\epsilon^3, \quad \tilde{v}(\epsilon) = 1/\epsilon.$$

We can see that within ϵ time, $\tilde{u}$ grows from its initial value of order ϵ to $O(\epsilon^{-3})$, a factor of ϵ^{-4} amplification.

The key ingredient in obtaining the global existence in Theorem 3.2 is that the coefficient on the right-hand side of (3.22) is less than -1. For this ODE system, there are two distinguished phases. In the first phase, if $\tilde{v}$ is negative and large in magnitude, but $\tilde{u}$ is small, then $\tilde{v}$ can experience

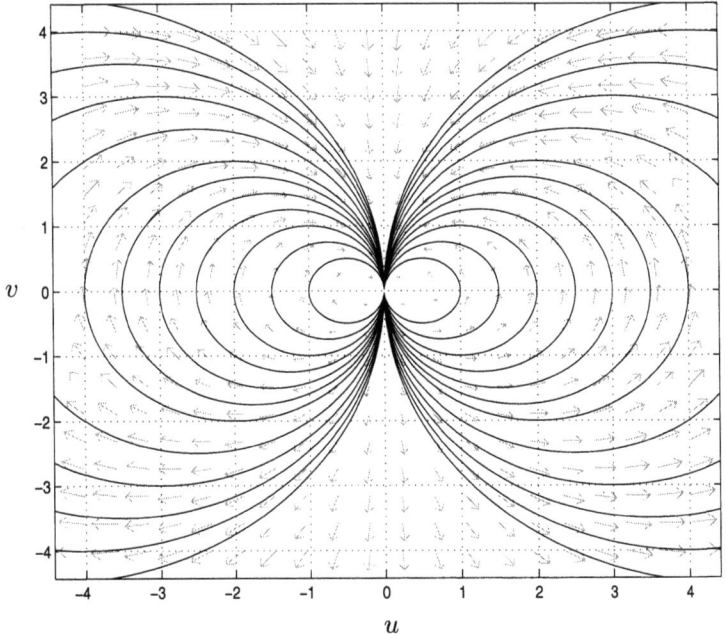

Figure 3.1. The phase diagram for the ODE system.

tremendous dynamic growth, which is essentially governed by

$$\tilde{v}_t = -\tilde{v}^2.$$

However, as $\tilde{v}$ becomes very large and negative, it will induce a rapid growth in $\tilde{u}$. The nonlinear structure of the ODE system is such that $\tilde{u}$ will eventually grow even faster than $\tilde{v}$ and force $(\tilde{u})^2 - (\tilde{v})^2 < 0$ in the second phase. From this time on, $\tilde{v}$ will increase in time and eventually become positive. Once $\tilde{v}$ becomes positive, the nonlinear term, $-\tilde{v}^2$, becomes stabilizing for $\tilde{v}$. Similarly, the nonlinear term, $-2\tilde{u}\tilde{v}$, becomes stabilizing for $\tilde{u}$. This subtle dynamic stability property of the ODE system can be best illustrated by the phase diagram in Figure 3.1.

The reaction–diffusion model

In this subsection, we consider the reaction–diffusion system:

$$(\tilde{u})_t = \nu \tilde{u}_{zz} - 2\tilde{v}\tilde{u}, \qquad (3.29)$$

$$(\tilde{v})_t = \nu \tilde{v}_{zz} + (\tilde{u})^2 - (\tilde{v})^2. \qquad (3.30)$$

As we can see for the corresponding ODE system, the structure of the nonlinearity plays an essential role in obtaining global existence. Intuitively, one may think that the diffusion term would help to stabilize the dynamic growth induced by the nonlinear terms. However, because the nonlinear ODE system in the absence of viscosity is very unstable, the diffusion term can actually have a destabilizing effect. Below we demonstrate this somewhat surprising fact through careful numerical experiments.

In Figures 3.2–3.4, we plot a time sequence of solutions for the above reaction–diffusion system with the following initial data:

$$\tilde{u}_0(z) = \epsilon(2 + \sin(2\pi z)), \quad \tilde{v}_0(z) = -\frac{1}{\epsilon} - \sin(2\pi z),$$

where $\epsilon = 0.001$. For this initial condition, the solution is periodic in z with period one. We use a pseudo-spectral method to discretize the coupled system (3.29)–(3.30) in space and use the simple forward Euler discretization for the nonlinear terms and the backward Euler discretization for the diffusion term. In order to resolve the nearly singular solution structure, we use $N = 32,768$ grid points with an adaptive time step satisfying

$$\Delta t_n (|\max\{\tilde{u}^n\}| + |\min\{\tilde{u}^n\}| + |\max\{\tilde{v}^n\}| + |\min\{\tilde{v}^n\}|) \leq 0.01,$$

where $\tilde{u}^n$ and $\tilde{v}^n$ are the numerical solution at time t_n and $t_n = t_{n-1} + \Delta t_{n-1}$ with the initial time stepsize $\Delta t_0 = 0.01\epsilon$. During the time iterations, the smallest time step is as small as $O(10^{-10})$.

From Figure 3.2, we can see that the magnitude of the solution $\tilde{v}$ increases rapidly by a factor of 150 within a very short time ($t = 0.00099817$). As the solution $\tilde{v}$ becomes large and negative, the solution $\tilde{u}$ increases much more rapidly than $\tilde{v}$. By time $t = 0.0010042$, $\tilde{u}$ has increased to about 2.5×10^8 from its initial condition, which is of magnitude 10^{-3}. This is a factor of 2.5×10^{11} increase. At this time, the minimum of $\tilde{v}$ has reached -2×10^8. Note that since $\tilde{u}$ has outgrown $\tilde{v}$ in magnitude, the nonlinear term, $\tilde{u}^2 - \tilde{v}^2$, on the right-hand side of the $\tilde{v}$-equation has changed sign. This causes the solution $\tilde{v}$ to split. By the time $t = 0.001004314$ (see Figure 3.3), both $\tilde{u}$ and $\tilde{v}$ have split and settled down to two relatively stable travelling wave solutions. The wave on the left will travel to the left while the wave on the right will travel to the right. Due to the periodicity in z, the two travelling waves approach each other from the right side of the domain. The 'collision' of these two travelling waves tends to annihilate each other. In particular, the negative part of $\tilde{v}$ is effectively eliminated during this nonlinear interaction. By the time $t = 0.00100603$ (see Figure 3.4), the solution $\tilde{v}$ becomes all positive. Once $\tilde{v}$ becomes positive, the effect of nonlinearity becomes stabilizing for both $\tilde{u}$ and $\tilde{v}$, as in the case of the ODE system. From then on, the solution decays rapidly. By $t = 0.2007$, the magnitude of $\tilde{u}$ is as small as 5.2×10^{-8}, and $\tilde{v}$ becomes almost a constant function with value close to 5. From this time on, $\tilde{u}$ is essentially decoupled from $\tilde{v}$ and will decay like $O(1/t)$.

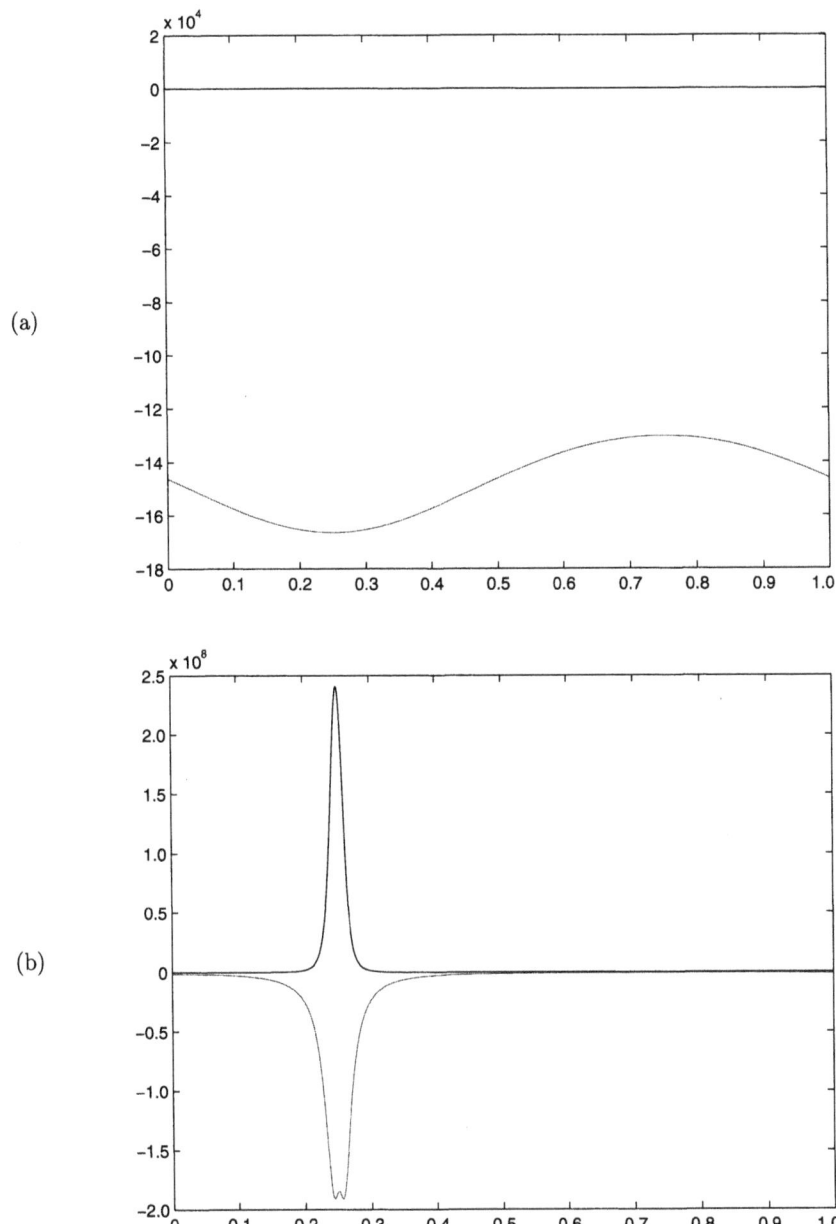

Figure 3.2. The solutions u (dark curve) and v (light curve) at
(a) $t = 0.00099817$, and (b) $t = 0.0010042$, respectively; $N = 32768$, $\nu = 1$.

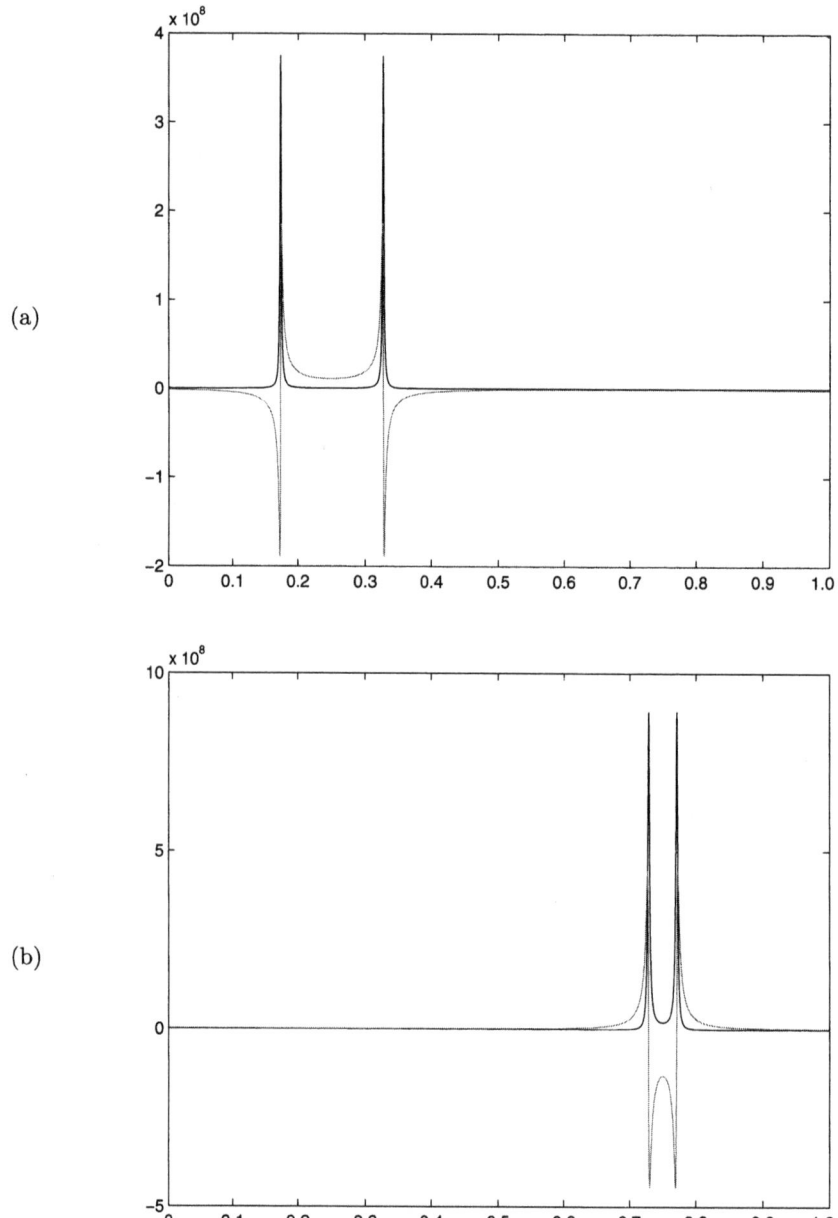

Figure 3.3. The solutions u (dark) and v (light) at (a) $t = 0.001004314$, and (b) $t = 0.001005862$, respectively; $N = 32768$, $\nu = 1$.

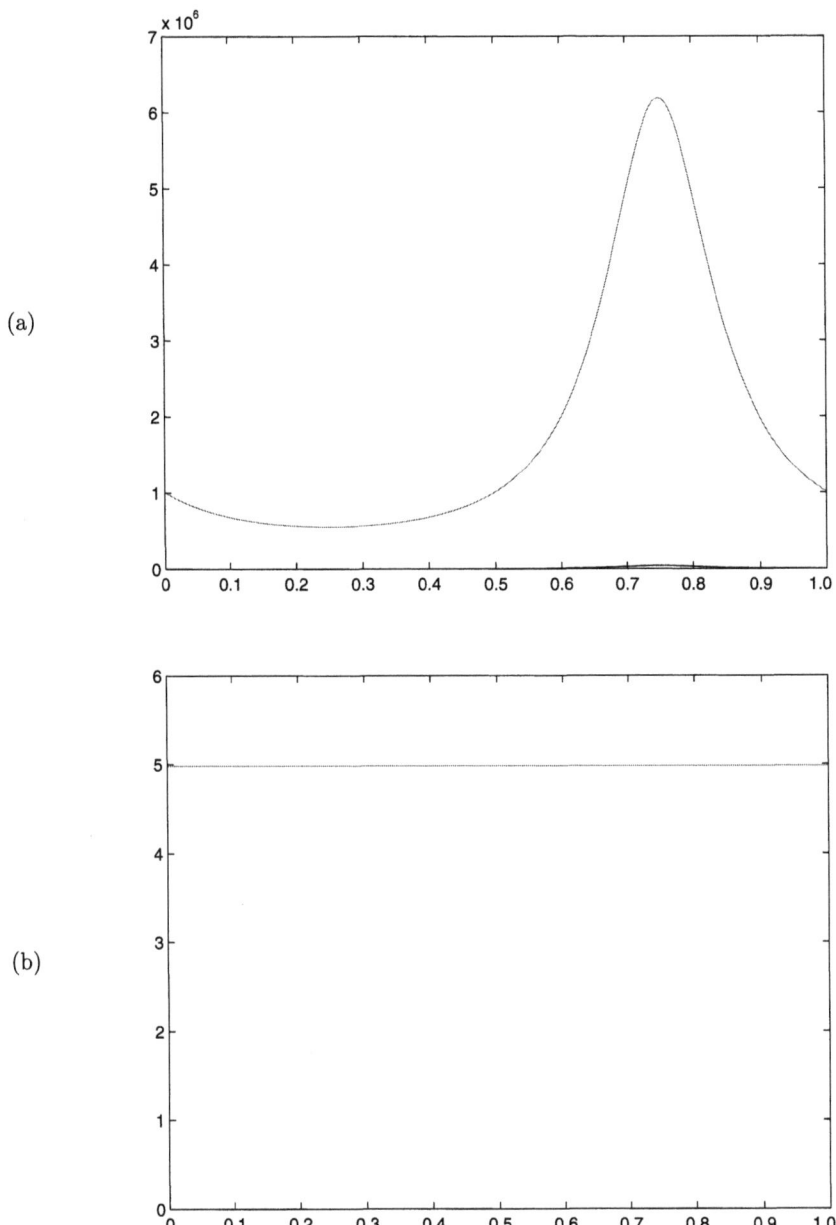

Figure 3.4. The solutions u (dark) and v (light) at (a) $t = 0.00100603$, and (b) $t = 0.2007$, respectively; $N = 32768$, $\nu = 1$. Note that at $t = 0.00100603$, the value of u becomes quite small and is very close to the x-axis. By $t = 0.2007$, the value of u is of the order 5.2×10^{-8} and is almost invisible.

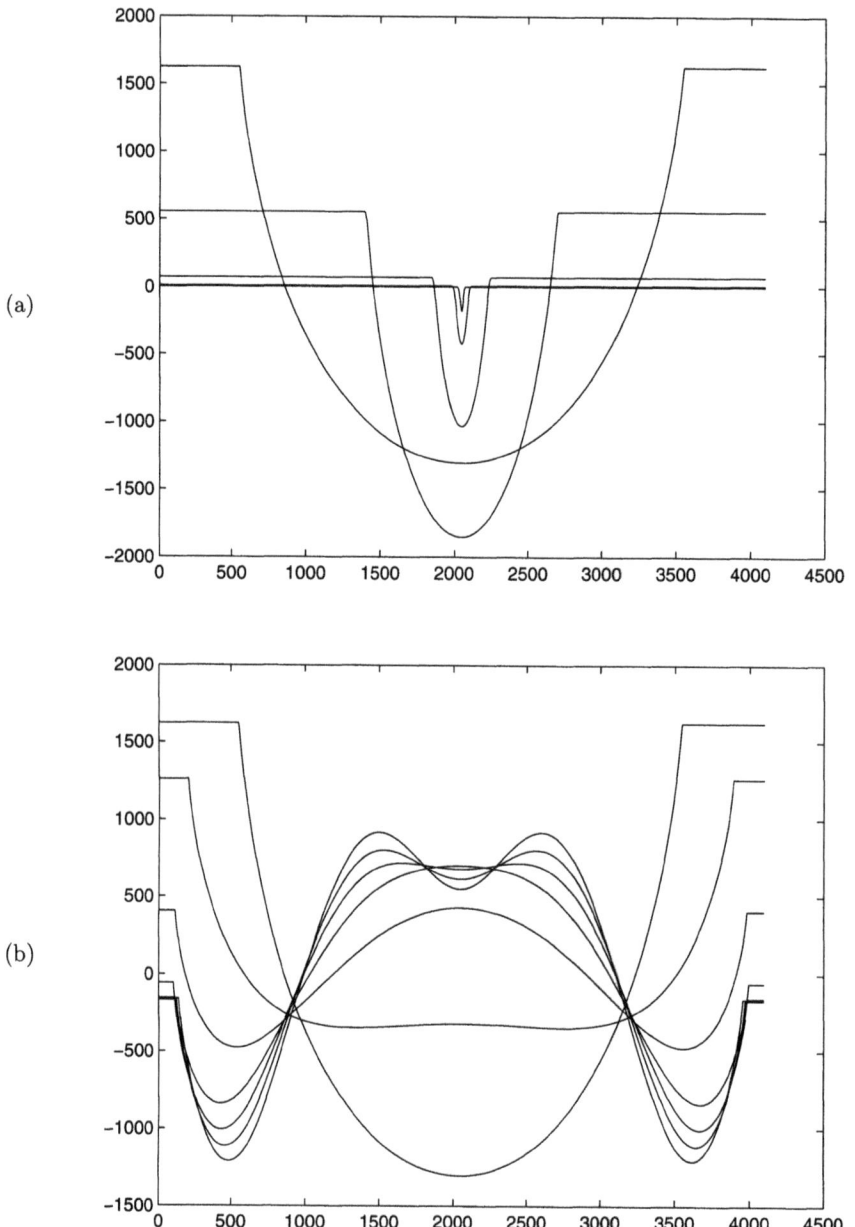

Figure 3.5. The time sequence of v in the Eulerian coordinate, $N = 4096$, $\nu = 0$. (a) $t = 0, 0.0033, 0.0048, 0.0055, 0.0059$, (b) $t = 0.0059, 0.0062, 0.0066, 0.007, 0.0074, 0.0078, 0.0081$. The solutions are plotted against the number of grid points corresponding to the range $[0, 1]$ in physical space.

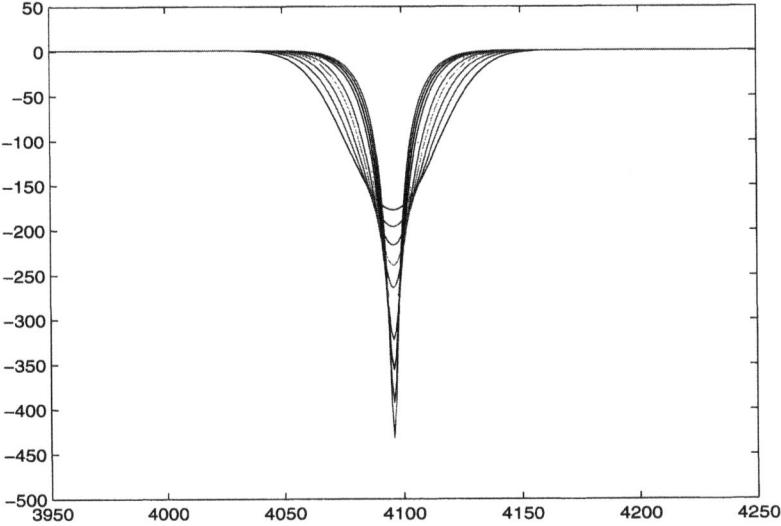

Figure 3.6. The time sequence of solution v in the Lagrangian coordinate by solving the model equation with the wrong sign, $N = 8192$, $\nu = 0$. The time sequence is from $t = 0$ to 0.0033 corresponding to a sequence of curves from the top to the bottom. The solutions are plotted against the number of grid points corresponding to the range $[0.482, 0.519]$ in physical space.

3.4. Global well-posedness of the full 1D model

We have also performed numerical studies of the full 1D model. We find that the solution behaviour of the full 1D model is completely different from the reaction–diffusion model. In particular, the convection term plays an essential role in regularizing the nearly singular behaviour of the reaction–diffusion model. In our numerical computations, we use a pseudo-spectral method to discretize in space and a second-order Runge–Kutta discretization in time with an adaptive time-stepping. The initial data are given by

$$u(\alpha, 0) = 1, \quad v(\alpha, 0) = 1 - \frac{1}{\delta} \exp^{-(x-0.5)^2/\epsilon},$$

with $\epsilon = 0.00001$ and $\delta = \sqrt{\epsilon \pi}$. In Figure 3.5, we plot a sequence of snapshots of the solution. We see that the solution experiences a similar splitting process as in the reaction–diffusion model. On the other hand, we observe that as the solution $\tilde{v}$ grows large and negative, the initial sharp profile of $\tilde{v}$ becomes defocused and smoother. This is a consequence of the incompressibility of the fluid flow. If we change the sign of the convection velocity from $2\tilde{\psi}$ to $-2\tilde{\psi}$, the profile of $\tilde{v}$ becomes focused dynamically and seems to evolve into a focusing finite-time blow-up: see Figure 3.6.

Based on our numerical studies, we become convinced that the solution of the full 1D model should be regular for all times. However, it is extremely difficult, if not impossible, to prove the global regularity of the 1D model by using an energy type of estimates. If we multiply the $\tilde{u}$-equation by $\tilde{u}$, and the $\tilde{v}$-equation by $\tilde{v}$, and integrate over z, we arrive at

$$\frac{1}{2}\frac{d}{dt}\int_0^1 \tilde{u}^2\,dz = -3\int_0^1 (\tilde{u})^2\tilde{v}\,dz - \nu\int_0^1 \tilde{u}_z^2\,dz, \tag{3.31}$$

$$\frac{1}{2}\frac{d}{dt}\int_0^1 \tilde{v}^2\,dz = \int_0^1 \tilde{u}^2\tilde{v}\,dz - 3\int_0^1 (\tilde{v})^3\,dz - \nu\int_0^1 \tilde{v}_z^2\,dz. \tag{3.32}$$

Even for this 1D model, the energy estimate shares the some essential difficulty as the 3D Navier–Stokes equations. It is not clear how to control the nonlinear vortex-stretching-like terms by the diffusion term. On the other hand, if we assume that

$$\int_0^T \|\tilde{v}\|_{L^\infty}\,dt < \infty,$$

similar to the Beale–Kato–Majda non-blow-up condition for vorticity (Beale, Kato and Majda 1984), then one can easily show that there is no blow-up before $t = T$.

In order to obtain the global regularity of the 1D model, we need to use a local estimate. The key is to obtain a pointwise estimate for a positive Lyapunov function. Convection is found to play an essential role in cancelling the destabilizing vortex stretching terms. Using this pointwise estimate, we can prove that if the initial conditions for $\tilde{u}$ and $\tilde{v}$ are in C^m with $m \geq 1$, then the solution remains in C^m for all times.

Theorem 3.3. (Hou and Li 2008a) Assume that $\tilde{u}(z,0)$ and $\tilde{v}(z,0)$ are in $C^m[0,1]$ with $m \geq 1$ and periodic with period 1. Then the solution $(\tilde{u}, \tilde{v})$ of the 1D model will be in $C^m[0,1]$ for all times and for $\nu \geq 0$.

Proof. The key is to obtain a *pointwise* estimate *a priori* for the positive Lyapunov function $\tilde{u}_z^2 + \tilde{v}_z^2$. Differentiating (3.20)–(3.21) with respect to z, we get

$$(\tilde{u}_z)_t + 2\tilde{\psi}(\tilde{u}_z)_z - 2\tilde{v}\tilde{u}_z = -2\tilde{v}\tilde{u}_z - 2\tilde{u}\tilde{v}_z + \nu(\tilde{u}_z)_{zz}, \tag{3.33}$$

$$(\tilde{v}_z)_t + 2\tilde{\psi}(\tilde{v}_z)_z - 2\tilde{v}\tilde{v}_z = 2\tilde{u}\tilde{u}_z - 2\tilde{v}\tilde{v}_z + \nu(\tilde{v}_z)_{zz}. \tag{3.34}$$

Note that the convection term contributes to stability by *cancelling one of the nonlinear terms* on the right-hand side. This gives

$$(\tilde{u}_z)_t + 2\tilde{\psi}(\tilde{u}_z)_z = -2\tilde{u}\tilde{v}_z + \nu(\tilde{u}_z)_{zz}, \tag{3.35}$$

$$(\tilde{v}_z)_t + 2\tilde{\psi}(\tilde{v}_z)_z = 2\tilde{u}\tilde{u}_z + \nu(\tilde{v}_z)_{zz}. \tag{3.36}$$

Multiplying (3.35) by $2\tilde{u}_z$ and (3.36) by $2\tilde{v}_z$, we obtain

$$(\tilde{u}_z^2)_t + 2\tilde{\psi}(\tilde{u}_z^2)_z = -4\tilde{u}\tilde{u}_z\tilde{v}_z + 2\nu\tilde{u}_z(\tilde{u}_z)_{zz}, \tag{3.37}$$

$$(\tilde{v}_z^2)_t + 2\tilde{\psi}(\tilde{v}_z^2)_z = 4\tilde{u}\tilde{u}_z\tilde{v}_z + 2\nu\tilde{v}_z(\tilde{v}_z)_{zz}. \tag{3.38}$$

Now, we add (3.37) to (3.38). *Surprisingly, the remaining nonlinear vortex stretching terms cancel each other exactly.* We get

$$(\tilde{u}_z^2 + \tilde{v}_z^2)_t + 2\tilde{\psi}(\tilde{u}_z^2 + \tilde{v}_z^2)_z = 2\nu(\tilde{u}_z(\tilde{u}_z)_{zz} + \tilde{v}_z(\tilde{v}_z)_{zz}). \tag{3.39}$$

Further, we can rewrite equation (3.39) as follows:

$$(\tilde{u}_z^2 + \tilde{v}_z^2)_t + 2\tilde{\psi}(\tilde{u}_z^2 + \tilde{v}_z^2)_z = \nu(\tilde{u}_z^2 + \tilde{v}_z^2)_{zz} - 2\nu[(\tilde{u}_{zz})^2 + (\tilde{v}_{zz})^2]. \tag{3.40}$$

Now it is easy to see that $(\tilde{u}_z^2 + \tilde{v}_z^2)$ satisfies a *maximum principle* for all $\nu \geq 0$:

$$\|\tilde{u}_z^2 + \tilde{v}_z^2\|_{L^\infty} \leq \|(\tilde{u}_0)_z^2 + (\tilde{v}_0)_z^2\|_{L^\infty}.$$

It is worth emphasizing that the cancellation between the convection term and the vortex stretching term takes place at the inviscid level. Viscosity does not play an essential role here. Since $\tilde{v}$ has zero mean, the Poincaré inequality implies that $\|\tilde{v}\|_{L^\infty} \leq C_0$, with C_0 defined by

$$C_0 = \|((\tilde{u}_0)_z^2 + (\tilde{v}_0)_z^2)^{\frac{1}{2}}\|_{L^\infty}.$$

The boundedness of $\tilde{u}$ follows from the bound on $\tilde{v}$, that is, $\|\tilde{u}(t)\|_{L^\infty} \leq \|\tilde{u}_0\|_{L^\infty} \exp(2C_0 t)$. The higher-order regularity follows from the standard estimates. This proves Theorem 3.3. □

3.5. Construction of a family of 3D globally smooth solutions

We can use the solution from the 1D model to construct a family of globally smooth solutions for the 3D axisymmetric Navier–Stokes equations with large initial data of finite energy. We remark that a special feature of this family of globally smooth solutions is that the solution can potentially develop very large dynamic growth and it violates the smallness condition required by classical global existence results (Constantin and Foias 1988, Temam 2001).

Theorem 3.4. (Hou and Li 2008a) Let $\phi(r)$ be a smooth cut-off function and u_1, ω_1 and ψ_1 be the solution of the 1D model. Define

$$u^\theta(r,z,t) = r u_1(z,t)\phi(r) + \tilde{u}(r,z,t),$$
$$\omega^\theta(r,z,t) = r \omega_1(z,t)\phi(r) + \tilde{\omega}(r,z,t),$$
$$\psi^\theta(r,z,t) = r \psi_1(z,t)\phi(r) + \tilde{\psi}(r,z,t).$$

Then there exists a family of globally smooth functions $\tilde{u}$, $\tilde{\omega}$ and $\tilde{\psi}$ such that u^θ, ω^θ and ψ^θ are globally smooth solutions of the 3D Navier–Stokes equations with finite energy.

4. Stabilizing effect of convection for 3D Navier–Stokes

Hou and Lei (2009b) studied the stabilizing effect of the convection term in the 3D incompressible Euler or Navier–Stokes equations using a new 3D model. This model was derived from the reformulated Navier–Stokes equations. It shares many properties with the 3D Euler or Navier–Stokes equations. First of all, it has the same nonlinear vortex stretching term. Secondly, it has the same type of *a priori* energy identity. Thirdly, almost all the existing non-blow-up criteria for the 3D Euler or Navier–Stokes equations are also valid for our model. A 3D model that satisfies all these properties seems hard to find in general. But in terms of the equations for the new variables, u_1, ω_1, and ψ_1, we obtain our 3D model equations by simply dropping the convective term from the reformulated Navier–Stokes equations (3.16):

$$\partial_t u_1 = \nu\left(\partial_r^2 + \frac{3}{r}\partial_r + \partial_z^2\right)u_1 + 2u_1\psi_{1z}, \tag{4.1a}$$

$$\partial_t \omega_1 = \nu\left(\partial_r^2 + \frac{3}{r}\partial_r + \partial_z^2\right)\omega_1 + (u_1^2)_z, \tag{4.1b}$$

$$-\left(\partial_r^2 + \frac{3}{r}\partial_r + \partial_z^2\right)\psi_1 = \omega_1. \tag{4.1c}$$

Note that (4.1) is already a closed system. The main difference between our 3D model and the Navier–Stokes equations is that we neglect the convection term in our model. If we add the convection term back to our 3D model, we will recover the Navier–Stokes equations.

Below we will summarize some important properties of the model equations (4.1).

4.1. Properties of the 3D model

This 3D model shares many important properties with the axisymmetric Navier–Stokes equations. First of all, one can define an incompressible velocity field for the 3D model,

$$\mathbf{u}(t, \mathbf{x}) = u^r(t, r, z)e_r + u^\theta(t, r, z)e_\theta + u^z(t, r, z)e_z, \tag{4.2}$$

$$u^\theta = ru_1, \quad u^r = -r\psi_{1z}, \quad u^z = 2\psi_1 + r\psi_{1r}, \tag{4.3}$$

where $\mathbf{x} = (x, y, z)$, $r = \sqrt{x^2 + y^2}$. It is easy to check that

$$\nabla \cdot \mathbf{u} = \partial_r u^r + \partial_z u^z + \frac{u^r}{r} = 0, \tag{4.4}$$

which is the same as the original Navier–Stokes equations.

Furthermore, Hou and Lei (2009b) proved the following energy identity for the 3D model.

Theorem 4.1. (Energy identity (Hou and Lei 2009b)) The strong solution of (4.1) satisfies

$$\frac{1}{2}\frac{d}{dt}\int (|u_1|^2 + 2|D\psi_1|^2)r^3\,dr\,dz + \nu\int (|Du_1|^2 + 2|D^2\psi_1|^2)r^3\,dr\,dz = 0, \tag{4.5}$$

Here D is the first-order derivative operator defined in $\mathbb{R}^5$.

This energy identity is equivalent to that of the Navier–Stokes equations, which has the form

$$\frac{1}{2}\frac{d}{dt}\int (|u_1|^2 + |D\psi_1|^2)r^3\,dr\,dz + \nu\int (|Du_1|^2 + |D^2\psi_1|^2)r^3\,dr\,dz = 0. \tag{4.6}$$

Another result obtained by Hou and Lei is a non-blow-up criterion of the 3D model equations (4.1), which is an analogue of the Beale–Kato–Majda (BKM) result for the 3D Euler and Navier–Stokes equations. For the 3D Euler and Navier–Stokes equations, the BKM non-blow-up criterion states that the solution **u** blows up at time $T < \infty$ if and only if the accumulation of vorticity $\int_0^T \|\nabla_x \times \mathbf{u}\|_{L^\infty(\mathbb{R}^3)}\,dt$ is infinite (Beale, Kato and Majda 1984). The BKM non-blow-up criterion was later improved by Kozono and Taniuchi (2000), who proved that the $\|\cdot\|_{L^\infty}$-norm can be replaced by the norm in the BMO space. This generalization is interesting because some crucial Sobolev embedding theorems can be applied to the BMO space, but not to the L^∞-space. A non-blow-up result formulated in terms of the BMO space has a broader range of applications.

Theorem 4.2. (A non-blow-up criterion of Beale–Kato–Majda type (Hou and Lei 2009b)) A smooth solution (u_1, ω_1, ψ_1) of the model (4.1) for $0 \le t < T$ blows up at time $t = T$ if and only if

$$\int_0^T \|\nabla \times \mathbf{u}\|_{\mathrm{BMO}(\mathbb{R}^3)}\,dt = \infty, \tag{4.7}$$

where **u** is defined in (4.2)–(4.3).

There have been many results on the global regularity of the solutions of the 3D Navier–Stokes equations under some additional conditions imposed on the solution. In particular, the papers of Prodi (1959) and Serrin (1963) gave the following non-blow-up criterion for the solution of the 3D Navier–Stokes equations: *Any Leray–Hopf solution u to the 3D Navier–Stokes equations on $[0,T]$ is smooth on $[0,T]$ if* $\|\mathbf{u}\|_{L_t^q L_x^p([0,T]\times\mathbb{R}^3)} < \infty$ *for* *some p, q satisfying* $(3/p) + (2/q) \leq 1$, $3 < p \leq \infty$. A local version was later established by Serrin (1962) for $(3/p) + (2/q) < 1$ and by Struwe (1988) for $(3/p) + (2/q) = 1$. The highly non-trivial end-point case of $p = 3$ was recently established by Iskauriaza, Seregin and Sverak (2003).

To demonstrate the similarity between the 3D model equations (4.1) and the axisymmetric Navier–Stokes equations, Hou and Lei proved a non-blow-up criterion of the Prodi–Serrin type for their model.

Theorem 4.3. (A non-blow-up criterion of Prodi–Serrin type (Hou and Lei 2009b)) A weak solution (u_1, ω_1, ψ_1) of the model (4.1) is smooth on $[0, T] \times \mathbb{R}^3$ provided that

$$\|u^\theta\|_{L^q_t L^p_x ([0,T] \times \mathbb{R}^3)} < \infty \tag{4.8}$$

for some p, q satisfying $\frac{3}{p} + \frac{2}{q} \leq 1$ with $3 < p \leq \infty$ and $2 \leq q < \infty$.

Finally, Hou and Lei (2009a) studied the local behaviour of the solutions to the 3D model equations and established an analogue of the Caffarelli–Kohn–Nirenberg partial regularity theory (Caffarelli et al. 1982) for their model. They proved that for any suitable weak solution of the 3D model in an open set in space-time, the one-dimensional Hausdorff measure of the associated singular set is zero. The proof of this partial regularity result is similar in spirit to that of Lin (1998), but there are some new technical difficulties associated with the 3D model. One of the difficulties is in handling the singularity induced by the cylindrical coordinates. This makes it difficult to analyse the partial regularity of the 3D model in $\mathbb{R} \times \mathbb{R}^3$. To overcome this difficulty, they performed their partial regularity analysis in $\mathbb{R} \times \mathbb{R}^5$. By working in $\mathbb{R}^5$, they avoided the difficulty associated with the coordinate singularity.

Another difficulty in obtaining our partial regularity result is that we do not have an evolution equation for the entire velocity field. We need to reformulate the model in terms of a new vector variable. This new variable can be considered as a 'generalized velocity field' in $\mathbb{R}^5$. We remark that the partial regularity theory for Navier–Stokes equations in $\mathbb{R}^5$ is still open due to the lack of certain compactness. When formulating the 3D model in $\mathbb{R} \times \mathbb{R}^5$, they found a 3D structure which has the same scaling as that of the 3D Navier–Stokes equations. This is why the partial regularity analysis can be carried out for the 3D model in $\mathbb{R} \times \mathbb{R}^5$ using a strategy similar to that of Lin (1998).

Theorem 4.4. (An analogue of Caffarelli–Kohn–Nirenberg partial regularity result (Hou and Lei 2009a)) For any suitable weak solution of the 3D model equations (4.1) on an open set in space-time, the one-dimensional Hausdorff measure of the associated singular set is zero.

4.2. Potential singularity formation of the 3D model

Despite the striking similarity at the theoretical level between the 3D model and the Navier–Stokes equations, the former displays a completely different behaviour from the full Navier–Stokes equations. In the next subsection, we will present numerical evidence which seems to support that the model may develop a potential finite-time singularity from smooth initial data with finite energy. Before we do that, we would like to gain some understanding at the theoretical level why the 3D model may develop a finite-time singularity. For this purpose, we consider the inviscid model by setting $\nu = 0$ in (4.1):

$$\partial_t u_1 = 2 u_1 \psi_{1z}, \qquad (4.9a)$$

$$\partial_t \omega_1 = (u_1^2)_z, \qquad (4.9b)$$

$$-\left(\partial_r^2 + \frac{3}{r}\partial_r + \partial_z^2\right)\psi_1 = \omega_1. \qquad (4.9c)$$

If we let $v = \log(u_1^2)$, then we can further reduce the 3D model to the following non-local nonlinear wave equation:

$$v_{tt} = 4\big((-\Delta_5)^{-1} e^v\big)_{zz}, \qquad (4.10)$$

where $-\Delta_5 = -\big(\partial_r^2 + \frac{3}{r}\partial_r + \partial_z^2\big)$, and $\int e^v r^3 \, dr \, dz \le C_0$. Note that $(-\Delta_5)^{-1}$ is a positive operator. If we were to omit $(-\Delta_5)^{-1}$ from (4.10), we would obtain a 1D nonlinear wave equation, $v_{tt} = 4(e^v)_{zz}$, or

$$v_{tt} = 4e^v v_{zz} + 4e^v (v_z)^2, \qquad (4.11)$$

which we expect to develop a finite-time singularity.

4.3. Special blow-up solutions of the 3D model

We can construct a special class of blow-up solutions by letting $u_1 = z\tilde{u}(r,t)$, $\omega_1 = z\tilde{\omega}(r,t)$, and $\psi_1 = z\tilde{\psi}(r,t)$. Then it is easy to derive the following system for $\tilde{u}(r,t)$, $\tilde{\omega}(r,t)$, and $\tilde{\psi}(r,t)$:

$$\partial_t \tilde{u} = 2\tilde{\psi}\tilde{u} + \nu(\partial_r^2 + \frac{3}{r}\partial_r)\tilde{u}, \qquad (4.12a)$$

$$\partial_t \tilde{\omega} = 2\tilde{u}^2 + \nu(\partial_r^2 + \frac{3}{r}\partial_r)\tilde{\omega}, \qquad (4.12b)$$

$$-\left(\partial_r^2 + \frac{3}{r}\partial_r\right)\tilde{\psi} = \tilde{\omega}. \qquad (4.12c)$$

Note that the nonlinear terms become local and quadratic. It is easy to show that if the initial data are positive, then the solution of (4.12) will remain positive for all times. Using this property, we can prove that the above system has finite-time blow-up solutions. However, such singular solutions have infinite energy unless we introduce a cut-off along the z-direction.

We remark that Constantin (1986) has constructed a family of finite-time blow-up solutions to the distorted Euler equations:

$$(\nabla \mathbf{u})_t + (\nabla \mathbf{u})^2 + \nabla\nabla p = 0, \qquad (4.13)$$

where $-\Delta p = Tr((\nabla \mathbf{u})^2)$. We note that $\mathbf{u}$ is no longer divergence-free (in fact, there is no evolution equation for $\mathbf{u}$), and the model does not conserve energy. Moreover, the blow-up solution has infinite energy.

4.4. Numerical evidence for a potential finite-time singularity

In this subsection, we present numerical evidence which seems to support that the model may develop a potential finite-time singularity from smooth initial data with finite energy. By exploiting the axisymmetric geometry of the problem, we obtain a very efficient adaptive solver with an optimal complexity which provides effective local resolutions of order 4096^3. With this level of resolution, we obtain an excellent fit for the asymptotic blow-up rate of maximum axial vorticity. If we denote by w^z the axial vorticity component along the z-direction, we find that $\|w^z\|_\infty(t) \approx C(T-t)^{-1}$ and the potential singularity approaches the symmetry axis (the z-axis) as $t \to T$. Moreover, our study seems to suggest that the potential singularity is locally self-similar and isotropic.

The initial condition considered in our numerical computations is given by

$$u_1(z, r, 0) = (1 + \sin(4\pi z))(r^2 - 1)^{20}(r^2 - 1.2)^{30}, \qquad (4.14)$$
$$\psi_1(z, r, 0) = 0, \qquad (4.15)$$
$$\omega_1(z, r, 0) = 0. \qquad (4.16)$$

A second-order finite-difference discretization is used in space, and the classical fourth-order Runge–Kutta method is used to discretize in time. Since we expect that the potential singularity will appear along the symmetry axis at $r = 0$, we use the following coordinate transformation along the r-direction to achieve the adaptivity by clustering the grid points near $r = 0$:

$$r = f(\alpha) \equiv \alpha - 0.9\sin(\pi\alpha)/\pi. \qquad (4.17)$$

With this change of variables, we can achieve an effective resolution up to 4096^3 for the corresponding 3D problem.

We now present numerical results which show that the solution of the viscous model becomes nearly singular. We choose the viscous coefficient to be $\nu = 0.001$ and perform a series of resolution studies using the adaptive method. We have used both uniform mesh and adaptive mesh with N_z ranging from 256 to 4096. Below we present the computational results obtained by using the adaptive mesh with the highest resolution $N_z = 4096$, $N_r = 400$, and $\Delta t = 2.5 \times 10^{-7}$. We will also perform a resolution study to demonstrate that our computations are well resolved.

From our analytical study of the 3D model, it follows by using a standard energy estimate that if u_1 is bounded, then the solution of the viscous 3D model cannot blow up in a finite time. Thus it is sufficient to monitor the growth of $\|u_1\|_\infty$ in time. We will present numerical evidence which seems to support that u_1 may develop a potential finite-time singularity for the initial condition we consider. The nature of this potential singularity and the mechanism for generating this potential singularity will be analysed in a later subsection.

In Figure 4.1, we plot the maximum of u_1 in time over the time interval $[0,\ 0.021]$ using the adaptive mesh method with $N_z = 4096$ and $N_r = 400$. The time step is chosen to be $\Delta t = 2.5 \times 10^{-7}$. We can see that $\|u_1\|_\infty$ experiences a very rapid growth in time after $t = 0.02$. In Figure 4.1(b), we also plot $\log(\log(\|u_1\|_\infty))$ as a function of time. We can see clearly that $\|u_1\|_\infty$ grows much more rapidly than double exponential in time, which implies that the solution of our model may develop a finite-time singularity. We will present more careful analysis of this potentially singular behaviour later.

In Figures 4.2–4.3, we show a sequence of contour plots for u_1 from $t = 0.014$ to $t = 0.021$. At early times, we observe that the solution forms two large focusing centres of u_1 which approach each other. As this occurs, these rather localized regions are squeezed and form a thin layer parallel to the r-axis and with large gradients along the z-direction. As these regions approach each other and develop a thin layer parallel to the r-axis, the solution becomes locally z-dominant near the region where u_1 achieves its maximum. In this region, the 3D model can be approximated to the leading order by the corresponding 1D model along the z-direction. Hou and Lei (2009b) proved that the solution of the 1D model cannot blow up. The solution survives this potential blow-up scenario. After $t = 0.0172$, the maximum of u_1 starts to decrease. The two focusing centres move away from each other and their supports become more isotropic. As time increases, we observe that there is a strong nonlinear interaction between u_1 and $(\psi_1)_z$, which is induced by the overlap between the support of maximum of u_1 and the support of maximum of $(\psi_1)_z$. By the support of maximum of u_1, we mean the region in which u_1 is comparable to its maximum. The strong alignment between u_1 and $(\psi_1)_z$ near the support of maximum of u_1 leads to a rapid growth of the solution which may become singular in a finite time.

Another important observation is that as time increases, the position at which u_1 achieves its maximum also moves towards the symmetry axis. This suggests that the potential singularity will be along the symmetry axis at the singularity time. We note that $\lim_{r\to 0+} u_1 = 0.5 \lim_{r\to 0+} \omega^z$. Thus, the blow-up of u_1 characterizes the blow-up of the axial vorticity, ω^z.

Next, we perform a detailed study for the 3D model and push our computations very close to the potential singularity time. We use a sequence of

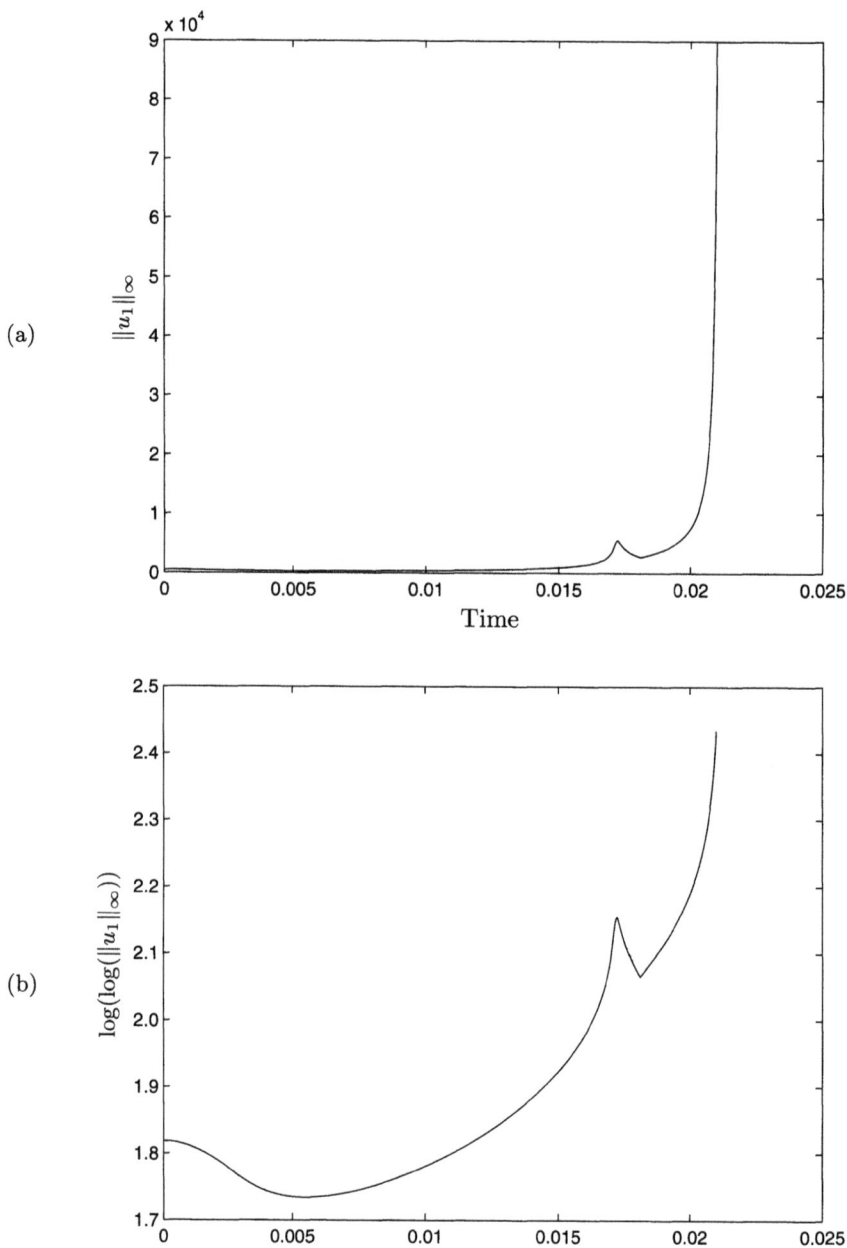

Figure 4.1. (a) $\|u_1\|_\infty$ as a function of time over the interval $[0, 0.021]$, (b) $\log(\log(\|u_1\|_\infty))$ as a function of time over the same interval. The solution is computed by an adaptive mesh with $N_z = 4096$, $N_r = 400$, $\Delta t = 2.5 \times 10^{-7}$, $\nu = 0.001$.

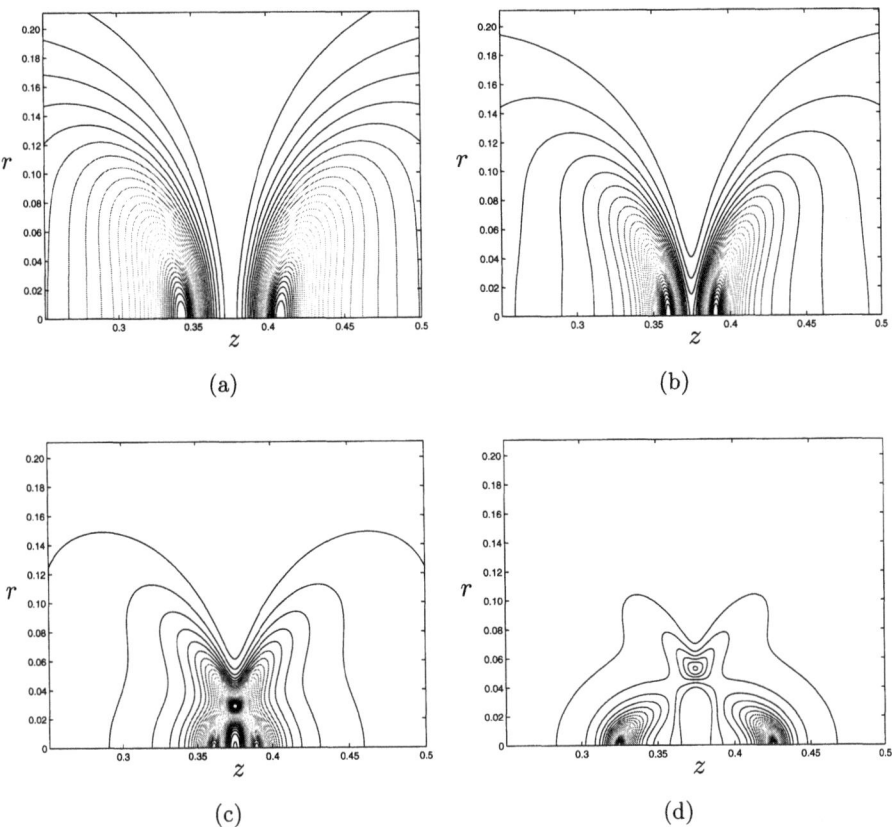

Figure 4.2. (a)–(d) The contour plots of u_1 for the viscous model at $t = 0.014$, 0.016, 0.018 and 0.02 respectively. Adaptive mesh computation with $N_z = 4096$, $N_r = 400$, $\Delta t = 2.5 \times 10^{-7}$, $\nu = 0.001$.

resolutions using both uniform and adaptive mesh. For the uniform mesh, we use resolutions for $N_z \times N_r$ ranging from 256×256 to 2048×2048 with time steps ranging from $\Delta t = 5 \times 10^{-6}$ to 5×10^{-7}. For the adaptive mesh, we use $N_z \times N_r = 2048 \times 256$, $N_z \times N_r = 3072 \times 328$ and $N_z \times N_r = 4096 \times 400$ respectively. The corresponding time steps for these computations are $\Delta t = 10^{-6}$, $\Delta t = 5 \times 10^{-7}$, and $\Delta t = 2.5 \times 10^{-7}$ respectively. With $N_z \times N_r = 4096 \times 400$, we achieve an effective resolution of 4000×4000 near the region of $r = 0$ where the solution is most singular.

To obtain further evidence for a potential finite-time singularity, we use a systematic singularity form fit procedure to obtain a good fit for the possible singularity of the solution. The procedure of our form fit is as

328   T. Y. HOU

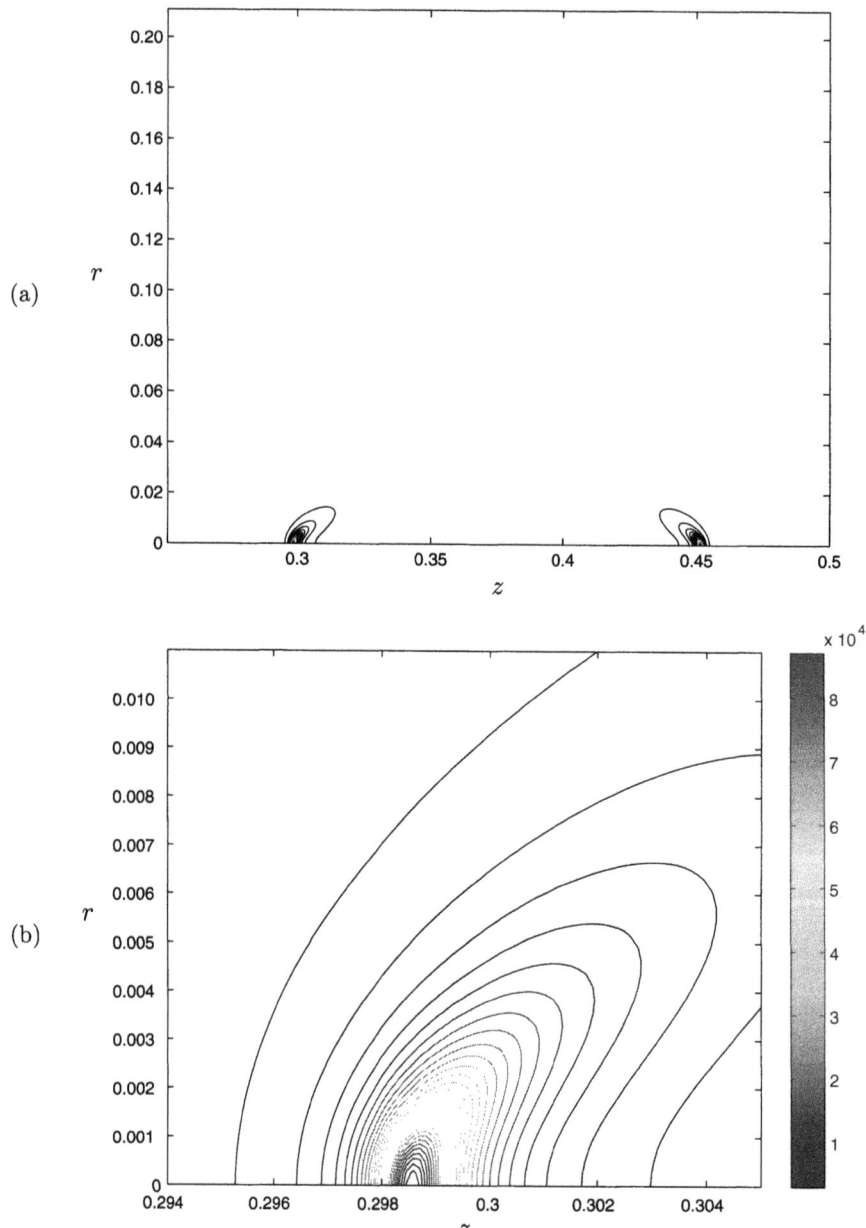

Figure 4.3. The contour of u_1 at $t = 0.021$ (a) and its close-up view (b) for the viscous model computed by the adaptive mesh with $N_z = 4096$, $N_r = 400$, $\Delta t = 2.5 \times 10^{-7}$, $\nu = 0.001$.

follows. We look for a finite-time singularity of the form

$$\|u_1\|_\infty \approx \frac{C}{(T-t)^\alpha}. \qquad (4.18)$$

We have tried several ways to determine the fitting parameters T, C and α. Ultimately, we find that the best way is to study the inverse of $\|u_1\|_\infty$ as a function of time using a sequence of numerical resolutions. For each resolution, we find that the inverse of $\|u_1\|_\infty$ is almost a perfect linear function of time: see Figures 4.4 and 4.5. By using a least-squares fit of the inverse of $\|u_1\|_\infty$, we find that $\alpha = 1$ gives the best fit. The same least-squares fit also determines the potential singularity time T and the constant C. We remark that the $O(1/(T-t))$ blow-up rate of u_1, which measures the axial vorticity, is consistent with the non-blow-up criterion of Beale–Kato–Majda type.

To confirm that the above procedure indeed gives a good fit for the potential singularity, we plot $\|u_1\|_\infty^{-1}$ as a function of time in Figure 4.4(a). We can see that the agreement between the computed solution with $N_z \times N_r = 4096 \times 400$ and the fitted solution is almost perfect. In Figure 4.4(b) we plot $\|u_1\|_\infty$ computed by our adaptive method against the form fit $C/(T-t)$ with $T = 0.02109$ and $C = 8.20348$. The two curves are almost indistinguishable during the final stage of the computation from $t = 0.018$ to $t = 0.021$.

We further investigate the potential singular behaviour of the solution by using a sequence of resolutions to study the limiting behaviour of the

Table 4.1. Resolution study of parameters T and C in the asymptotic fit for the viscous model: $\|u_1\|_\infty^{-1} \approx \frac{(T-t)}{C}$ using different resolutions $h_z = 1/(2N_z)$. The resolutions we use in our adaptive computations are $N_z \times N_r = 1024 \times 128$, 2048×256, 3072×328 and 4096×400 respectively. The corresponding time steps are $\Delta t = 10^{-6}$, 5×10^{-7}, 3.625×10^{-7} and 2.5×10^{-7} respectively. The last row is obtained by extrapolating the second-order polynomial that interpolates the data obtained using $h_z = 1/4096$, $1/6144$ and $1/8192$.

h_z	T	C
1/2048	0.02114	8.409
1/4096	0.0211	8.2237
1/6144	0.021093	8.20946
1/8192	0.02109	8.20348
extrapolation to $h_z = 0$	0.021083	8.1901

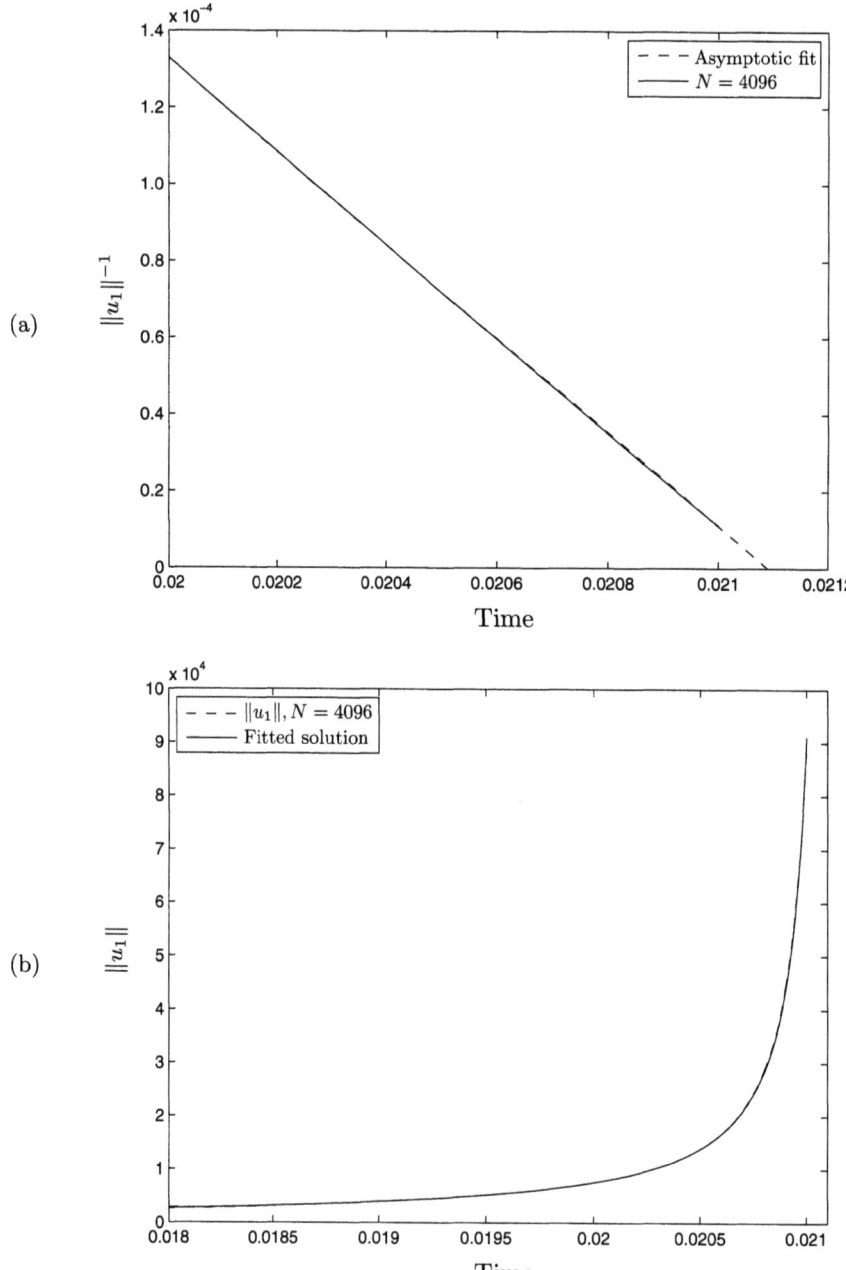

Figure 4.4. (a) The inverse of $\|u_1\|_\infty$ (solid line) versus the asymptotic fit (dashed line) for the viscous model; (b) $\|u_1\|_\infty$ (solid line) versus the asymptotic fit (dashed line). The asymptotic fit is of the form $\|u_1\|_\infty^{-1} \approx \frac{(T-t)}{C}$ with $T = 0.02109$ and $C = 8.20348$. The solution is computed by adaptive mesh with $N_z = 4096$, $N_r = 400$, $\Delta t = 2.5 \times 10^{-7}$; $\nu = 0.001$.

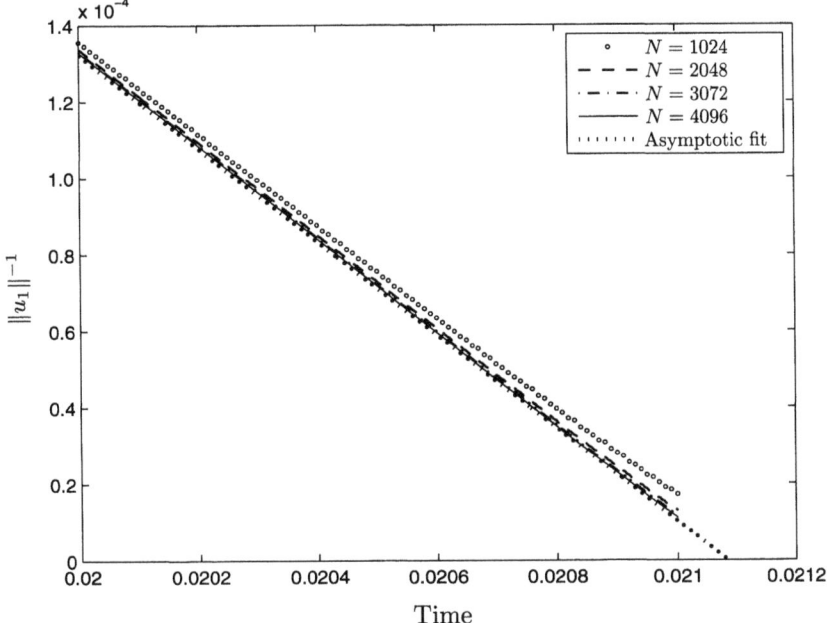

Figure 4.5. The inverse of $\|u_1\|_\infty$ in time for the viscous model. The solution is computed by adaptive mesh with $N_z = 1024$, 2048, 3072 and 4096 respectively (ordering from top to bottom in the figure), $\Delta t = 10^{-6}$, 5×10^{-7}, 3.625×10^{-7}, and 2.5×10^{-7} respectively. The last curve is the singularity fit by extrapolating the computational results obtained by $N_z = 2048$, 3072 and 4096 to infinite resolution $N_z = \infty$. The fitted curve is of the form $\|u_1\|_\infty^{-1} \approx (T-t)/C$, with $T = 0.021083$ and $C = 8.1901$; $\nu = 0.001$.

computed solution as we refine our resolutions. The space resolutions we use are $N_z \times N_r = 1024 \times 128$, 2048×256, 3072×328 and 4096×400 respectively. The corresponding time steps are $\Delta t = 10^{-6}$, 5×10^{-7}, 3.625×10^{-7} and 2.5×10^{-7} respectively. For each resolution, we obtain an optimal least-squares fit of the singularity of the form $\|u_1\|_\infty^{-1} \approx (T-t)/C$. The results are summarized in Table 4.1. Based on the fitted parameters T and C from the three largest resolutions, we construct a second-order polynomial that interpolates T and C through these three data points. We then use the polynomial to extrapolate the values of T and C to the infinite resolution limit. The extrapolated values at $h_z = 0$ are $T = 0.021083$ and $C = 8.1901$ respectively. In Figure 4.5, we plot the inverse of $\|u_1\|_\infty$ as a function of time using four different resolutions. We can see that as we refine the resolution, the computed solution converges to the extrapolated singularity limiting profile.

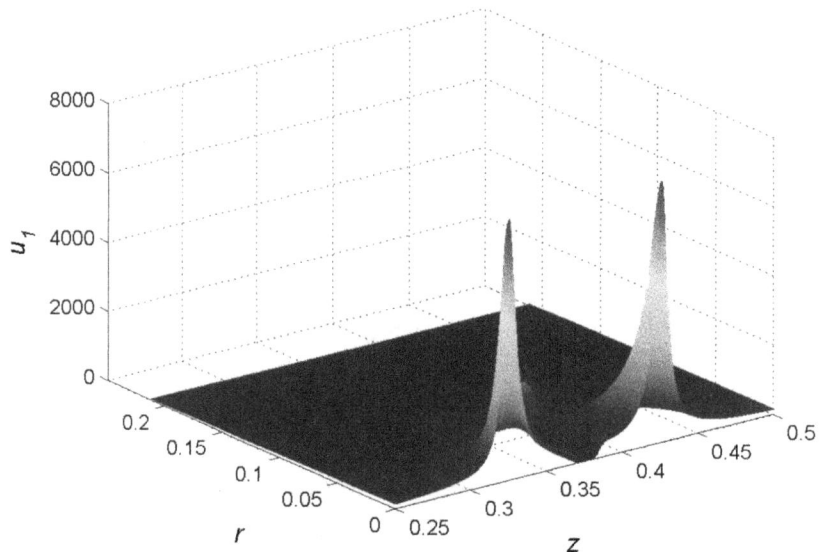

Figure 4.6. The 3D view of u_1 at $t = 0.02$ for the viscous model computed by the adaptive mesh with $N_z = 4096$, $N_r = 400$, $\Delta t = 2.5 \times 10^{-7}$, $\nu = 0.001$.

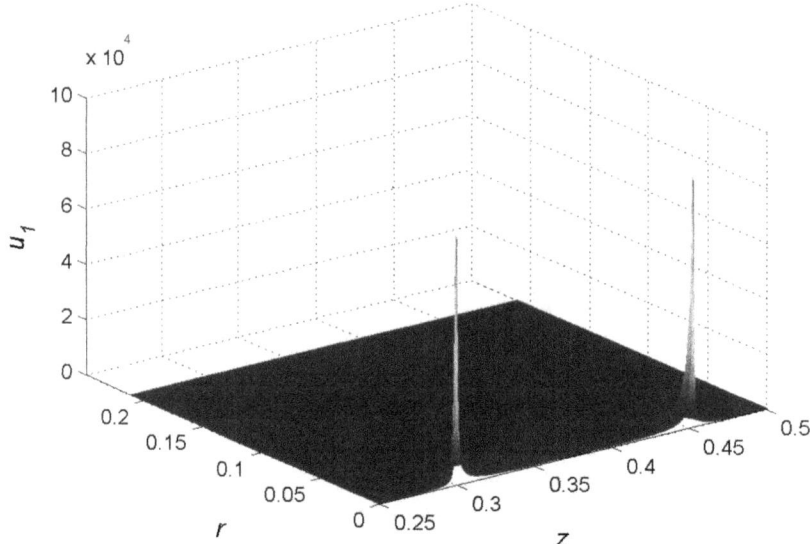

Figure 4.7. The 3D view of u_1 at $t = 0.021$ for the viscous model computed by the adaptive mesh with $N_z = 4096$, $N_r = 400$, $\Delta t = 2.5 \times 10^{-7}$, $\nu = 0.001$.

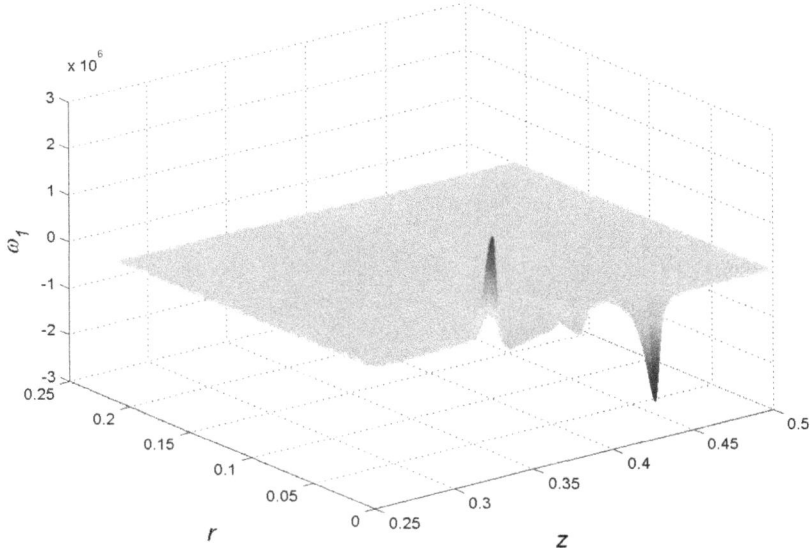

Figure 4.8. The 3D view of w_1 at $t = 0.02$ for the viscous model computed by the adaptive mesh with $N_z = 4096$, $N_r = 400$, $\Delta t = 2.5 \times 10^{-7}$, $\nu = 0.001$.

To illustrate the nature of the nearly singular solution, we show the 3D view of u_1 as a function of r and z in Figures 4.6 and 4.7. We also show the 3D view of w_1 as a function of r and z in Figure 4.8. While u_1 is symmetric with respect to $z = 0.375$, w_1 is anti-symmetric with respect to $z = 0.375$. We can see that the support of the solution u_1 in the most singular region is isotropic and appears to be locally self-similar (Hou and Lei 2009b).

Resolution study
Finally, we perform a resolution study for our computations by comparing the computation obtained by three different resolutions, which are $N_z \times N_r = 2048 \times 256$, $N_z \times N_r = 3072 \times 328$, and $N_z \times N_r = 4096 \times 400$. In Figure 4.9, we plot $\|u_1\|_\infty$ as a function of time using these three resolutions $N_z \times N_r = 2048 \times 256$, $N_z \times N_r = 3072 \times 328$, and $N_z \times N_r = 4096 \times 400$ over the time interval $[0, 0.021]$. We can see that while the computation with $N_z = 2048$ under-resolves the solution near the end of the computation, the solution obtained by using $N_z = 3072$ gives an excellent agreement with that obtained by using $N_z = 4096$.

We also compare the solution of u_1 at $r = 0$ using three different resolutions. Using the partial regularity theory for the 3D model, any singularity of our 3D model must lie on the symmetry axis, $r = 0$. Thus it makes

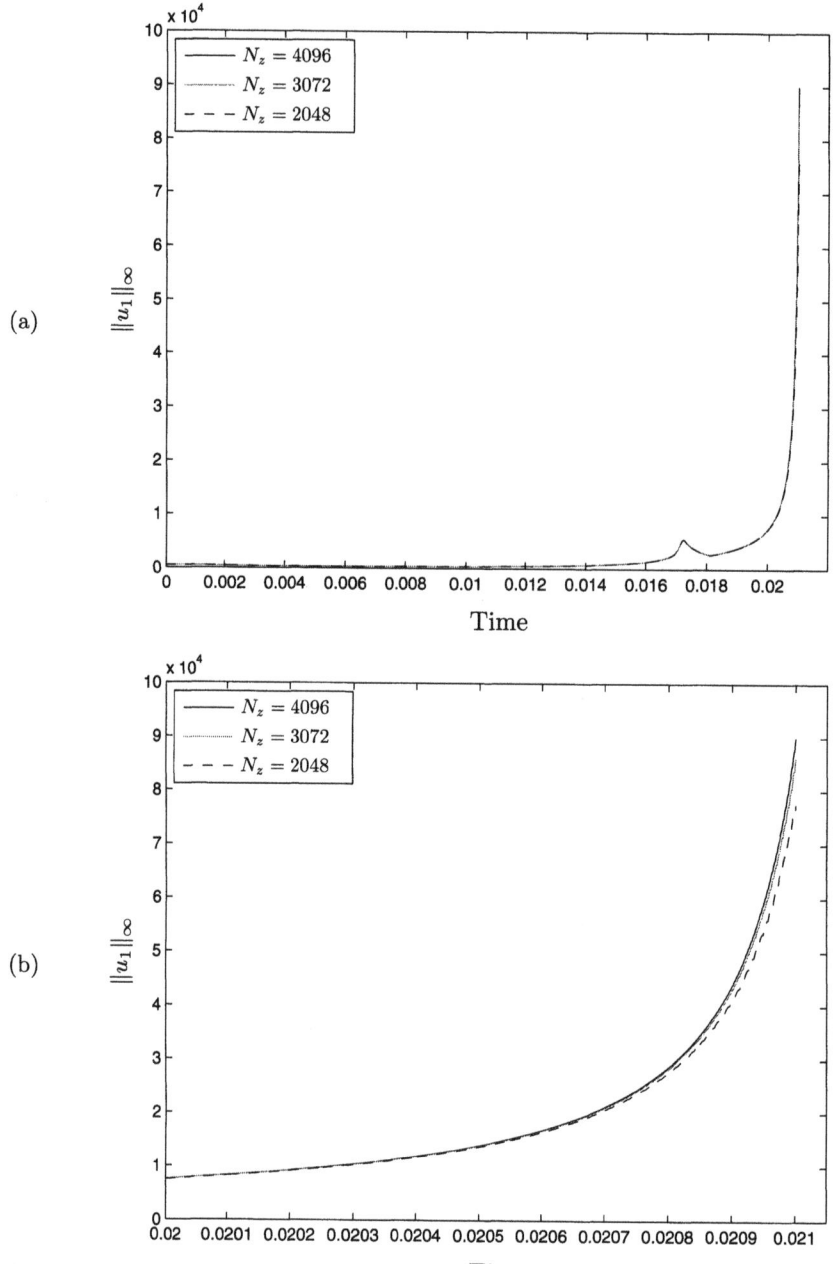

Figure 4.9. Convergence study for $\|u_1\|_\infty$ in time for the viscous model with three resolutions: $N_z \times N_r = 2048 \times 256$, $\Delta t = 5 \times 10^{-7}$ (dashed line), $N_z \times N_r = 3072 \times 328$, $\Delta t = 3.625 \times 10^{-7}$ (grey line), $N_z \times N_r = 4096 \times 400$, $\Delta t = 2.5 \times 10^{-7}$ (solid line). Figure (a) is over the time interval $[0, 0.021]$; (b) is a close-up view over the time interval $[0.02, 0.021]$; $\nu = 0.001$.

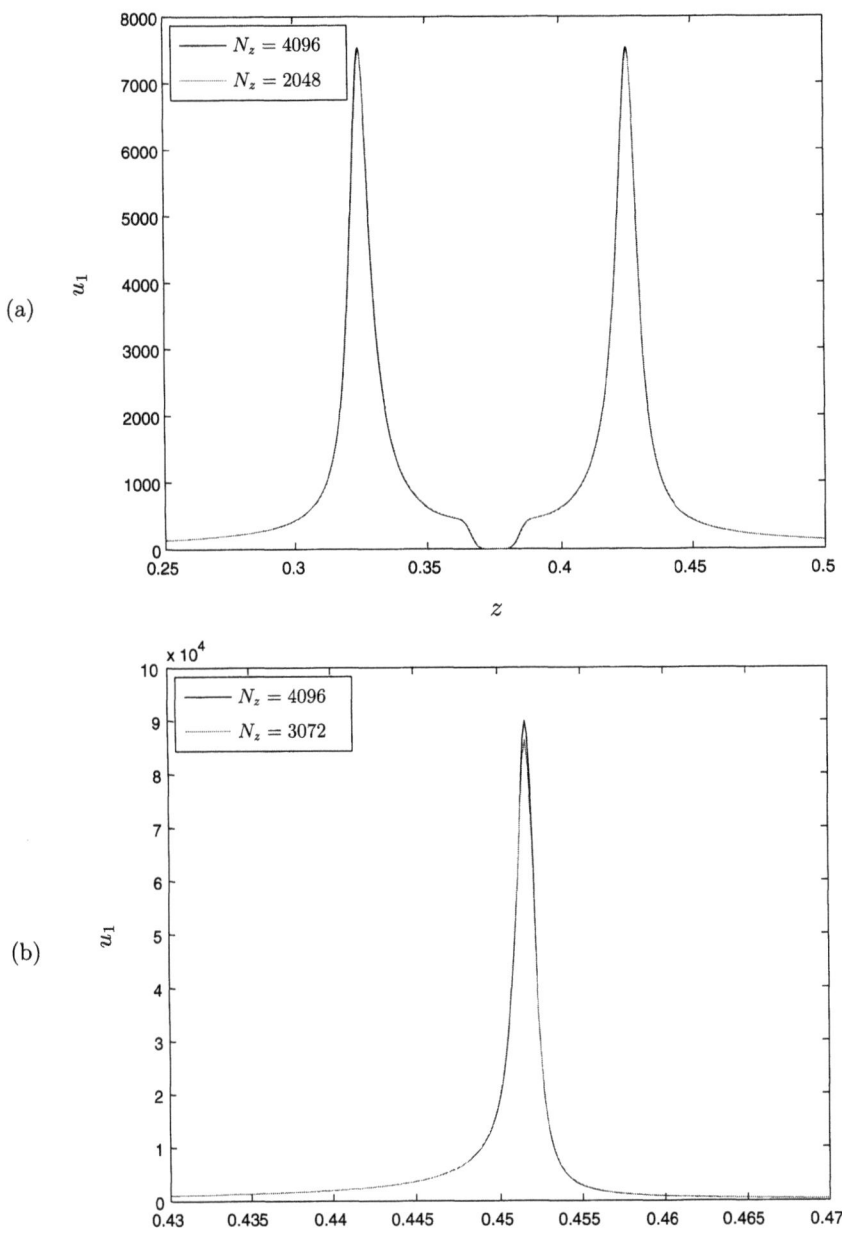

Figure 4.10. Convergence study for u_1 at $r = 0$ and $t = 0.02$ and $t = 0.021$ for the viscous model with different resolutions. Figure (a) is the comparison between $N_z \times N_r = 2048 \times 256$ (solid line) and $N_z \times N_r = 4096 \times 400$ (grey line); (b) is the comparison between $N_z \times N_r = 3072 \times 328$ (solid line) and $N_z \times N_r = 4096 \times 400$ (grey line); $\nu = 0.001$.

sense to perform a resolution study for the solution along the symmetry axis which is the most singular region of the solution. In Figure 4.10(a), we plot the solutions obtained by two resolutions using $N_z \times N_r = 2048 \times 256$ ($\Delta t = 5 \times 10^{-7}$) and $N_z \times N_r = 4096 \times 400$ ($\Delta t = 2.5 \times 10^{-7}$) on top of each other at $t = 0.02$. The two solutions are almost indistinguishable. However, the computation with $N_z \times N_r = 2048 \times 256$ is not sufficient to resolve the nearly singular behaviour of the solution at $t = 0.021$. On the other hand, the computation with $N_z \times N_r = 3072 \times 328$ ($\Delta t = 3.625 \times 10^{-7}$) gives much improved resolution. In Figure 4.10(b) we compare the solution obtained by using $N_z \times N_r = 3072 \times 328$ with that obtained by using $N_z \times N_r = 4096 \times 400$ at $t = 0.021$. We observe that the agreement of the two solutions is very good except near the points where u_1 attains its maximum.

4.5. Mechanism for a finite-time blow-up

To understand the mechanism for the potential blow-up of the viscous model, we plot the solution u_1 on top of $(\psi_1)_z$ along the symmetry axis $r = 0$ at t=0.021 in Figure 4.11. We see that there is a significant overlap between the supports of the maximum of u_1 and of the maximum of $(\psi_1)_z$. Moreover, the solution u_1 has a strong alignment with $(\psi_1)_z$ near the region of maximum of u_1. The local alignment between u_1 and $(\psi_1)_z$ induces a strong nonlinearity on the right-hand side of the u_1-equation, which has the form $2(\psi_1)_z u_1$. This strong alignment between u_1 and $(\psi_1)_z$ is the main mechanism for the potential finite-time blow-up of the 3D model. Similar alignment between u_1 and $(\psi_1)_z$ near the region of maximum u_1 is also observed for the inviscid model (Hou and Lei 2009b).

It is interesting to note that the position at which u_1 attains its maximum does not coincide with that at which $(\psi_1)_z$ attains its maximum. In fact, at the point where u_1 reaches its maximum, the value of $(\psi_1)_z$ is relatively small, or even negative. This misalignment between the position at which u_1 attains its maximum and the position at which $(\psi_1)_z$ attains its maximum induces a dynamic motion which pushes the two focusing centres of u_1 to move away from each other. This dynamics reinforces the local alignment between u_1 and $(\psi_1)_z$. We remark that this wave-like behaviour of the solution along the z-direction is consistent with the nonlinear non-local wave equation (4.10) that we derived for $v = \log(u_1^2)$ for the inviscid model.

As we see in the next subsection, the inclusion of the convection term forces the two focusing centres to travel towards each other. Moreover, the local alignment between u_1 and $(\psi_1)_z$ is destroyed. As a result, the solution becomes defocused and smoother along the symmetry axis. There is no evidence that the solution of the full Navier–Stokes equations would develop a finite-time singularity, at least for the time interval considered here.

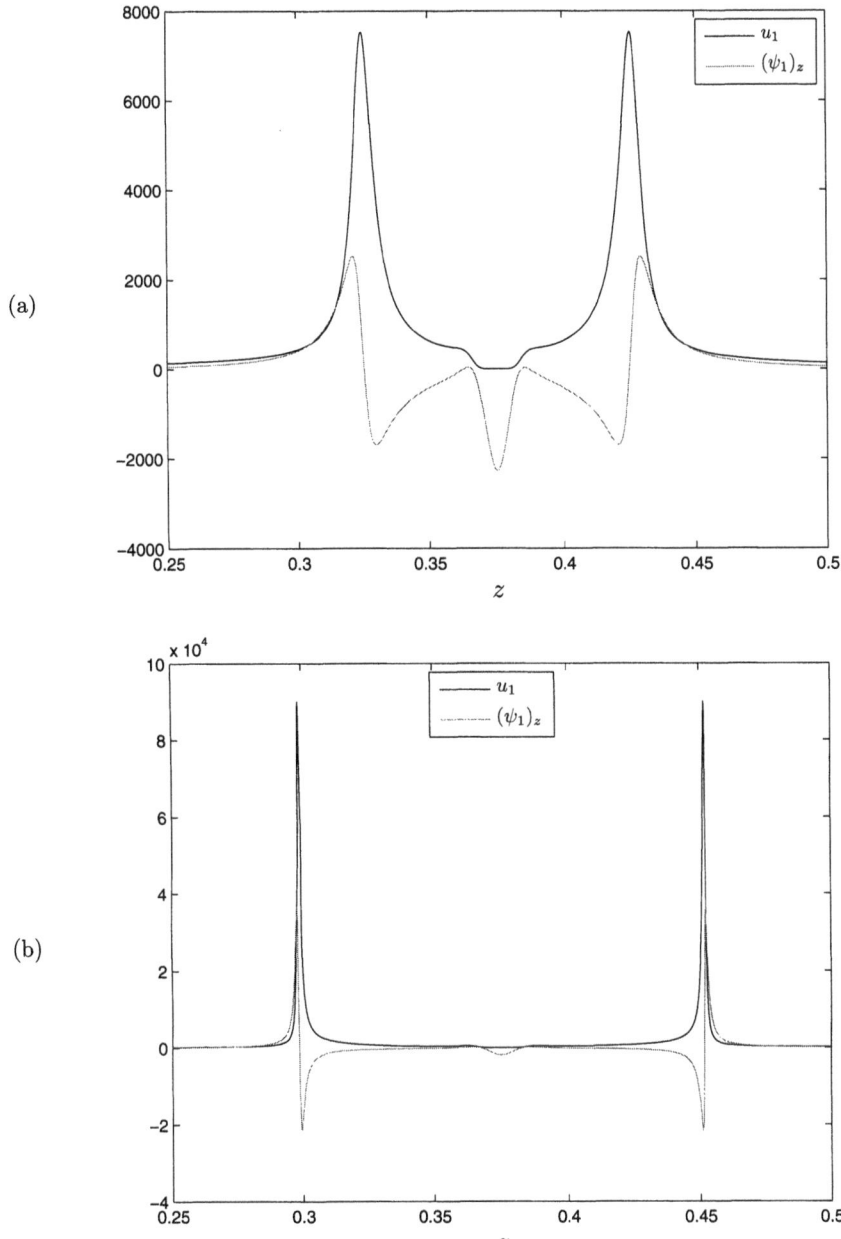

Figure 4.11. u_1 (solid line) versus $(\psi_1)_z$ (grey line) of the viscous model along the symmetry axis $r = 0$. (a) $t = 0.02$; (b) $t = 0.021$. Adaptive mesh computation with $N_z = 4096$, $N_r = 400$, $\Delta t = 2.5 \times 10^{-7}$, $\nu = 0.001$.

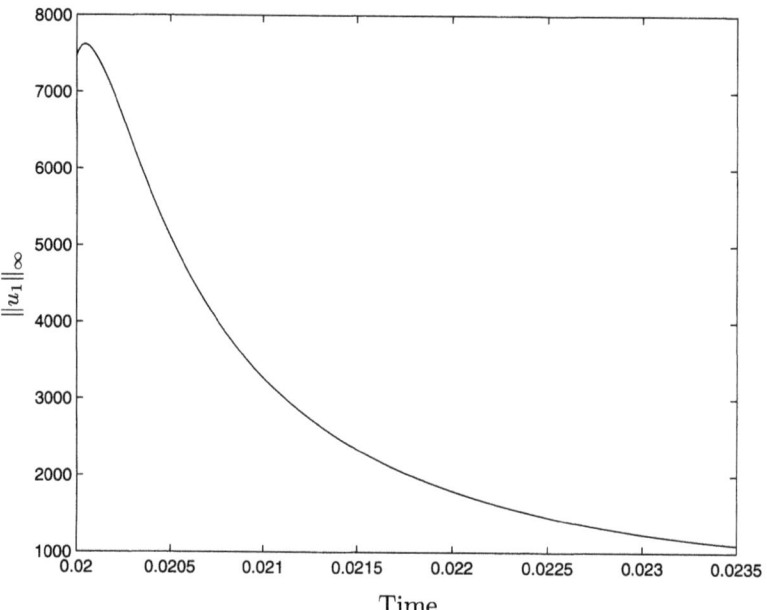

Figure 4.12. $\|u_1\|_\infty$ in time, the full Navier–Stokes computation using the solution of the 3D viscous model at $t = 0.02$ as the initial condition. Adaptive mesh computation with $N_z = 2048$, $N_r = 1024$, $\nu = 0.001$.

4.6. The stabilizing effect of the convection term

In this subsection, we show that by adding back the convection term to the 3D model, which recovers the reformulated Navier–Stokes equations, the solution behaves completely differently. The mechanism for generating the potential finite time singularity for the 3D model is destroyed. Even if we start with the nearly singular solution obtained by the 3D model at $t = 0.02$ and use it as the initial condition for the full Navier–Stokes equations, we observe that the maximum of u_1 soon decreases in time: see Figure 4.12. It is easy to see that the 3D axisymmetric Navier–Stokes equations with swirl cannot develop a finite-time singularity if u_1 is bounded. Thus the fact that $\|u_1\|_\infty$ is decreasing in time is a clear indication that the solution does not develop a finite-time singularity, at least over the time interval considered here.

We also observe that the local alignment between u_1 and $(\psi_1)_z$ near the region of maximum u_1 is destroyed by including the convection term (see Figure 4.13), as is the focusing mechanism. The solution becomes defocused (see Figure 4.14). As time evolves, the two focusing centres approach each other. This process creates a strong internal layer orthogonal to the z-axis

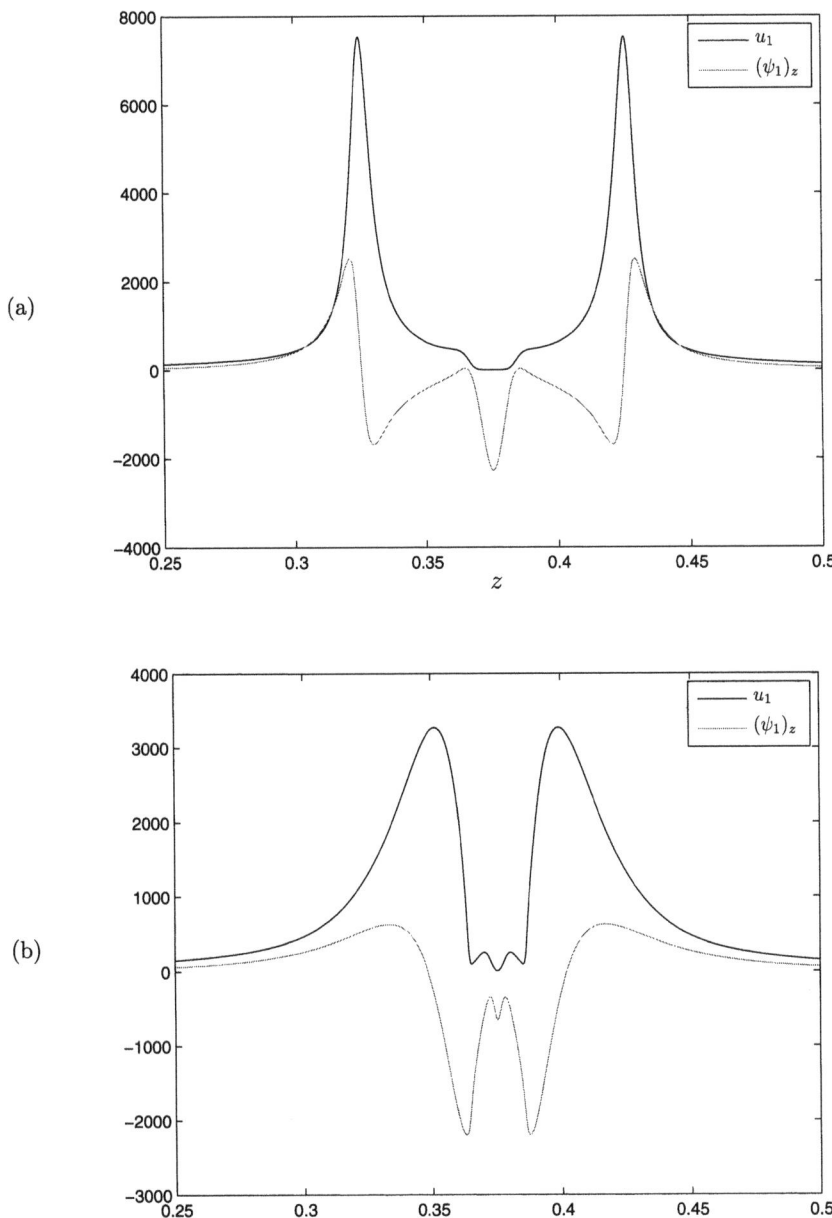

Figure 4.13. u_1 (solid line) versus $(\psi_1)_z$ (grey line) along the symmetry axis $r = 0$. Figure (a) corresponds to $t = 0.02$ (the solution from the 3D viscous model); (b) corresponds to $t = 0.021$ obtained by solving the full Navier–Stokes equations. Adaptive mesh computation with $N_z = 2048$, $N_r = 1024$, $\nu = 0.001$.

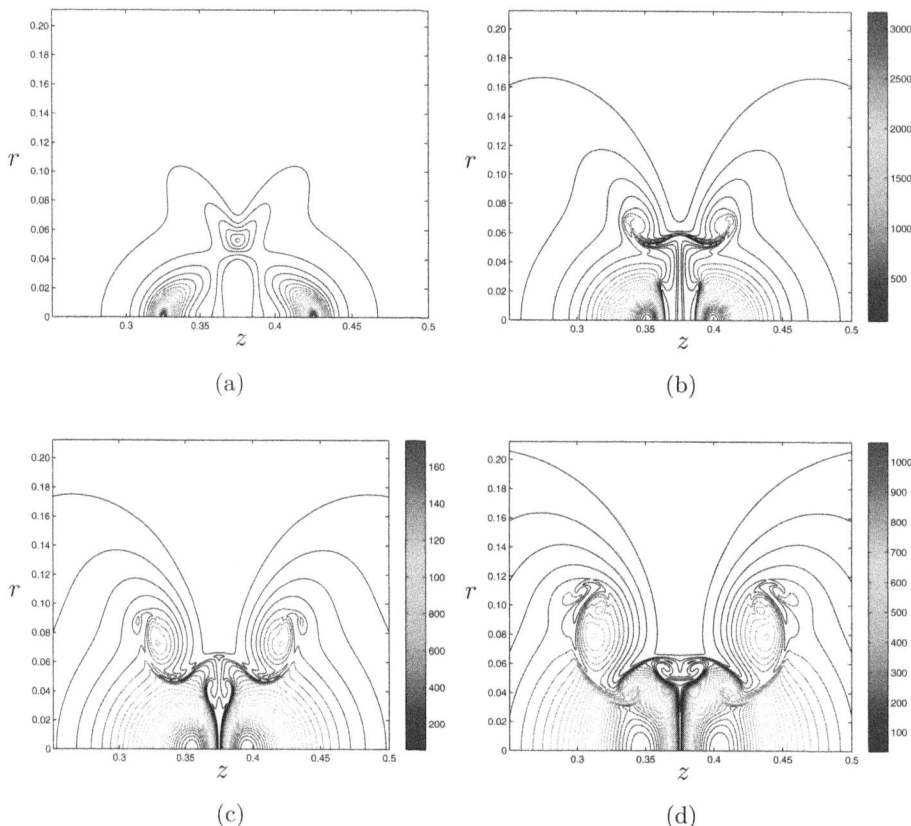

Figure 4.14. (a) The contour of u_1 at $t = 0.02$ obtained from the 3D viscous model which serves as the initial condition for the full Navier–Stokes equations. (b) The contour of u_1 at $t = 0.021$ obtained by solving the full Navier–Stokes equations. (c), (d) The contours of u_1 at $t = 0.022$ and $t = 0.0235$, respectively, by solving the full Navier–Stokes equations. Adaptive mesh computation with $N_z = 2048$, $N_r = 1024$, $\nu = 0.001$.

and forms a jet that moves away from the symmetry axis (the z-axis). The jet further generates some interesting vortex structures. This is illustrated in Figure 4.14. Since the most singular part of the solution of the Navier–Stokes equations moves away from the symmetry axis, we use a higher-resolution adaptive mesh along the r-direction with $N_r = 1024$ to better resolve the layered structure along the r-axis.

By the Caffarelli–Kohn–Nirenberg theory, the singularity of the 3D axisymmetric Navier–Stokes equations, if there is any, must be along the symmetry axis. The fact that the most singular part of the solution moves away from the symmetry axis indicates that the full Navier–Stokes equations will not form a finite-time singularity, at least not for the initial condition we consider here over the time interval for which we compute the solution. On the other hand, the solution of the 3D model with the same initial condition seems to develop a potential finite-time singularity in an earlier time. This confirms that convection plays an essential role in depleting the destabilizing effect induced by vortex stretching.

5. Concluding remarks

Our analysis and computations revealed a subtle dynamic depletion of vortex stretching. Sufficient numerical resolution is essential to capture this dynamic depletion. Our computations for the two antiparallel vortex tubes initial data showed that the velocity is bounded and that the vortex stretching term is bounded by $C\|\omega\|_{L^\infty} \log(\|\omega\|_{L^\infty})$. In Hou and Li (2008b), we also repeated the computation of R. Pelz using highly symmetric initial data (Pelz 1997). We found that while Pelz's vortex filament model indeed produces a finite-time self-similar singularity, the solution of the full 3D Euler equation with the same initial data gives only very modest growth dynamically. No evidence of finite-time singularities was found. Pelz's vortex filament computation was inspired by his earlier computation of the 3D Navier–Stokes equations (Boratav and Pelz 1994). However, our computation showed that the rapid growth of vorticity observed by Boratav and Pelz (1994) was due to under-resolution of his numerical solution (Hou and Li 2008b). The actual growth of maximum vorticity was only exponential in the time interval when the solution was still well resolved. It is natural to ask if the dynamic depletion that we observed is generic, and to consider the driving mechanism for this depletion of vortex stretching. Some recent progress has been made in analysing the dynamic depletion of vortex stretching and nonlinear stability for 3D axisymmetric flows with swirl (Hou and Li 2008a, Hou, Lei and Li 2008). A related study for the 2D quasi-geostrophic model can be also found in Deng, Hou, Li and Yu (2006b). The local geometric structure of the solution near the region of maximum vorticity and the anisotropic scaling of the support of maximum vorticity seem to play a key role in the dynamic depletion of vortex stretching.

We also studied the dynamic stability of the 3D Navier–Stokes equations via an exact 1D model. This 1D model is an exact reduction of the 3D Navier–Stokes equations along the symmetry axis for a special class of initial data. It retains some essential nonlinear features of the 3D Navier–Stokes equations. We proved the global regularity of this 1D model by using a pointwise estimate. The key was to show that a positive Lyapunov function satisfies a new maximum principle. Here convection played an essential role in cancelling the destabilizing vortex stretching terms. Using the solution of the 1D model as a building block, we constructed a family of solutions of the 3D Navier–Stokes equations which experience interesting dynamic growth but remain smooth for all times.

To gain further understanding of the stabilizing effect of convection, we constructed a new 3D model by neglecting the convection term from the reformulated Navier–Stokes equations. This 3D model shares almost all properties of the Navier–Stokes equations, including an equivalent energy identity and a partial regularity result. Our numerical results seemed to support the conclusion that the solution of the 3D model develops locally self-similar isotropic singularities. But when we added the convection term back to the 3D model, the mechanism for generating the finite-time singularity in the 3D model was destroyed.

The results presented in this paper may have some important implication to the global regularity of the 3D Navier–Stokes equations. Our studies indicate that a successful strategy in analysing the global regularity of the 3D Navier–Stokes equations need to take advantage of the stabilizing effect of the convection term in an essential way. So far most of the regularity analysis for the 3D Navier–Stokes equations has not used the stabilizing effect of the convection term. In many cases, the same results can also be obtained for our 3D model. We are currently working to prove that the 3D model develops finite-time singularities from smooth initial data with finite energy. Such a theoretical result would show convincingly that traditional energy estimates are inadequate to prove global regularity of the 3D Navier–Stokes equations. New analytical tools that exploit the local geometric structure of the solution and the stabilizing effect of convection would be needed.

We also investigated the performance of pseudo-spectral methods in computing nearly singular solutions of fluid dynamics equations. In particular, we proposed a novel pseudo-spectral method with a high (36th)-order Fourier smoothing which retains a significant portion of the Fourier modes beyond the 2/3 cut-off point. We demonstrated that the pseudo-spectral method with the high-order Fourier smoothing gives a much better performance than the pseudo-spectral method with the 2/3 de-aliasing rule. Moreover, we showed that the high-order Fourier smoothing method captures about $12 \sim 15\%$ more effective Fourier modes in each dimension than the 2/3 de-aliasing method. For the 3D Euler equations, the gain in the effective Fourier codes for the high-order Fourier smoothing method can be as large as 20% over the 2/3 de-aliasing method. Another interesting observation was that the error produced by the high-order Fourier smoothing method is highly localized near the region where the solution is most singular, while the 2/3 de-aliasing method tends to produce oscillations in the entire domain. The high-order Fourier smoothing method was found be very stable dynamically. No high-frequency instability was observed.

Acknowledgements

I would like to thank my collaborators, Drs Jian Deng, Zhen Lei, Congming Li, Ruo Li, and Xinwei Yu, who have contributed significantly to the results presented in this survey paper. We would like to thank Prof. Lin-Bo Zhang from the Institute of Computational Mathematics in Chinese Academy of Sciences (CAS) for providing us with the computing resources to perform this large-scale computational project. Additional computing resources were provided by the Center of High Performance Computing in CAS. We also thank Prof. Robert Kerr for providing us with his Fortran subroutine generating his initial data. This work was in part supported by the NSF under the NSF grant DMS-0713670.

REFERENCES

A. Babin, A. Mahalov and B. Nicolaenko (2001), '3D Navier–Stokes and Euler equations with initial data characterized by uniformly large vorticity', *Indiana Univ. Math. J.* **50**, 1–35.

J. T. Beale, T. Kato and A. Majda (1984), 'Remarks on the breakdown of smooth solutions of the 3-D Euler equations', *Comm. Math. Phys.* **96**, 61–66.

O. N. Boratav and R. B. Pelz (1994), 'Direct numerical simulation of transition to turbulence from a high-symmetry initial condition', *Phys. Fluids* **6**, 2757–2784.

J. P. Boyd (2000), *Chebyshev and Fourier Spectral Methods*, 2nd edn, Dover.

L. Caffarelli, R. Kohn and L. Nirenberg (1982), 'Partial regularity of suitable weak solutions of the Navier–Stokes equations', *Commun. Pure Appl. Math.* **35**, 771–831.

R. Caflisch (1993), 'Singularity formation for complex solutions of the 3D incompressible Euler equations', *Physica D* **67**, 1–18.

C. Cao and E. S. Titi (2007), 'Global well-posedness of the three-dimensional primitive equations of large scale ocean and atmosphere dynamics', *Ann. of Math.* **166**, 245–267.

D. Chae (2006), 'Global regularity of the 2d Boussinesq equation with partial viscous terms', *Adv. Math.* **203**, 497–513.

D. Chae and J. Lee (2002), 'On the regularity of the axisymmetric solutions of the Navier–Stokes equations', *Math. Z.* **239**, 645–671.

C. Chen, R. M. Strain, T. Tsai and H. T. Yau (2008), 'Lower bound on the blow-up rate of the axisymmetric Navier–Stokes equations', *Int. Math. Res. Not.*, article ID rnn016.

C. Chen, R. M. Strain, T. Tsai and H. T. Yau (2009), 'Lower bound on the blow-up rate of the axisymmetric Navier–Stokes equations II', to appear in *Comm. Partial Differential Equations*.

A. Chorin (1982), 'The evolution of a turbulent vortex', *Comm. Math. Phys.* **83**, 517.

A. J. Chorin and J. E. Marsden (1993), *A Mathematical Introduction to Fluid Mechanics*, 3rd edn, Springer.

P. Constantin (1986), 'Note on loss of regularity for solutions of the 3-D incompressible Euler and related equations', *Comm. Math. Phys.* **104**, 311–326.

P. Constantin (1994), 'Geometric statistics in turbulence', *SIAM Review* **36**, 73.

P. Constantin and C. Foias (1988), *Navier–Stokes Equations*, Chicago University Press.

P. Constantin, C. Fefferman and A. Majda (1996), 'Geometric constraints on potentially singular solutions for the 3-D Euler equation', *Comm. Partial Differential Equations* **21**, 559–571.

P. Constantin, P. D. Lax and A. Majda (1985), 'A simple one-dimensional model for the three-dimensional vorticity equations', *Commun. Pure Appl. Math.* **38**, 715–724.

J. Deng, T. Y. Hou and X. Yu (2005), 'Geometric properties and non-blowup of 3-D incompressible Euler flow', *Comm. Partial Differential Equations* **30**, 225–243.

J. Deng, T. Y. Hou and X. Yu (2006a), 'Improved geometric conditions for non-blowup of 3D incompressible Euler equation', *Comm. Partial Differential Equations* **31**, 293–306.

J. Deng, T. Y. Hou, R. Li and X. Yu (2006b), 'Level set dynamics and the non-blow-up of the quasi-geostrophic equation', *Methods Appl. Anal.* **13**, 157–180.

V. M. Fernandez, N. J. Zabusky and V. M. Gryanik (1995), 'Vortex intensification and collapse of the Lissajous-elliptic ring: Single and multi-filament Biot–Savart simulations and visiometrics', *J. Fluid Mech.* **299**, 289–331.

J. Goodman, T. Y. Hou and E. Tadmor (1994), 'On the stability of the unsmoothed Fourier method for hyperbolic equations', *Numer. Math.* **67**, 93–129.

R. Grauer and T. Sideris (1991), 'Numerical computation of three dimensional incompressible ideal fluids with swirl', *Phys. Rev. Lett.* **67**, 3511.

R. Grauer, C. Marliani and K. Germaschewski (1998), 'Adaptive mesh refinement for singular solutions of the incompressible Euler equations', *Phys. Rev. Lett.* **80**, 19.

T. Y. Hou and Z. Lei (2009a), 'On the partial regularity of a 3D model of the Navier–Stokes equations', *Comm. Math. Phys.* **287**, 589–612.

T. Y. Hou and Z. Lei (2009b), 'On the stabilizing effect of convection for three dimensional incompressible flows', *Commun. Pure Appl. Math.* **62**, 501–564

T. Y. Hou and C. Li (2005), 'Global well-posedness of the viscous Boussinesq equations', *Discrete Contin. Dyn. Syst.* **12**, 1–12.

T. Y. Hou and C. Li (2008a), 'Dynamic stability of the 3D axi-symmetric Navier–Stokes equations with swirl', *Commun. Pure Appl. Math.* **61**, 661–697.

T. Y. Hou and R. Li (2006), 'Dynamic depletion of vortex stretching and non-blowup of the 3-D incompressible Euler equations', *J. Nonlinear Sci.* **16**, 639–664.

T. Y. Hou and R. Li (2007), 'Computing nearly singular solutions using pseudo-spectral methods', *J. Comput. Phys.* **226**, 379–397.

T. Y. Hou and R. Li (2008b), 'Blowup or no blowup? The interplay between theory and numerics', *Physica D* **237**, 1937–1944.

T. Y. Hou, Z. Lei and C. Li (2008), 'Global regularity of the 3D axi-symmetric Navier–Stokes equations with anisotropic data', *Comm. Partial Differential Equations* **33**, 1622–1637.

L. Iskauriaza, G. Seregin and V. Sverak (2003), '$L_{3,\infty}$-solutions of Navier–Stokes equations and backward uniqueness', *Uspekhi Mat. Nauk* **58**, 3–44.

R. M. Kerr (1993), 'Evidence for a singularity of the three dimensional, incompressible Euler equations', *Phys. Fluids* **5**, 1725–1746.

R. M. Kerr (2005), 'Velocity and scaling of collapsing Euler vortices', *Phys. Fluids* **17**, 075103–114.

R. M. Kerr and F. Hussain (1989), 'Simulation of vortex reconnection', *Physica D* **37**, 474.

G. Koch, N. Nadirashvili, G. Seregin and V. Sverak (2009), 'Liouville theorems for the Navier–Stokes equations and applications', to appear in *Acta Mathematica*.

H. Kozono and Y. Taniuchi (2000), 'Bilinear estimates in BMO and Navier–Stokes equations', *Math Z.* **235**, 173–194.

O. Ladyzhenskaya (1970), *Mathematical Problems of the Dynamics of Viscous Incompressible Fluids*, Nauka, Moscow.

R. J. LeVeque (1992), *Numerical Method for Conservation Laws*, Birkhäuser.

F. H. Lin (1998), 'A new proof of the Caffarelli–Kohn–Nirenberg theorem', *Commun. Pure Appl. Math.* **51**, 241–257.

J. G. Liu and W. C. Wang (2006), 'Convergence analysis of the energy and helicity preserving scheme for axisymmetric flows', *SIAM J. Numer. Anal.* **44**, 2456–2480.

A. J. Majda and A. L. Bertozzi (2002), *Vorticity and Incompressible Flow*, Cambridge University Press.

R. B. Pelz (1997), 'Locally self-similar, finite-time collapse in a high-symmetry vortex filament model', *Phys. Rev. E* **55**, 1617–1626.

G. Prodi (1959), 'Un teorema di unicità per el equazioni di Navier–Stokes', *Ann. Mat. Pura Appl.* **48**, 173–182.

A. Pumir and E. E. Siggia (1990), 'Collapsing solutions to the 3-D Euler equations', *Phys. Fluids A* **2**, 220–241.

G. Raugel and G. Sell (1993a), 'Navier–Stokes equations on thin 3D domains I: Global attractors and global regularity of solutions', *J. Amer. Math. Soc.* **6**, 503–568.

G. Raugel and G. Sell (1993b), Navier–Stokes equations on thin 3D domains III: Global and local attractors, in *Turbulence in Fluid Flows: A Dynamical Systems Approach*, Vol. 55 of *IMA Volumes in Mathematics and its Applications*, Springer, pp. 137–163.

G. Raugel and G. Sell (1994), Navier–Stokes equations on thin 3D domains II: Global regularity of spatially periodic solutions, in *Nonlinear Partial Differential Equations and their Applications, Collège de France Seminar*, Vol. XI of *Pitman Research Notes Math, Series 299*, Longman, pp. 205–247.

J. Serrin (1962), 'On the interior regularity of weak solutions of the Navier–Stokes equations', *Arch. Ration. Mech. Anal.* **9**, 187–195.

J. Serrin (1963), The initial value problem for the Navier–Stokes equations, in *Nonlinear Problems* (R. Langer, ed.), University of Wisconsin Press, pp. 69–98.

M. J. Shelley, D. I. Meiron and S. A. Orszag (1993), 'Dynamical aspects of vortex reconnection of perturbed anti-parallel vortex tubes', *J. Fluid Mech.* **246**, 613–652.

M. Struwe (1988), 'On partial regularity results for the Navier–Stokes equations', *Commun. Pure Appl. Math.* **41**, 437–458.

R. Temam (2001), *Navier–Stokes Equations*, AMS, Providence, RI.

Online reference

C. Fefferman, Clay Mathematical Institute, Millennium Open Problems:
www.claymath.org/millennium/Navier–Stokes_Equations/

For EU product safety concerns, contact us at Calle de José Abascal, 56–1°, 28003 Madrid, Spain or eugpsr@cambridge.org.

www.ingramcontent.com/pod-product-compliance
Ingram Content Group UK Ltd.
Pitfield, Milton Keynes, MK11 3LW, UK
UKHW050107230326
469255UK00016B/236